AF307915

Environmental Engineering

Series Editors: U. Förstner, R. J. Murphy, W. H. Rulkens

Springer

Berlin
Heidelberg
New York
Barcelona
Budapest
Hong Kong
London
Milan
Paris
Santa Clara
Singapore
Tokyo

Charles G. Gunnerson
Jonathan A. French (Eds.)

Wastewater Management for Coastal Cities

The Ocean Disposal Option

With contributions by Qian Ming Lu, Jørgen Færch Knudsen, Søren K. Eskesen, J. T. Powers, Frederick Shremp, Douglas A. Segar, Elaine Stamman and Zhou Yucheng

Second revised Edition

Originally Published in 1988 as Technical Paper 77 by The World Bank, Washington, DC

Springer

Series Editors

Prof. Dr. U. Förstner

Arbeitsbereich Umweltschutztechnik
Technische Universität Hamburg-Harburg
Eißendorfer Straße 40
D-21073 Hamburg, Germany

Prof. Robert J. Murphy

Dept. of Civil Engineering and Mechanics
College of Engineering
University of South Florida
4202 East Fowler Avenue, ENG 118
Tampa, FL 33620-5350, USA

Prof. Dr. ir. W. H. Rulkens

Wageningen Agricultural University
Dept. of Environmental Technology
Bomenweg 2, P.O. Box 8129
NL-6700 EV Wageningen, The Netherlands

Editors

Charles G. Gunnerson

Environmental Engineering and Policy,
Laguna Hills, California, USA

Jonathan A. French

Camp, Dresser and McKee,
Cambridge, Massachusetts, USA

ISBN-13:978-3-642-79731-6

Cataloging-in-Publication Data applied for

Die Deutsche Bibliothek – Cip-Einheitsaufnahme
Wastewater management for coastal cities : the ocean disposal option ; originally published
in 1988 as Technical paper 77 by The World Bank, Washington, DC / Charles G. Gunnerson ;
Jonathan A. French (ed.). - 2., rev. ed. - Berlin ; Heidelberg ; New York ; Barcelona ; Budapest ;
Hong Kong ; London ; Milan ; Paris ; Santa Clara ; Singapore ; Tokyo : Springer, 1996
  (Environmental engineering)
  ISBN-13:978-3-642-79731-6   e-ISBN-13:978-3-642-79729-3
  DOI: 10.1007/ 978-3-642-79729-3

NE: Gunnerson, Charles G. [Hrsg.]

Typesetting: Camera-ready by editors
SPIN:10465317   61/3020-5 4 3 2 1 0 - Printed on acid -free paper

Foreword

Some three-quarters of the earth's surface is covered by the oceans and in the not too distant past many city officials, engineers, and planners mistakenly believed the oceans could provide an infinitely great and endless repository for the liquid and solids wastes of coastal cities. This practice has at times threatened the well-being of bathers, harmed tourism, filled marine food resources with pathogens and poisons, and destroyed delicate marine ecosystems.

The truth of the matter is, that while the oceans are indeed great, the very limited coastal waters nearest to and affected by centers of human population make up less than one-tenth of one percent of the seas. These relatively circumscribed areas of the seas have only limited pollution assimilation capacity and are the areas most used by millions of people for bathing and recreation. They also serve as one of the richest reservoirs of shellfish and other seafood resources that concentrate human pathogens and chemical pollution from seawater to epidemic levels . Thus the protection of coastal waters from direct pollution from coastal cities is a vital task in preserving marine ecosystems and promoting human health. While our understanding of the ecological and oceanographic constraints imposed on disposal of urban wastes into coastal waters is incomplete, enough is known in most cases to provide a very high level of protection to human health, sea food resources, and delicate marine ecosystems. There are, however, particularly sensitive enclosed seas and marine ecosystems where appropriate treatment followed by land disposal is demonstrably the most rational means for providing environmental protection.

This book, edited by two leading experts on wastewater management for coastal cities is in many ways unique in both the depth and breadth of it's coverage of the subject. It delves deeply into ecological and oceanographic fundamentals that are essential to an understanding of the scientific foundations of the impacts of wastes in the complex marine environment. These fundamentals provide for a fuller understanding of the requirements for rational engineering design and operation of the physical and institutional components of coastal city wastewater management. To this end, guidelines for hydraulic design, ocean outfall construction, monitoring, cost recovery, and other economic aspects are provided to the manager, designer, and engineer. The book is enriched by including case studies drawn on the basis of world-wide field experience of the editors.

Hillel I. Shuval, Kunin-Lunenfeld
Professor of Environmental Sciences
The Hebrew University, Jerusalem, Israel.

Preface

This second edition of Wastewater Management for Coastal Cities is for students as well as the practitioners to whom the first edition was directed. Its scope has been expanded to include costing and other non-structural aspects of engineering. Additional empirical and theoretical background information along with worked examples are included. The book is about the ways in which scientific, engineering, economic, and institutional frameworks interact and from which investment and assessment decisions in wastewater management for coastal cities are made. The long-range goal of the editors and contributors is the same as that of others committed to the practice of international engineering in support of economic development: it is to make investments in coastal city water sector infrastructures, particularly those for wastewater management, more efficient.

This work is constructed on three premises. First, that sustaining coastal cities and the lives of their peoples is essential; second, that while the oceans are large, deep, and stable, coastal waters are not, and sustaining them is also important; and third, that these goals are most efficiently approached in the environmental sector when informed cooperation and consensus are the instruments of choice in both industrial and developing countries.

Damage to coastal waters from municipal wastewaters is everywhere accompanied by other corollaries of industrial and commercial development. Local and national problems of overfishing eventually disappear along with the fish. These effects are increasing on global scales where problems are aggregating, intermediate scales where cooperative monitoring can measure rates of degradation and recovery, and local scales where problems are solved Meanwhile, coastal city and megacity populations and their economic activities and impacts are increasing at combined rates of, say, 6 to 8 percent per annum worldwide. It will take much money and ingenuity to ensure environmental sustainability.

Our second edition comes from some fifteen years of the editors' collaboration in environmental engineering practice along parallel paths of civil service and private enterprise that began with the preparation of first edition. Our individual ventures have been made possible by our collegial and counterpart engineers, scientists, planners, economists, and managers in domestic and foreign coastal cities and megacities, and in the World Bank and other international organizations. These are the people with whom we have had the good fortune to work, learn each others' skills and cultures, and depend on for our achievements in the United States and throughout the world.

All of this is with due recognition that the original work was made possible by the interest, institutional, and technical support of many World Bank colleagues including John M. Kalbermatten, Shaul Arlosoroff, and James K. Feather and their staffs; and by U.S. Environmental Protection Agency and National Oceanic and Atmospheric Administration advisors and consultants Victor J. Cabelli, Joel B. O'Connor, and William F. Garber.

x

We are especially indebted for manuscript reviews, comments, updated information, and coordination of the chapters in this book to Lars Frendrup Jensen, Jørgen Færch Knudsen, and Lee Lund Mathiasen, COWIconsult, Denmark; to Qian-ming Lu, Jacob Steen Møller, Wagner Jacobsen, and Peter V. Nielsen, Danish Hydraulic Institute;   Hanne Bach, Water Quality Institute, Denmark; L.D. Jones, National Rivers Authority, Thames Region, U.K.: M.R. Hoffman, Thames Water Pic, London; Paul Woodcock, Anglian Water Services, Ltd., Cambridge; M.T. George Taylor, Consultants in Environmental Sciences, Ltd., Beckenham, Kent, UK; Zhou Yucheng, Shanghai Sewerage Project Construction Company;   John H. Dorsey, City of Los Angeles Bureau of Sanitation; Charles C. Carey, Thomas LeBrun and Frances Garrett, Los Angeles County Sanitation Districts; Jeffrey Cross, Southern California Coastal Water Research Project; Terry Fleming, USEPA Region IX, San Francisco: Joel O'Connor, USEPA Region II, New York; Irwin Haydock, Sanitation Districts of Orange County, California; James E. Foxworthy, Loyola Marymount University; R. Lawrence Swanson, State University of New York, Stony Brook, Philip J. Valent, U.S. Naval Research Laboratory; and Geoffrey Read, World Bank.

We also are grateful for the otherwise unacknowledged debt to those upper division and graduate students in environmental engineering and sciences whose questions evolved from the first edition and from the complementary works that distinguish sanitary and environmental engineering.

The editors

Contents

1 Overview

It is the certain fate of domestic, municipal, and surface drainage waters in coastal cities to flow to the ocean, either over the land or as submarine springs, somewhat diminished in volume but always enriched in fertilizer and other foreign materials.

Coastal cities are there because water, arable flood plains, fisheries, communications, transport, and opportunities for regional and international trade were there in the first place. As cities prospered, increasing amounts of fresh water were taken from one convenient place, used, and then dumped into another.

Chapters of this book describe frameworks and alternatives for planning and implementing waste water management in low, middle, and high income coastal cities in an international environmental engineering context. It is an introduction to oceanography for engineers, to engineering for oceanographers and to comparative costing and cost recovery for planners from both disciplines, all with due regard for the fact that there are some coastal waters into which even the most highly treated effluents should not be purposefully discharged.

Chapter 2. Oceanography at the Margin, introduces the physical, chemical, and geological framework of the unstable land-sea boundary (which is why the shoreline is where it is). Local diurnal and seasonal currents, tides, and wave characteristics in this high energy zone contribute to stirring and mixing of drainage in seawater, and to sedimentation processes. Sea level changes due to long-term glacial melting and other global warming trends, and to local tectonic or isostatic movement are factors in hydraulic design. Observation and theory of how shoaling wave energy is dissipated by bottom friction, sediment movement, breakers and surf, reflection and refraction are design elements in coastal projects. Energy differences between tides and waves determine whether beaches or tidal flats are found. Every few years, *El Niño* appears as an oceanic-scale perturbation in Pacific Rim land-air-sea interactions and currents that affect marine life and reduce coastal fisheries, many of which are already over-exploited.

Mixing in sea waters at the molecular scale is the last phase of dilution that begins with mechanical stirring imposed by waves, boundary conditions, and outfall discharges. Chemical and physical properties include the non-linear dependence of density upon temperature and salinity, where mixing of two water masses of the same density produces a third with slightly higher density. Under the influence of the earth's rotation, these density differences cause currents. The web of life in the sea always responds to local effects of wastewaters, some toxic and some fertilizing, that are superimposed on secular trends so that understanding evolution, adaptation, and recovery of damaged ecosystems requires epidemiological approaches to marine pollution monitoring (see Chapter 10). .

Chapter 3. Ecological Design. Public health has long been the principal factor in ecological design for water supply and wastewater management. In high income countries, this has been accompanied by successions in design criteria from receiving water quality standards to technology-based standards and, in the United States, on to more dependable risk-based standards. The latter begin with assessments of risk of exposure that by themselves occlude benefit-cost analyses. but that favor a 1972 legislative policy of achieving zero pollutant discharge, a culturally attractive but thermodynamically flawed concept. Credible risk-assessments also require site-specific epidemiological studies to quantify risk of infection. Conceptual and operational difficulties in the zero-discharge approach have been demonstrated in environmental protection reports by using such terms as "virtual elimination."

Meanwhile, limiting standards for coliform bacteria and measurements of overall first-order decay rates or times for 90 percent reductions (T$_{90}$s) in receiving waters continue to be used for design Finely scaled field studies of currents are useful in designing for initial dilution but less so for estimating subsequent reductions of concentrations because of the dominant effects of sedimentation There are no verified models for aggregating separate components of coliform decay coefficients from laboratory and computational studies of diffusion, sunlight, phage, chemical and DNA speciation, elutriation, flocculation, etc., into their combined effects.

Much less is known about marine ecosystems than about public health ones. Ecological data collection is expensive, particularly for unbounded environmental impact assessments, but it is conceptually simple and, with computer graphics, to render in fine detail. Data analysis and understanding are difficult, often even more expensive, and tend to be postponed when data and displays are voluminous. A rule of thumb is that, if its analysis is to be delayed, leave it for the archaeologists. We conclude that bounded system studies of recoveries of damaged ecosystems are more useful than open system compliance monitoring (see Chapter 10)

Chapter 4. Hydraulic Design. This chapter presents engineering principles and evolving practices in making first estimates and numerical models in outfall siting, field surveys, remote sensing, physical hydraulic models, and equations for estimating transport and turbulent diffusion along the ocean margin. Topics and worked examples include outlet design, open ends vs. multiport diffusers, initial dilution, information needs and collection, and worked examples. Other design factors are pipe diameter, thrust at bends, hydraulic transients, excess hydrostatic head, drop structures, tunneled conduits with buried risers for diffusers, and provision for pigging.

Chapters 5 through 8 summarize selected aspects of the structural design and sustained operation of ocean outfalls. Construction methods and materials are interrelated with pipe diameters and stresses during and after construction. For example, plastic pipe jacking requires a specially designed thick-walled or temporarily lined plastic pipe. Other materials include cast iron, wrought iron, reinforced concrete, and concrete coated steel pipe. Vertical and lateral stability of buried and unburied pipelines may be provided by anchoring, blanketing, or other engineering responses to hydrodynamic forces in the nearshore zone.

Stress analysis follows standard engineering practice, with particular attention to anchoring, unsupported spans, and installation conditions leading to collapse or buckling of the pipe. The shore approach is the most vulnerable part of the outfall because of wave battering and erosion during and after construction and as modified later by other coastal works.

Although marine borers can infest both temporary and permanent wood and concrete structures, it is the corrosion of steel in seawater that is most damaging. Principles for selection of cathodes, sacrificial anodes, external coating systems, and internal linings are described.

Chapter 9. Ocean Outfall Construction. Construction usually accounts for 60 to 80 percent of the costs of an ocean outfall Methods used in the shore approach or other shallow waters include preparing the pipe bed by pre-trenching with mechanical or hydraulic dredges followed by backfilling, post-trenching with hydraulic cutterhead dredges or jet sleds, cofferdams, and trestles. Offshore installation can be by bottom assembly from a jack-up platform similar to an oil drilling rig, lay barge. reel barge, bottom pull, conventional surface pull sometimes modified by using floats and chains, and tunneling. The latter is increasingly being used for ocean outfalls and includes conventional shield or pipe jacking methods for larger diameters, and remote controlled directional drilling and pipe jacking for smaller diameters. Practical limits to offshore tunneling technologies are continually being extended by the offshore oil and gas, and the utilities construction industries. Higher unit costs for onshore directional drilling tend to be offset by savings in time and surface activity disruption from that for trenching and backfilling

Chapter 10. Ocean Outfall System Performance Monitoring. Ocean outfalls are components of the larger systems described in Chapter 12. They include physical works, environmental forces, financial and institutional factors, fates and effects of discharge components, and people.

Although the two functions are complementary, monitoring is more likely to be funded than research since the former term implies to decision makers that specific actions will follow. Monitoring provides more job and program security than research and, when properly scaled, can be an essential element in treatment plant operations, in receiving water evaluation and response, in post-audits, in identifying research needs, and with appropriate data resolution, in regional and larger scale monitoring

Methods for collecting monitoring data are published in various sources and sometimes mandated by regulatory authority. This chapter focuses on monitoring data utilization with due regard for available resources, legislative mandates, and for local, national, and international environmental policies. Monitoring technologies can be exported from one country to another, but their regulatory aspects are based on national economic and cultural factors that cannot.

Chapter 11. Case Studies. Representative findings from case studies are:
- Shanghai is in the early phases of the Yangtze River wastewater management program with cooperative leadership shared by the Municipality of Shanghai and the World Bank. Bilateral assistance is provided by official development agencies and consultants from Australia, Canada, Denmark, France, Norway, and the U.K. Section 11.2 is a status report on state-of-the-art modeling of the scientific and engineering studies and their institutional frameworks as of the end of 1995.
- The Thames estuary story is one of continuing recovery of an ecosystem long damaged by wastewater discharges. A 2-dimensional mathematical model described in 1965 was intended only to assure enough dissolved oxygen for fish. In the hands of careful observers and committed institutions, the damaged ecosystem below London Bridge has been largely recovered.
- The unique hydrographic regime of the Bosporus and Sea of Marmara has a lower layer flowing north from the Mediterranean and Aegean Seas and a brackish upper layer carrying runoff from the Danube and other rivers entering the Black Sea. Sewage discharged into the lower layer thus goes mostly into the lower layer of the Black Sea. Since the late 1960s, institutional factors in environmental, engineering, and economic decisions have provided important secondary benefits.
- The Boston Harbor project uses state-of-the-art technologies with estimated system costs of $5.5 billion (1995 dollars) of which $334 million is for a 15.2 km long, 8.1 m diameter effluent tunnel and diffusers. The program is an example of how engineering ingenuity and state-of-the-art technology requires even more social than financial innovation as projects increase in size and elegance.
- The oceanographic edge of southern California's development is yielding new insights into natural land-sea interactions as they are changed by importing people and water into a "Cadillac Desert". Predictive models of these changes, rates of recovery of damaged ecosystems, and realization of secondary benefits from wastewater management abound.

Chapter 12. Cost and Sustainability factors. This chapter looks at non-structural engineering components in the urban water sector where the costs of sanitation always exceed the costs of water supply by ratios of 1.3:1 to at least 15:1 as the supply increases from a village level of 20 liters per capita per day to a large urban supply of, say, 700 lcd. This requires using the same demand projections and unified cost recovery principles for both water and sanitation. Operationally, the final selection of wastewater management technologies and instruments for a coastal city depends upon four interrelated sets of planning and implementation elements. These include: (1) what costs? who benefits? who pays? (counterpart and public participation, cost recovery, relative costs of water and sanitation, outfall costs, and shared benefits), (2) system options (scales in water supply, disposal, reclamation, and conservation, comparative costing), (3) water, sanitation and public health service levels, and (4) cooperation, competition, and water conflict resolution from local to international scales (capacity building, community participation, the prisoner's dilemma paradigm, and terms of reference for information and technology transfer).

2 Oceanography at the Margin

Ocean science is concerned with global navigation, fisheries, climate, geology and geophysics, water and energy balances, national security, and the origin of life. Most economic values of the ocean lie along its margin as fisheries, recreation, tourism, transportation, placer deposits, cooling water, and waste assimilative capacity. To make the best of these ocean resources, peoples and their governments must define, appreciate, allocate, and efficiently conserve them. This chapter identifies the non living and living frameworks of the ocean margin into which marine outfalls discharge. These areas have developed over millions of years in dynamic equilibrium with geologic and climatic change to be taken into account in engineering, environmental, and investment decisions on wastewater management.

2.1 The Physical, Geological, and Chemical Framework

A major portion of the world's peoples live within 100 km of the ocean, which provides them with food, transport, and communication. The oceans cover nearly 71 percent of the earth's surface and contain 86.5 percent of the earth's water. Their total volume of water is approximately 1,350 million km and their average depth almost 4 km. The geography, principal surface currents, and continental shelf and margin areas to 1,000 m depths are shown in Figures A1 and A2 of Appendix A.)

2.1.1 Oceanic Waters

Solar energy heats the ocean, which supplies energy for atmospheric circulation. Much of this energy is used to evaporate water, which is returned to the earth as rain. Approximately 10 million tons/sec of fresh water flow into the oceans, mostly from river runoff and melting ice (71). Shapes of the major ocean basins are listed in Table 2.1.

Temperature, salinity, and light are the primary variables that control biological and chemical interactions in the ocean. Density is a non-linear function of the temperature and salinity parameters that define oceanic water masses Water is incompressible except at depths greater than 4500 meters where adiabatic heating is significant. Figure 2.1 shows typical temperature, salinity, and density depth distributions in the mid-Pacific. Non-linearity of the density function is shown by the concave curvature of the isopycnals so that when two water masses mix

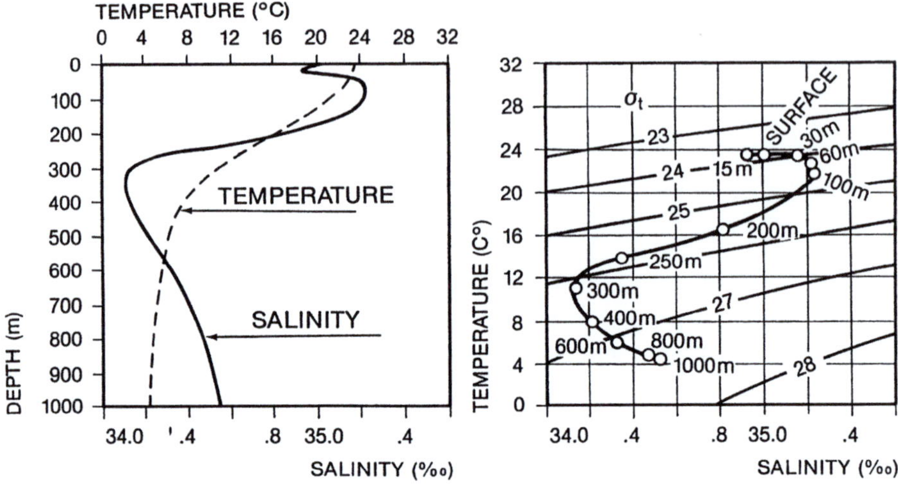

Figure 2.1. Mid-Pacific Ocean temperature and salinity near the Hawaiian Islands. The left-hand diagram shows a thermocline between about 100 and 400 meters and a halocline between about 60 and 300 meters. The right-hand diagram shows upper and lower water masses separated by a pycnocline between 60 and 300 meters. Density (ρ) is shown as σ_t where $\rho = (\sigma_t - 1) \times 1000$. Mixing takes place between waters of equal density. σ_t is a non-linear function of temperature (ζ) and salinity (s) and lines of equal σ_t are concave downward, indicating that a mixture of waters of the same density are slightly heavier than either of the source waters. The difference is sufficient to establish density-driven differential currents between large water masses. Source: Smith and Brown (62).

horizontally among layers of equal density, the resulting mixture always slightly heavier than either of the sources. The increase in density as σ_t is small, about 0.02 to 0.04 but because of the size of oceanic masses it is enough to be cause a slight fall in the surface elevation, and there will be a geostrophic flow that responds to the effects of the earth's rotation on currents. In the surface layer, temperature, salinity, and density are determined by heating, cooling, evaporation, and precipitation. The permanent pycnocline (there is also a shallower, seasonal one) is where the water density changes rapidly with depth at the base of the surface layer. This layer is stable and prevents deep ocean waters from mixing with surface waters. Most ocean water lies below the pycnocline, is cold, receives no light; and almost all of the organisms that live there depend on food sources that have settled into this layer from the upper layers.

Table 2.1.   Areas and depths of the oceans (65)

Ocean Area	Water Area (10^4 km^2)	Land Area (10^4 km^2)	Ratio Water / Land	Mean Depth (km)
Pacific	180	18	10	3.94
Atlantic	107	67	1.6	3.31
Indian	74	17	4.3	3.84
World Ocean	361	102	3.6	3.73

Source: Sverdrup, Johnson, and Fleming (65)

Surface waters are driven by winds, tides, density differences and the rotation of the earth (Coriolis effect). Bottom currents are due to density differences and the Coriolis effect Wind forces mix shallow coastal waters, which move to the bottom during afternoon sea-breezes or other on-shore winds Solar energy drives the principal surface currents shown in Figure A-1 through heating and evaporation. The large-scale air-sea interactions between weather, wind, and currents are shown in Figure 2.2.

El Niño is an example of perturbations in oceanic-scale air-sea interactions that affect nearshore oceanographic conditions and continental weather. Originally applied only to Peruvian waters, the term, "El Niño," (Christ Child) was coined long ago by local marine fishermen to describe warm ocean surface temperatures and related weather conditions that temporarily eliminate the local anchovy fishery, populations of sea birds that feed on the fish, production of guano needed by Peruvian farmers as fertilizer, and unusual or severe marine and continental weather conditions that begin aperiodically every few years during northern winters around Christmas time. Related observations have been recorded by ships captains as early as 1726. The interactions are so tightly drawn that it is not known whether the events begin in the atmosphere or in the water, although progress is being reported in predicting an event's intensity once it is being observed (80, 89, 96).

The warm ocean surface layer that identifies an El Niño may last for a few months or several years. Its effects on fisheries can be widespread and severe along the eastern boundary of the central Pacific Ocean. Recent research has shown El Niños are elements of large-scale Pacific air-sea interactions formally known as El Niño-Southern Oscillation (ENSO) events of which the 1982-83 is the most remarkable and best studied. There are latitudinal and longitudinal changes in equatorial and trade winds in both hemispheres, reduced rainfall over Indonesia, increased rainfall over the central Pacific, thickening of the warm upper layer of water in the central and eastern Pacific, flooding in the Andean foothills, and (arguably) droughts as far away as Australia and sub-Saharan Africa (80, 88). Along with overfishing, it masks all but the most local small-scale oceanographic effects of river and waste discharges. El Niño's economic and cultural effects include acceptance as evidence of global warming, semi-popular (in California) blame for a variety of human behavioral and economic aberrations, and calls for government interventions such as eliminating foreign fishermen from the 200-mile extended economic zone claimed by maritime nations.

2.1.2 Coastal Waters

Many oceanographic processes in coastal waters are similar to those in the open ocean, except that in coastal waters they take place more quickly. Much of the circulation of coastal waters is driven by trade winds, diurnal land and sea breezes, wave movement, storm surges, coastal currents, and upwelling that affect the water column and coastal sediment transport .

Water temperatures control the physical, chemical, and biological conditions in the coastal ocean (6, 14). Whereas bottom waters in deeper ocean areas generally undergo very little temperature change, coastal water masses are subject to seasonal and diurnal temperature fluctuations. Thus the highest and lowest ocean surface temperatures are found in coastal waters. During the day, surface waters are warmed by the sun at depths up to 10 m. These waters reach their highest temperature in mid afternoon and their lowest at dawn.

A thermocline is a layer of water with a large vertical temperature gradient that separates a warmer, upper layer from a cooler, bottom layer (see Figure 2.1). Permanent thermoclines are resistant to change and are the major stabilizing feature in the water column. Shallower, seasonal thermoclines are less stable and can disappear for short periods following major storms.

Haloclines are layers in the water column characterized by a strong vertical salinity gradient. Salinity varies greatly in coastal waters because of the freshwater coming in from rivers, melting ice, and dispersed land runoff. Strong haloclines develop where river discharges of low salinity flow out over heavier, more saline coastal ocean waters. Since density is a function of both temperature and salinity, the pycnocline may be a thermocline, a halocline, or a combination of both. Thermoclines in coastal waters are often seasonal, developing in spring as waters are warmed, becoming stronger in the summer, and disappearing in the fall and winter as the waters become thoroughly mixed. This seasonal temperature cycle of waters is strongest at mid latitudes where changes in air temperatures are also greatest. Salinity changes are also seasonal.

Currents.   Currents in open coastal waters are a result of winds, tides, and geostrophic forces. Geostrophic forces are generated by the earth's rotation so that they are deflected to the right in the northern hemisphere. This is accompanied by a redistribution of mass so that the lighter (warmer) water lies to the right with corresponding adjustments in the lateral slope of the sea surface (29, 65, 86).

Currents may be local, regional, or planetary and are generally permanent, seasonal, or diurnal. Near the sea surface and in shallow water, currents are directly related to geostrophic winds (such as trade winds), to seasonal (monsoon) winds, and to diurnal land breeze and sea breeze systems. When winds blow the surface waters, high and low points are formed on the sea surface. As the sea surface slopes downward, water gravitates downhill and flows in a direction influenced by the Coriolis effect of the earth's rotation. As a result, water is deflected up to 45° to the right of the wind direction in the northern hemisphere and to the left in the southern hemisphere (Figure 2.2). The Coriolis effect increases with increasing latitude, but is constrained in coastal waters by the shoreline and other factors.

Geostrophic currents are strongest when river runoff is large and when strong winds parallel the coast. However, even weak geostrophic currents will move

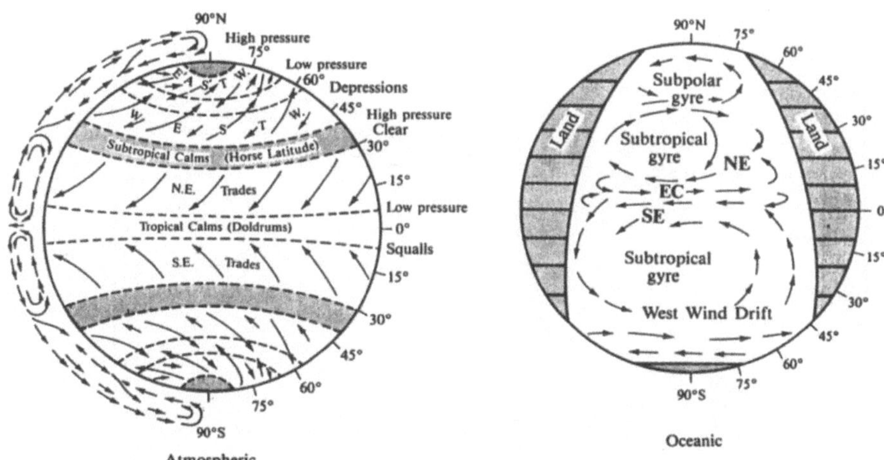

Figure 2.2. Planetary atmospheric and oceanic circulation. NE indicates North Equatorial Current; EC indicates Equatorial Current; and SE indicates South Equatorial Current. Sources: adapted from Fleming (23) and Gross (29).

wastewaters counterclockwise along the right bank of an embayment in the northern hemisphere (clockwise in the southern), a factor to be considered in outfall siting. Geostrophic forces also control the large, near-coastal currents such as the Gulf Stream and the Peru Current, and are reflected in the counterclockwise circulation of the Mediterranean and Black Seas).

Surface waters moved away from the coast by winds or by the Coriolis effect are replaced by subsurface waters that upwell to replace them (Figure A-1). Upwelled waters usually form small, cold, nutrient-rich surface water masses notably along arid coasts in subtropical areas, such as those of Peru and Ecuador. Since freshwater runoff is not present in such areas, there is no low-salinity surface layer, and cool, upwelled waters readily rise to the surface. In areas with large freshwater discharges, the low-salinity surface layer inhibits wind-induced upwelling.

Tidal currents are caused by the pull of the sun and the moon on the earth. They normally move in a rotary pattern in the open ocean, but, in the coastal ocean, the typical pattern is elliptical (see Figure 2.3). On essentially open coastlines, currents generated by small tidal ranges (≤ 0.2 m) are only a minor component of the total currents. In semi-enclosed basins such as Long Island Sound, even small ranges result in locally strong currents. In either situation, large local tidal ranges (≥ 1m) result in significant components of coastal currents.

2.1.3 Waves, Tides, and Sea Level

Waves and wave energy. Waves are described by what they do, which can be measured, rather than by what they are, which cannot. This has led to a large number of explanations that are disarmingly and elegantly developed by Kinsman

8 meters above bottom

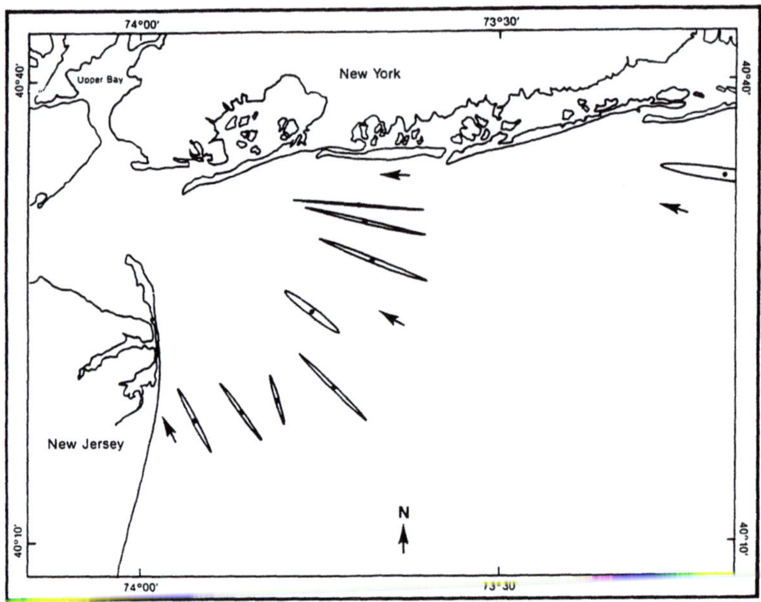

3 meters or less above bottom

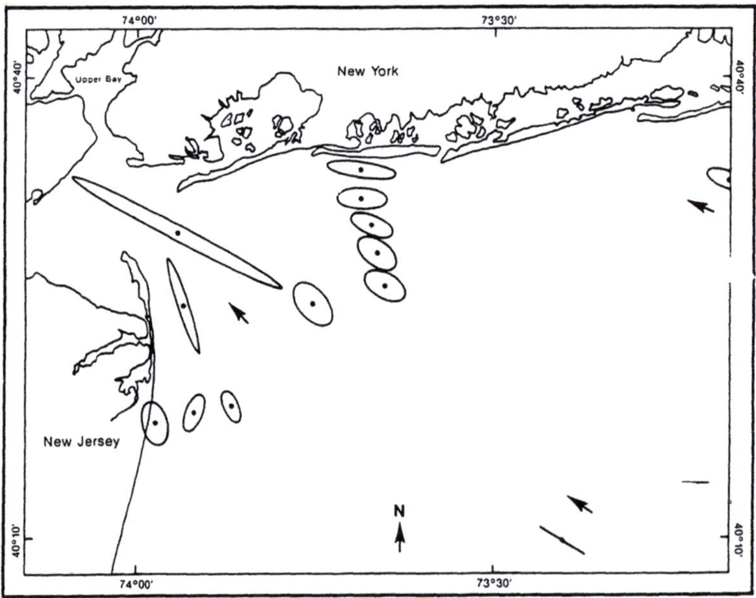

Figure 2.3. Tidal currents in the New York Bight. Rotation is predominantly clockwise at 8m and counterclockwise at ≤3m above bottom. Vectors show maximum lunar semidiurnal flood currents. Scale 1:650,000. Source: Hansen (33)

(41) as an infinite number of accelerations subject to the restoring forces of surface tension, gravity, and the Coriolis effect. These can be rationalized in a collection of zero to fourth order approximations and assumptions about boundary conditions, vorticity, conservation of momentum, stream functions, and continuity. To the great relief of designers of outfalls and other coastal works, the linear approximations by Airy in 1845 of simple harmonic, small amplitude waves are adequate for estimating wave forces and particle velocities for the design and construction of coastal structures (13, 41). These are summarized in Table 2.2..

Progressive ocean waves transmit energy from sources to sinks and to new equilibria determined by gravity and surface tension. The principal energy sources include gravitational attraction of the moon and sun, surface winds, and (for tsunamis) earthquakes. and turbidity currents. Energy sinks include white-capping, breakers and surf, bottom sediment movement, and shoreline erosion.. Most of the ocean surface wave energy is carried by the common 15 to 30 second gravity waves familiar to beach patrons (see Figure 2..4).

The principal descriptors of waves are their height measured from crest to crest (L), height measured from to trough to crest, and period (T) usually measured in seconds or minutes extending to hours, days, or years for astronomic tides (87). Their steepness is H/L and their frequency is 1/T.   Equations 2.1 to 2.3 relate wave speed (celerity) to depth and wave length. Note that for x<0.5, tanh x $\approx$ x, and for x>π, tanh x $\approx$ 1.

Wave speed (celerity) is $c = \left((gL/2\pi) \ \tanh 2\pi d/L \right)^{1/2}$ (2.1)

In deep water where the depth d > L/2, $c = \left(gL/2\pi \right)^{1/2}$ (2.2)

In shallow water where the depth, d > L/20 according to oceanographers

(65, 87)) or > 25 according to engineers (13) , $c = \left(gd \right)^{1/2}$ (2.3)

It is in the intermediate range of a shoaling bottom that only the general equation 2.1 applies. As the depth decreases, waves steepen from a sinusoidal shape to a trochoidal one because their speed, $c = \sqrt{gd}$, is greater at the crest than in the trough, and they expend their energy by moving bottom sediment and by breaking. As the bottom slope increases, the waves change to spilling, plunging, or collapsing breakers, then to swash and, against a vertical slope, to surge. The equations of wave geometry and their effects as they evolve from circular functions in deep water through intermediate stage hyperbolic functions and return to circular functions in shallow water are listed on Table 2.2.

Figure 2.5 shows (Eulerian) streamlines and (Lagrangian) trajectories of accelerating and decelerating water particles that participate in wave forming and energy transfer. Particles converge to form wave crests, diverge to form the next trough and change from circular orbits at the surface to back-and-forth movement at the bottom. Wave mass and movement impart the total kinetic and potential energy (E) expressed by

Wave energy, $E = \frac{1}{8}\left(\rho g H^2 \right)$. (2.4)

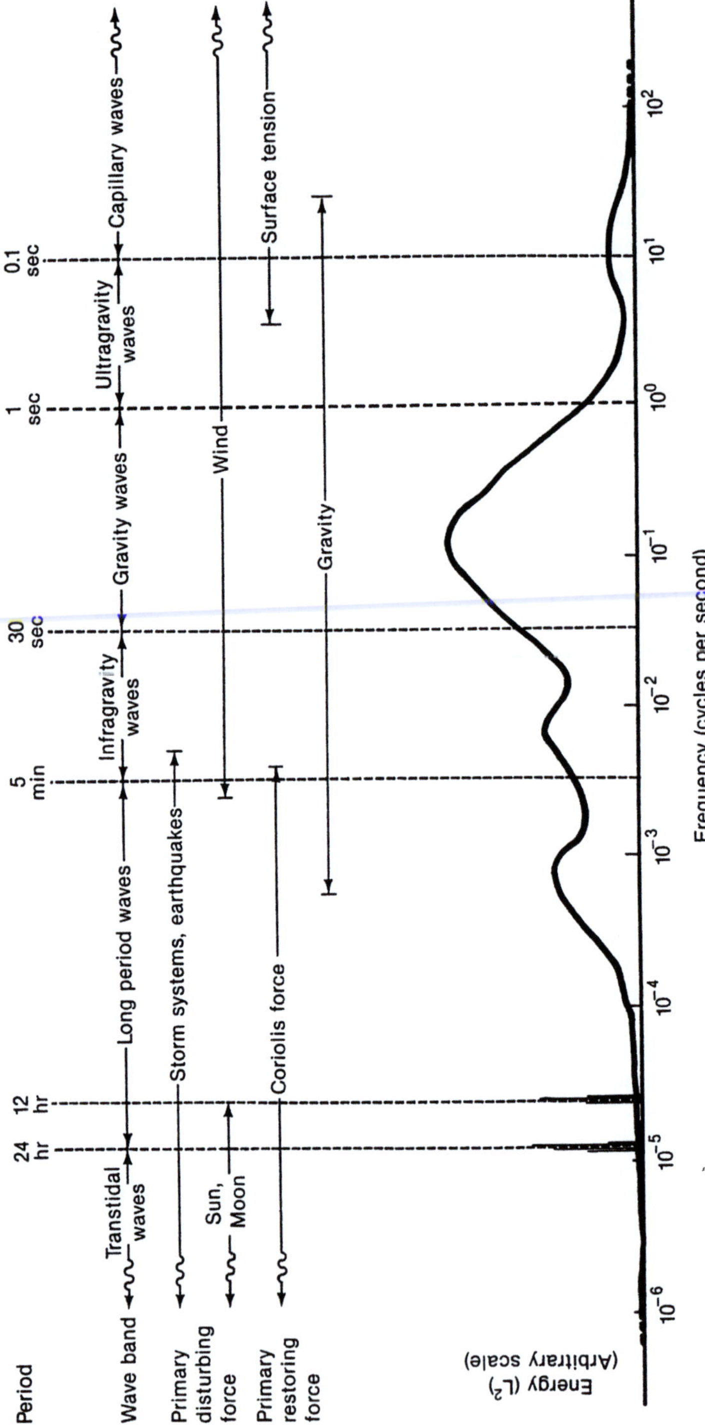

Figure 2.4. Estimated energy spectrum for ocean surface waves. Source after Kinsman (41).

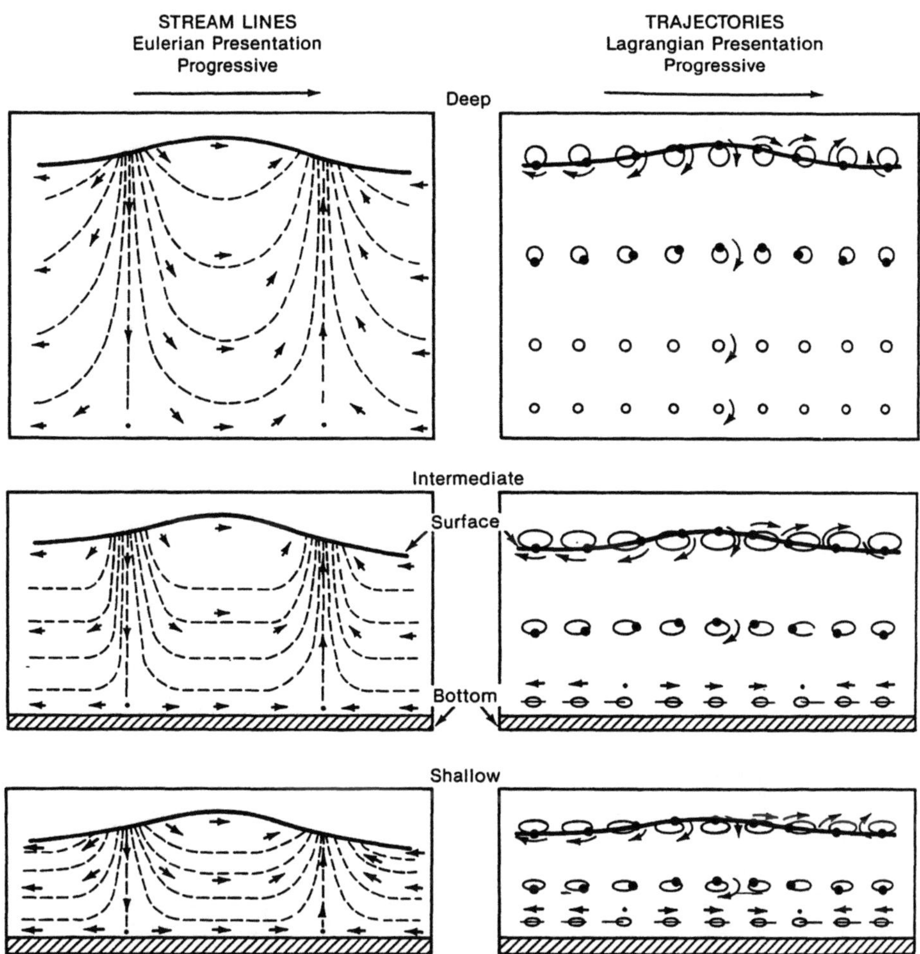

Figure 2.5. Water particle paths (Lagrangian viewpoint) and streamlines (Eulerian viewpoint) for a sinusoidal gravity wave. Source: after Kinsman (41)

Table 2.2. Summary of wave characteristics from linear (Airy) wave theory.-Source: CERC (13).

RELATIVE DEPTH	SHALLOW WATER $\frac{d}{L} < \frac{1}{25}$	TRANSITIONAL WATER $\frac{1}{25} < \frac{d}{L} < \frac{1}{2}$	DEEP WATER $\frac{d}{L} > \frac{1}{2}$
1. Wave profile	Same As $\longrightarrow$	$\eta = \frac{H}{2}\cos\left[\frac{2\pi x}{L} - \frac{2\pi t}{T}\right] = \frac{H}{2}\cos\theta$	Same As $\longleftarrow$
2. Wave celerity	$C = \frac{L}{T} = \sqrt{gd}$	$C = \frac{L}{T} = \frac{gT}{2\pi}\tanh\left(\frac{2\pi d}{L}\right)$	$C = C_0 = \frac{L}{T} = \frac{gT}{2\pi}$
3. Wavelength	$L = T\sqrt{gd} = CT$	$L = \frac{gT^2}{2\pi}\tanh\left(\frac{2\pi d}{L}\right)$	$L = L_0 = \frac{gT^2}{2\pi} = C_0 T$
4. Group velocity	$C_g = C = \sqrt{gd}$	$C_g = nC = \frac{1}{2}\left[1 + \frac{4\pi d/L}{\sinh(4\pi d/L)}\right]\cdot C$	$C_g = \frac{1}{2}C = \frac{gT}{4\pi}$
5. Water Particle Velocity (a) Horizontal	$u = \frac{H}{2}\sqrt{\frac{g}{d}}\cos\theta$	$u = \frac{H}{2}\frac{gT}{L}\frac{\cosh[2\pi(z+d)/L]}{\cosh(2\pi d/L)}\cos\theta$	$u = \frac{\pi H}{T}e^{\frac{2\pi z}{L}}\cos\theta$
(b) Vertical	$w = \frac{H\pi}{T}\left(1 + \frac{z}{d}\right)\sin\theta$	$w = \frac{H}{2}\frac{gT}{L}\frac{\sinh[2\pi(z+d)/L]}{\cosh(2\pi d/L)}\sin\theta$	$w = \frac{\pi H}{T}e^{\frac{2\pi z}{L}}\sin\theta$
6. Water Particle Accelerations (a) Horizontal	$a_x = \frac{H\pi}{T}\sqrt{\frac{g}{d}}\sin\theta$	$a_x = \frac{g\pi H}{L}\frac{\cosh 2\pi(z+d)/L]}{\cosh(2\pi d/L)}\sin\theta$	$a_x = 2H\left(\frac{\pi}{T}\right)^2 e^{\frac{2\pi z}{L}}\sin\theta$
(b) Vertical	$a_z = -2H\left(\frac{\pi}{T}\right)^2\left(1 + \frac{z}{d}\right)\cos\theta$	$a_z = -\frac{g\pi H}{L}\frac{\sinh[2\pi(z+d)/L]}{\cosh(2\pi d/L)}\cos\theta$	$a_z = -2H\left(\frac{\pi}{T}\right)^2 e^{\frac{2\pi z}{L}}\cos\theta$
7. Water Particle Displacements (a) Horizontal	$\xi = -\frac{HT}{4\pi}\sqrt{\frac{g}{d}}\sin\theta$	$\xi = -\frac{H}{2}\frac{\cosh 2\pi(z+d)/L]}{\sinh(2\pi d/L)}\sin\theta$	$\xi = -\frac{H}{2}e^{\frac{2\pi z}{L}}\sin\theta$
(b) Vertical	$\zeta = \frac{H}{2}\left(1 + \frac{z}{d}\right)\cos\theta$	$\zeta = \frac{H}{2}\frac{\sinh[2\pi(z+d)/L]}{\sinh(2\pi d/L)}\cos\theta$	$\zeta = \frac{H}{2}e^{\frac{2\pi z}{L}}\cos\theta$
8. Subsurface Pressure	$p = \rho g(\eta - z)$	$p = \rho g\eta\frac{\cosh[2\pi(z+d)/L]}{\cosh 2\pi d/L)} - \rho g z$	$p = \rho g\eta e^{\frac{2\pi z}{L}} - \rho g z$

Although waves move such small amounts of water that the Coriolis effect is negligible, they effectively transmit energy from winds to the nearshore areas. Gravity wave energy (see Figure 2.4) ranges from 1.4 megawatts/kilometer (mw/km) during relative calms with 0.6 m waves to 40 mw/km during storms. Wave heights during storms depend upon wind velocity, duration, and fetch (distance) and upon interference with other waves or opposing currents. Wave heights during storms in the open ocean have reported at up to 35m (4), Waves move more quickly over long distances than do the storms that created them and are forced by gravity to lower heights and longer lengths known as swell.

The small numbers of water particles that ride the crests of waves up onto the shore build up sufficient height (head) to create rip currents with velocities up to 2 m/s in small embayments creating hazards to swimmers in the surf zone. Rips can be recognized from above by their calming effect on breakers.

As waves enter depths less than about half their length, their energy remains constant. Their speed and length decrease, their steepness and height increase, and, when the water depth is about four-thirds to twice their height, the waves begin to break. Here, spring dynamometers have measured pressures up to 70 tons/m^2 to which must be added wave momentum (16, 41).

Along the shoreline, winter storm surf removes beach and nearshore bottom sand and stores it further offshore from where it is returned during the summer months. Unusually heavy storm surf has broken off and moved reinforced concrete harbor sections weighing over 1200 tons. Bigelow and Edmondson (4) and Kinsman (41) report that wave action has been known to move rocks weighing over 100 kg at depths of 30 m , and that even much less noticeable swells have moved 0.5 kg rocks into lobster pots at depths of 55 m.

Wave refraction. Nearshore waves respond to changes in depth due to sudden shoaling or submarine canyons. Since shallow wave speed is controlled by depth, the alignment of the crests along orthoganals and wave direction along rays that approach the coast are changed (refracted). Wave energy continues to be conserved, so that concentration or dispersion of the rays result in increased or decreased shoreline erosion, respectively. The long-term effect of wave refraction and littoral drift is to approach equilibrium by eroding headlands and filling in embayments

Long waves.   Long waves include standing waves or seiches generated by tidal and other forces in   bathtubs, lakes, embayments , canals, straits, estuaries, harbors, and larger basins.   Water levels are constant at nodes and variable at antinodes. Water motion is maximum at nodes and minimum at antinodes.

For a closed basin with a single node, the period is $T = 2L/\sqrt{gd}$. (2.5)

For a basin open to the sea, where the node is at the open end, $T = 4L/\sqrt{gd}$ (2.6)

For a canal, channel, or strait, and a node at each end, $T = (\sqrt{gd})?L$ (2.7)

Seiches due to heavy winds have been observed on open coasts such as the Atlantic between Cape Hatteras and Nantucket Island, a distance of some 900 km, with a period of about 15 minutes (87).

Tides. Ocean tides are long waves due to gravitational forces, angular velocities of the earth relative to the stars, the moon around the earth, the long axis of the elliptic of the moon, and the earth around the sun, all modified by boundary geometry and latitude. They may be predominantly diurnal, semidiurnal, or mixed

Table 2.3. Tidal periods

Semidiuranl	hours	Diurnal	hours	Long-period	hours
Principal lunar	12.42	Luni-solar	23.93	Lunar fortnightly	227.86
Principal solar	12.00	Principal lunar	25.82	Lunar monthly	661.30
Larger lunar elliptic	12.66	Principal solar	24.07	Lunar semi-annual	2191.43
Luni-solar	11.97				

Source: adapted from Sverdrup, et al (65)

and are modified by ocean and land geometry and latitude as shown on Annex A, Figure A-8. Table 2.3 lists semidiurnal, diurnal, and longer period tides each in order of relative importance from harmonic analysis of tide-produced forces. The differences are revealed in 2-hour current measurements over two months. Where channels narrow or slopes increase, tidal wave height and speed ($c=\sqrt{gh}$) increase to form tidal bores notably in the Amazon, Colorado, Elbe, and Yangtze Rivers.

Internal waves. Like surface waves, internal waves are formed by disturbances at the interface of two dissimilar fluids. For the same wave energy, the height of an internal wave will be greater than that of a surface wave by the ratio of the density differences across the interface, $\rho_o/(\rho_d - \rho_o)$. Internal waves with tidal periods and heights to 120 m affect measurements being used for diffuser system design. Internal waves along the thermocline shoal, break, surge, and form bores when they contact the shoreline and trigger turbidity currents that remove solids from continental slopes 84) and outfall areas (see section 2.2.4).   Energy from small vessel propellers at 1 or 2 meters can be transformed to internal waves along a diurnal thermocline to create a condition called "sticky water." During the late summer in nearshore water depths of 50 to 100 m. when the seasonal thermocline at, say 20-30 m depth is at its strongest. for an upper layer $\sigma_t = 24.0$, the lower layer $\sigma_t = 26.0$ ($\rho=1.0240$ and 1.0260), the density difference is 0.002. Internal waves are damped vertically to a few centimeters at the surface, and to zero at the bottom where they form internal tidal bores along the shore (65, 97).

Tsunamis. These highly destructive long waves with periods of 15 to 60 minutes and speeds of 700 to 800 km/h are caused by volcanic explosions.or by submarine landslides associated with earthquakes around the Pacific Rim. Their location marks the collision of oceanic and continental tectonic plates in which one plate is slowly subducted and buried for geologic time scales This had lead to proposals for placing nuclear wastes in areas where they will be subducted and sequestered long enough for their radioactivity to be reduced to background levels.

The Lisbon earthquake of November 1, 1755, for example, caused a locally destructive wave that reached a height of 12 m (40 ft) and that continued across the Atlantic and reached the West Indies with a still disastrous height of 4 to 6 m (65). In 1958, a rockfall triggered by an earthquake caused a local surge up to 530 m on the opposite side of Lituya Bay, Alaska (59). Volcanic explosions that were responsible for the formation of the caldera at Thera (Santorini) in the Aegean Sea during the thirteenth or fourteenth century BC. are credited with causing a major and perhaps fatal disruption of Minoan civilization, the Exodus, and a series of destructive waves from Minoan Crete to the Phoenician ports of the eastern Mediterranean. A similar but smaller explosion on Mount Krakatoa in Indonesia in 1883 caused a destructive wave that traveled from the Sunda Strait across the Indian Ocean into the Atlantic Ocean as far as the English Channel (44, 51, 65).

Sea level. Long-term changes in sea level relative to land level due to global warming, isostatic adjustment, and tectonic movement are shown in Figure 2.6, which shows the short-term variability in the record, and Table 2.4 which shows the spatial variability. The table includes 35 from the list of 635 selected by Emery and Aubrey (80) for their elucidation of global geology; it contains a greater proportion of developing country stations. Record lengths range from 7 to 114 years. Negative sea level changes in Scandinavia are due to isostatic rebound following melting of ice sheets. The largest vertical displacements, both positive and negative, are in Japan and are caused by seismic and volcanic activity. Other major subsidence is due to compaction of deltaic muds or to ground water or petroleum extraction. Collectively, the records show that the land-ocean boundaries are unstable, which is after all the reason that coastlines are where they are

Both long-term tide-gauge measurements and archaeological records are used in coastal engineering. Many of these displacements vary widely over short distances. According to Fleming (24), archaeological evidence from ancient harbors indicates that the average uplift at the western end of Crete over the last 2,000 years has been 50 cm per century and average sinking has been 20 cm per century at the eastern end. Such movements are large enough to affect gravity sewers and interceptors with design periods of over 20 years.

There is an international scientific consensus that global warming first predicted by Arrhenius in 1896 (77), due to the greenhouse effect of carbon dioxide from fossil fuel combustion has increased surface temperatures by about $0.5^{\circ}C$ (91). Earlier increases are generally synchronic with the industrial revolution, and recent ones began about 1950 when global energy use started to escalate. Different estimates have been made of a corresponding sea level rise due to thermal expansion of the upper ocean water layer. For design of infrastructures a conservative estimate is 30 cm/century sea level due to global warming that is expected to increase as fossil fuel use continues (80). The effects of these movements are magnified by slopes of deltaic or coastal plain areas.   Pilkey (55) has calculated that each meter of sea level rise along the U.S. East Coast will result in a 100 to 10,000 (or average 1,000) meter shoreward advance of the surf zone.   Similar ratios apply to other mid-latitude coastal plains within about 20 km of the shoreline. One design criterion is to assume flooding for structures within 1m of mean higher high water (MHHW) (81).

The design wave. Particle movements in gravity waves that move bottom sediments are determined from observations at specific points (Eulerian) or from their trajectory (Lagrangian) (see Figure 2.5). The datum for the design wave includes effects of tides and storm surges due to wind set-up and lowered

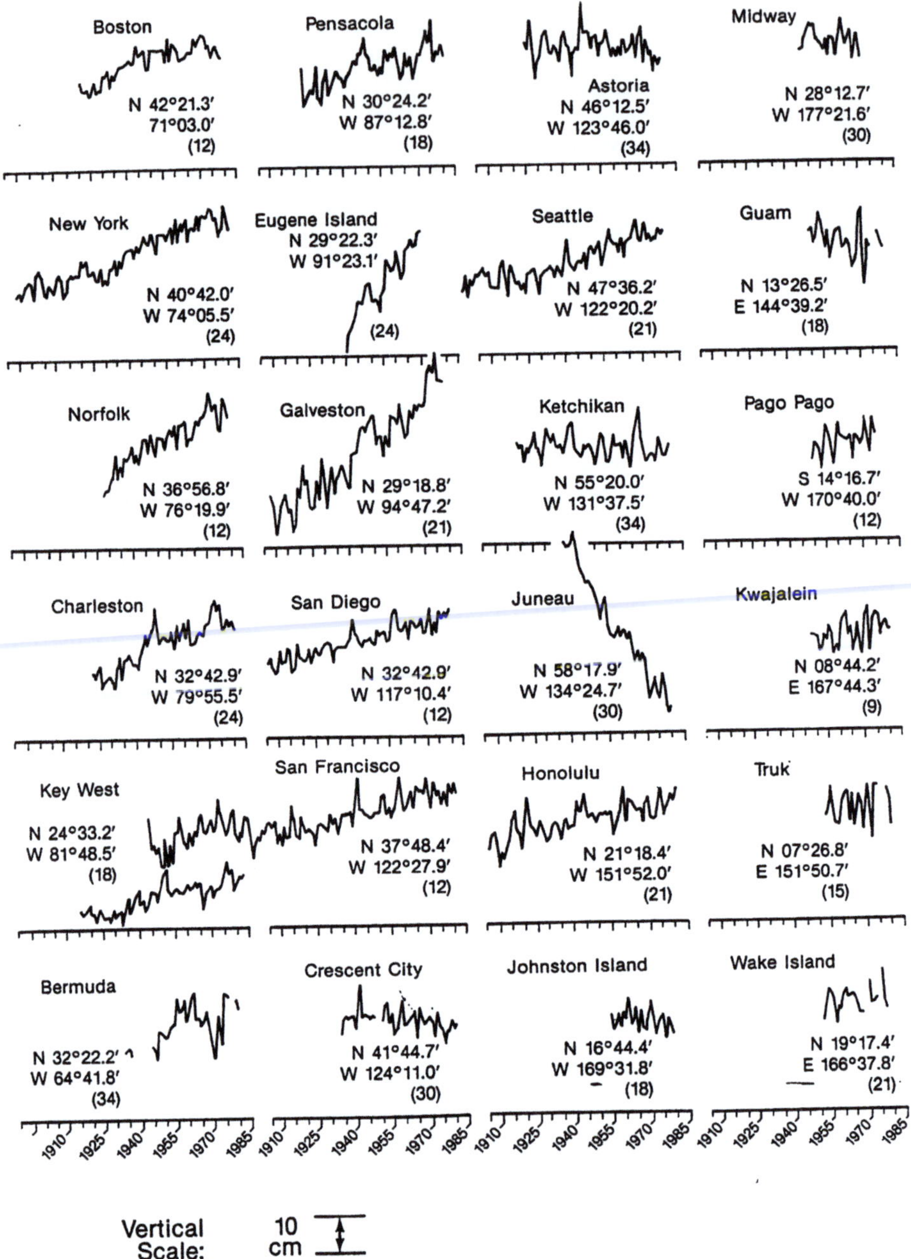

Figure 2.6. Mean annual sea level changes for U.S. tide gauge stations for period of record from 1850. Source: adapted from Hicks (35).

Table 2.4. Mean annual sea level changes for global tide gauge stations

Country and Tide Station	Latitude	Longitude	Record years	Sea Level Change, mm/y	t-confidence
Argentina, B. Aires	34°36'S	58°22'W	25	1.4	0.93
Australia, Sydney	33°51'S	151°14'E	90	0.6	1.00
Brazil, Recife	08°03'S	34°52'W	21	- 0.2	0.98
Canada, Vancouver	49°17'N	123°07'W	54	0.0	1.00
Halifax	44°40'N	63°35'W	62	3.7	1.00
China, Huludao	44°40'N	121°00'E	13	5.5	0.76
Shanghai	31°21'N	121°30"E	23	2.6	0.89
Wei Chou Tao	21°03'N	109°08'E	13	11.4	0.90
Chile, Valparaiso	33°02'S	71°38'W	13	1.7	0.88
Cuba, Guantanamo	19°54'N	75°09'W	31	1.9	1.00
France, Brest	48°23'N	04°30'W	114	0.7	1.00
Gibralter, Gibraltar	37°07'N	05°21'W	27	- 1.7	0.93
Haiti, Port au Prince	18°34'N	72°21'W	13	12.3	0.89
India, Bombay	18°55'N	72°50'E	104	0.9	1.00
Calcutta	22°33'N	88°18'E	50	6.9	0.97
Cochin	09°58'N	76°15'E	43	2.0	1.00
Indonesia, Jakarta	06°06'N	106°54'E	7	8.7	<0.8
Japan- Hokkaido	44°21'N	143°22'E	24	1.1	0.98
- Honshu	39°16'N	141°54'E	12	15.7	0.80
- Honshu	34°41'N	135°11'E	27	6.4	0.90
- Honshu	32°42'N	128°51'E	21	- 87.9	0.53
Korea, Pusan	35°06'N	129°02'E	26	0.6	1.00
Mexico, Veracruz	19°11'N	96°01'W	33	1.6	1.00
Myanmar, Rangoon	16°46'N	96°10'E	35	3.4	0.97
Namibia,Walvis Bay	22°57'S	14°30'E	22	2.2	0.95
Norway, Norvik	64°32'N	11°15'E	14	- 113.7	0.51
Pakistan, Karachi	24°48'N	66°58'E	17	0.3	1.00
Panama, Balboa	08°58'N	79°34'W	62	1.6	1.00
Philippines,Manila	14°35'N	120°58'E	73	5.1	1.00
Senegal,Dakar	14°40'N	17°33'N	15	1.4	0.99
Thailand, Frachula	13°33'N	100°35'E	45	11.5	0.99
Turkey, Antalya	36°53'N	30°42'E	36	- 3.8	0.95
U.K., Aberdeen	57°09'N	02°05'W	104	0.6	1.00
U.S.A., Charleston	33°47'N	79°56'W	67	3.4	1.00
San Francisco	37°48'N	122°28'W	114	1.1	1.00

Source: Emery and Aubrey (78). Note: t-confidence values from eigenanalysis of spatial and temporal variability between gauge locations and record lengths.

atmospheric pressure at the center of major storms. Where historical records exist, they can be used to validate a design wave based on predicted conditions. Table 2.5 indicates that total ranges of water surface elevations, including waves, depend upon both latitude and ocean boundary orientation.

Table 2.5. Maximum recorded departures from mean high water and mean low water along U.S. coastlines.

Location	Meters above MHW	Meters below MLW
North Atlantic (Maine - No. Carolina)	1.04 to 3.99	0.52 to 2.04
South Atlantic (So. Carolina to Florida)	0.91 to 2.35	0.49 to 1.34
Gulf Coast (Florida to Texas)	0.67 to 3.08	0.46 to 1.62
Pacific Coast (California to Washington)	0.46 to 0.88	0.79 to 1.43
Alaska	1.01 to 5.09	0.73 to 1.83

Source: CERC (13).

The accepted approach to selecting a design wave for coastal engineering works at present is to base it on predicted rather than historical records since the latter require observers to be present at the time. An authoritative working basis for design is the set of simplifying assumptions and worked examples using linear wave theory and the storm surge predictions of the U.S. Army Corps of Engineers and the National Oceanic and Atmospheric Administration, respectively (13).

2.1.4 Marine Geology and Sedimentation

The principal bathymetric features of the continental border and ocean floor are shown in Figure 2.7 from Smith and Brown (62). Liquid wastes and sludges discharged onto the continental shelf or barged further offshore are integrated into the hydrographic and sedimentary regimes of the sea.

Approximately 70 percent of continental shelf areas consist of terrestrial deposits laid down between 10,000 and 25,000 years ago on river and coastal plains (10). These deposits were subsequently overlain by more recent marine deposits, most of which are river borne bedload and suspended sediments that flocculate, settle, and add to the bedload when the salinity reaches about 4 parts per thousand ($\permil$) in estuaries (72). The usually polluted bedload is dredged from navigation channels and harbors and is often dumped offshore together with sewage sludge, building and construction debris, and industrial wastes.

Occasionally, bottom sediments of nepheloid suspensions accumulate forming unstable slopes that slump under seismic or even static conditions to forms a fast-moving turbidity current along the bottom. Along the continental slope, these heavy, sediment-laden flows can reach velocities of some 90 km/h, and are credited with creating and sustaining submarine canyons. They also carry suspended sediments from rivers down slope along the bottom or spreading in a water layer of the same density (30, 37, 69).

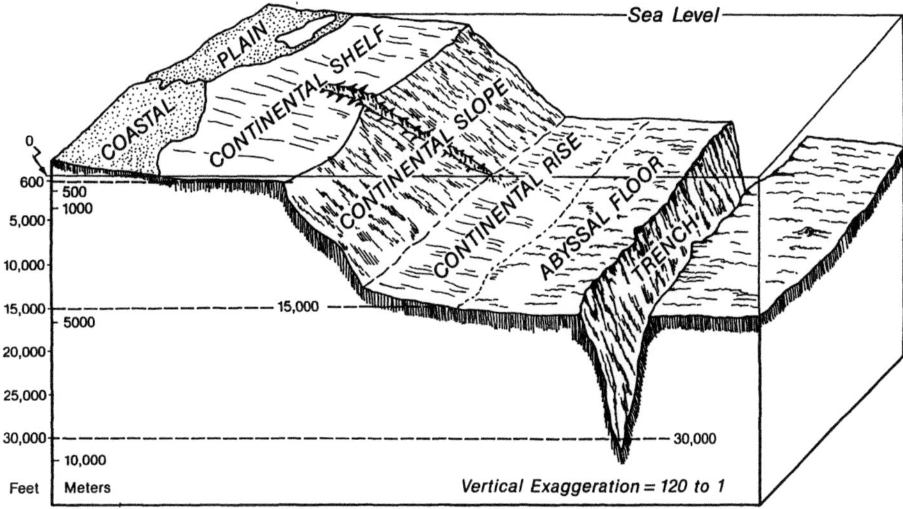

Figure 2.7. Nomenclature of continental borderland and ocean floor. Source: Note vertical exaggeration. Gradients are less than one degree for the continental shelf about four degrees for the slope, and somewhat less for the continental rise. Smith and brown (42)).

The seafloor is composed primarily of unconsolidated sediments of organic and inorganic origin. The inorganic fraction of continental shelf sediments originates predominantly from the land and is introduced by a variety of means: rivers and streams carrying both particulate and dissolved material; sheet runoff; shoreline erosion; glaciers and sea ice; winds; volcanoes; and biological activity (65).

There are three classifications of natural marine particles: lithogenous, biogenic, and authigenic. Lithogenous particles are those derived from the weathering of rock and soil. These may be sand and silt sediments formed by the disintegration and mechanical breakdown of rock into smaller fragments, or they may result from the chemical decomposition of rock particles by air and water. Biogenic particles are derived from marine plants and animals. Various organisms remove calcium carbonate and silica from seawater to build their skeletons which, upon the organism's death, become the principal constituents in sediments known as oozes, defined as sediment containing more than 30% biogenous constituents by volume (29). Insoluble fragments of bones, teeth, and shells account for the larger particles in biogenic sediments. Authigenic minerals such as manganese or phosphorite particles or nodules are formed by chemical precipitation that occurs in seawater or within the interstitial waters of sediments. Flocculation, which occurs when very fine clay particles come in contact with the dissolved salts in seawater, is responsible for much of the sediment deposited near river mouths.

Sediments are generally classified according to their grain size either as clay, silt, sand, gravel, cobbles, or boulders. If their densities are uniform, size and shape determine where they will be found in the water column during transportation and the distance that they will travel before settling to the bottom. In addition, size determines the degree of sorting that can be accomplished by the physical processes involved in their transport, such as waves and currents. A well-sorted sediment such as beach sand contains particles of a limited size range since the other sizes have been removed, usually by physical means. Poorly sorted sediments indicate that little mechanical energy was available to separate the different sizes.

The distribution of sediments throughout the ocean is related to their origin and to currents. Terrestrial sediments accumulate near the continents and cover about 25 percent of the ocean bottom. The remainder of the ocean floor is covered by pelagic deposits consisting largely of biogenous particles. Most land-derived sediments are not transported to deep ocean basins. Sewage effluents contain organic particles of varying density and size, mostly with low settling velocities. Effluents from different sources and treatment plants can vary substantially in their distribution of particle settling rates (see Chapter 3). The speed at which particles settle depends on their specific gravity, size, and shape, and on the specific gravity and viscosity of the water (65). Occasionally, bottom sediments of nepheloid suspensions accumulate forming unstable slopes that slump under seismic or even static conditions to forms a fast-moving turbidity current along the bottom. Along the continental slope, these heavy, sediment-laden flows can reach velocities of some 90 km/h, and are credited

Waves and currents are the primary mechanisms for sediment transport in coastal waters, either in suspension or as bed load carried by traction (rolling and sliding) and saltation (bouncing). When current velocities are no longer able to maintain the particles in suspension, sedimentation takes place. Silt and clay will occur in suspension near the seafloor, with larger particles near the bottom, depending upon the current velocity and the roughness of the bottom (84, 85).

Land-derived sediments are introduced both by runoff and by wave action along the coasts, which causes slumping, sliding, grinding, and breaking of rock along the shoreline. Durable rocks are slowly eroded to sedimentary material, whereas sandstones and glacial drift are readily eroded in large quantities. Along limestone coasts, the skeletal remains of calcareous marine organisms abound.

Coarse sediment is dragged out to sea by rip currents, backwash, and undertow and is deposited in shallow depths offshore. As each wave crest passes a given reference line on the bottom, the forward drag of the wave crest moves sand grains shoreward; with each passing wave trough, sand is dragged seaward (see Figure 2.5 and Table 2.2). This back-and-forth motion causes sand ripples to be formed.

The degree of erosion (scour) and deposition (fill) of sediments near river mouths varies according to season. As on land, scouring and flushing are more important during increasing discharge; filling dominates during decreasing flows. Offshore deposition of sediments occurs most often after storms, whereas inshore deposition is most common during periods of low wave energy.

Seafloor topography is another factor. Topographic highs are marked by coarser material transported by waves and currents. Depressions and basins contain fine-grained material from the continental shelf to deeper water (65, 84).

Beaches, Tidal Flats, and Sediment Movement. Figure 2.4 shows that the energy in gravity waves in the open ocean or along the open coast greatly exceeds that in tides.. Gravity waves are attenuated when they enter relatively enclosed areas formed by offshore bars or breakwaters. Over time, the sediments in these areas respond to the dominant force. Beaches are found where wave energy exceeds tidal energy, and tidal flats are formed when tidal energy exceeds wave energy.

After deposition, sediments can be resuspended by wave action or tidal currents with competent velocities as shown on Figure 2.8. For preliminary design purposes, rough estimates of these velocities can be made from grain size distributions of the sediments (16, 63, 69).

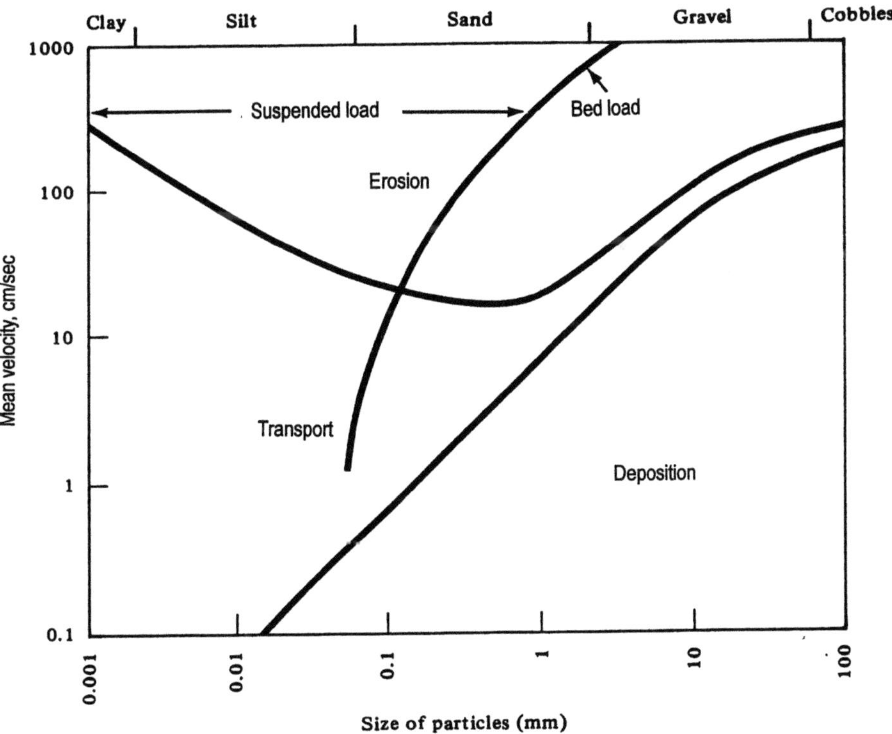

Figure 2.8. Effects of Current Velocity on Sediments. Source: adapted from Hjulstrøm, in Trask (69) and Sternberg (63)

2.1.5   Chemistry of Seawater and Suspended Sediments

Seawater is a thin soup of elements, organic and inorganic ions and compounds, and dissolved gases. Total dissolved inorganic salts (salinity) are conventionally expressed in parts per thousand, ‰ or ppt. Prior to the mid-1960s, salinity was calculated from its chlorinity determined by $AgNO_3$ titration against a standard seawater sample with Cl = 19.386‰ where S, ‰, = 1.8050 x Cl ‰ + 0.03, which is still the formal definition of salinity. Since then, salinity has been determined from the ratio of sample conductivity at 15° C and 1 atmosphere and a standard 32.4356‰ KCl solution. The precision of the measurements is an expression of the constancy of seawater composition of conservative elements, meaning those which are not involved in biological processes. Standard seawater has a salinity of 35.00 ‰, which is typical for the open ocean. Locally, salinity varies from a few parts per thousand near major rivers or melting ice to about 37 ‰ in the eastern Mediterranean and Aegean seas and 40 ‰ in the Red Sea.

Seawater density (ρ) is shown in Figure 2.1 as a nonlinear function of its salinity (S) and temperature (T), and is conventionally expressed as sigma-T, where for ρ= 1.02468, σ_t= 24.68. (See section 2.2.1 and Figure 2.1 and nomographs for determining density are in Appendix D.)

Carbon dioxide, oxygen, and nitrogen are the principal dissolved gases in seawater in concentrations according to their partial pressures in the atmosphere due to constant stirring and mixing of winds and waves at the air-sea interface. Temperature and salinity limit the amount of gas that can be dissolved in surface seawater: when either of these increase, the amount of gas that can be dissolved decreases. Ocean surface waters above the oxygen compensation depth at the bottom of the photic zone are supersaturated with oxygen (82). Standard Methods (1) lists oxygen solubility values from 0 to 50° C and 0 to 20‰ Cl (36 ‰ S).

Seawater contains trace amounts of the majority of elements known to man (see Figure 2.8 and Table 2.6). Although most elements occur in very small concentrations in seawater, these concentrations can vary, depending on natural and cultural inputs. The residence times of different elements in the water column depend upon reactions between the element and other physical, chemical, and biological factors in the marine environment.

The dissolved organic matter in seawater comes from decaying plants and animals and from alimentary wastes from animals, people, cities, agriculture, and industries. Dissolved organics are present in seawater in small but variable concentrations, ranging from 0 to 6 mg/l (58). Concentrations are higher in areas of high biological productivity or waste discharges. Dissolved organic matter includes organic carbon, carbohydrates, amino acids, organic acids, proteins, vitamins, and nutrients (nitrogen and phosphorus), which are later oxidized.

Suspended particulate matter in seawater includes organic detritus, some complexes of organic material, and fine-grained minerals. The amount of suspended material in any particular location varies greatly because it is influenced by local geography, biological productivity, and atmospheric conditions. Suspended materials are unevenly distributed throughout the water column, but are especially prevalent at the surface layer. Essentially all of the heavy metals, toxic hydrocarbons, and microorganisms associated with waste disposal are adsorbed onto the suspended solids. Very fine sand, silt, clays and the bits of organic detritus that

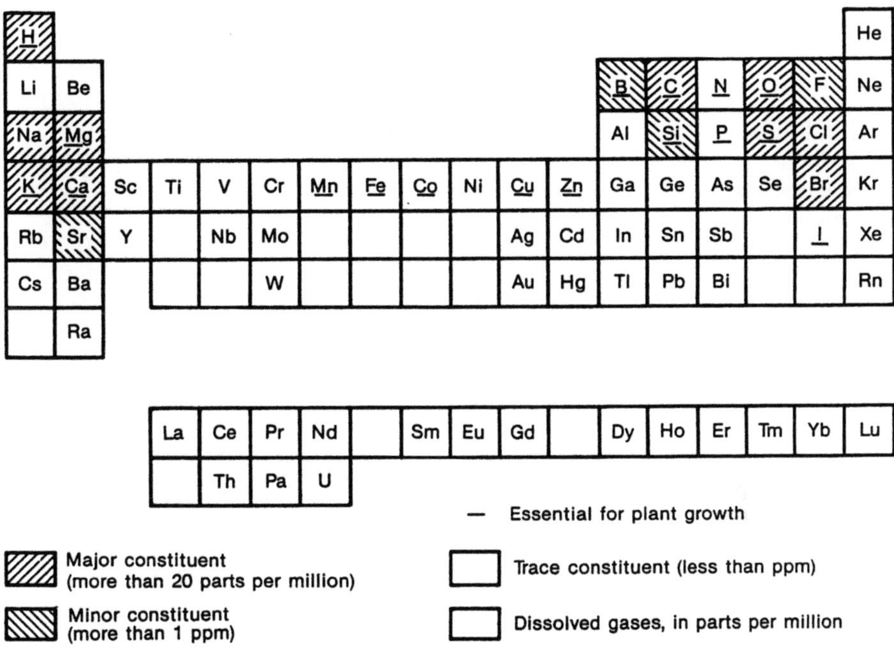

Figure 2.9. Elements detected in sea salts. Most non-conservative elements (underlined) are involved in biological processes. Source: Gross (29)

sink to the ocean floor as marine snow often remain suspended near the bottom in a nepheloid layer where they are occasionally disturbed by abyssal storms (83).

Seawater also contains precipitated materials, principally calcium carbonate, which tends to precipitate when salinity and temperature increase and carbon dioxide content decreases. This reaction is most likely to occur in areas of active photosynthesis and high or rising temperatures, as in coastal tropical waters. Conversely, low photosynthesis and low temperatures favor the solution of these elements. Other precipitation reactions are less common.

Chemical interactions that occur between the seawater and settled or suspended sediments can be summarized as follows: (1) dissolution of certain constituents of the sedimentary particles; (2) adsorption of seawater constituents onto the sediments; (3) ionic exchange; and (4) reactions that form new substances.

Local variations in the minor and trace elements and organic chemical composition of coastal ocean waters depend upon stream runoff from weathered rocks and soils, plant and animal life, and anthropogenic inputs. Over time, oceanic averages are more important. Dissolved nutrient elements are taken up by

Table 2.6. Constituents in solution in ocean water at a salinity of 35.00 ‰

	Grams per kilogram,	Grams/liter at 20° C (sp. g. 1.025)
Total salts	35.1	36.0
Sodium	10.77	11.1
Magnesium	1.30	1.33
Calcium	0.409	0.42
Potassium	0.388	0.39
Strontium	0.010	0.01
Chloride	19.37	19.8
Sulfate as SO_4	2.71	2.76
Bromide	0.065	0.066
Boric acid as H_3BO_3	0.026	0.026
Carbon:		
Present as bicarbonate, carbonate, and molecular carbon dioxide	0.023 g at pH 8.4 0.025 g at pH 8.2 0.026 g at pH 8.0 0.027 g at pH 7.8	
As dissolved organic matter	0.001 - 0.0025 g	
Oxygen (where in equilibrium with the atmosphere at 15° C)	0.008 g. = 5.8 cm^3 per 1	
Nitrogen (where in equilibrium with the atmosphere at 15° C)	0.013 g. = 10.5 cm3 N2 per 1 + 0.28 cm^3 argon, etc.	
Other elements	0.005	

Note: Calculated from chlorinity = 19.37 ‰ and S = 1.805 Cl - 0.03. Essentially all of the heavy metals, toxic hydrocarbons, and microorganisms associated with waste disposal are adsorbed onto the suspended solids. Very fine sand, silt, and clays are often suspended in a nepheloid layer near the bottom. Source: Harvey (34). See also Sections 2.3.4, 3.2.2 and 3.3.3 for more information on mercury..

mixed marine plankton communities consisting of both phytoplankton and zoo plankton in weight ratios of approximately 41:7:1 and molar ratios of 105:15:1 (the Redfield ratio) for organic C:N:P, respectively (65, 85, 88). Carbon may be locally limiting but in the sea is available as bicarbonate. Calcium and dissolved silica for skeletal growth are available in concentrations that are consistent with those in deep ocean sediments. The molar N:P ratio in mixed plankton is essentially the same as that for dissolved nitrate and phosphate in sea water, indicating dynamic equilibrium conditions that support arguments for the Gaia hypothesis of a self-regenerating earth.

2.1.6   Stirring and Mixing

Eckert's (18) explanation of dilution and dispersion in the sea--which he compares to adding cream to a cup of coffee--is particularly applicable to ocean outfalls. Three stages are observed when one fluid (cream) is introduced into another fluid of differing density (coffee). In the first stage, large volumes of the two fluids are distinctly visible, and there is a sharp gradient at the interface. If no motion is introduced, the boundaries persist for some time. The concentration gradient is high but, since the interfacial area is small, the average concentration gradient throughout the cup is small. The second, or intermediate, stage begins with stirring. The two fluids are distorted and the interfacial areas are increased. Since the concentration gradients are still visible and high, the average concentration gradient throughout the cup is high. In the final stage, the concentration gradient disappears almost instantaneously and reduces to zero. This is mixing.

Space and time scales for stirring and mixing in the ocean are much larger than those in a coffee cup. Nonetheless, Eckert's coffee cup gives an idea of the difference between near-field (mostly advective stirring) and far field dispersion and diffusion phenomena (mostly mixing that extends down to the molecular scale). Density stratification and vertical stability in the sea being what they are (see Figure 2.1), horizontal diffusion coefficients are many orders of magnitude greater than vertical ones, except near an outfall where stirring takes place and establishes the thickness and vertical stability of the initial sewage field. These factors are discussed in Chapter 4, along with examples of outfall design and calculations whose complexity depends upon the assumptions used. However, for small open-end outfalls of, say, less than 1 m diameter, Brooks' analysis (8) shows that initial (stirring) dilution is essentially the depth divided by the diameter. Subsequent dilution takes place slowly, and site-specific rates for the disappearance of bacteria and other non-conservative constituents become the principal design parameters. Costs are always a factor, particularly in developing countries where domestic and industrial wastewaters along with drainage reach the shoreline through natural channels where they are dispersed by stirring in the surf zone.

In recent years, our understanding of and ability to predict physical dilution processes in the coastal and shallower areas of the sea have improved considerably. Garret (27) points out that empirical observations and dimensional analysis of mixing in the English Channel the Gulf of Maine, and the Bay of Fundy reveal that a critical value for H/U^3 (where H is the depth and U the velocity) of about 70 $m^{-2}s^3$ and its equivalent log(depth/tidal dissipation) $\approx$ 1.9 marks the transition between well-stratified and vertically mixed conditions. The H/U^3 parameter, originally suggested by Simpson and Hunter (61), appears to be applicable to waste discharge problems in estuarine and coastal waters, although the numerical threshold values are expensively determined to be site-specific. This specificity, part of that discussed in Section 3.2.1, has most recently been confirmed in connection with studies of the effects of currents and stratification throughout the water column on Honolulu effluent plumes in Mamala Bay (90) These studies included measuring currents by acoustic Doppler current profilers at vertical intervals of 2 meters and time intervals of 30 minutes. It was found that the principal variance in the record was due to tidal currents parallel to the bottom contours as to be expected from Figure 2.2. Maximum currents were 30 cm/s.

2.1.7   Bays, Estuaries, and Straits

A bay is ".. a well-marked indentation whose penetration is in such proportion to the width of its mouth as to contain land-locked waters and constitute more than a mere curvature of the coast . . (whose) . . area is as large as, or larger than, that of the semicircle whose diameter is a line drawn across the mouth of that indentation." (70, 86). For defining territorial waters, the closing line across the mouth of the bay cannot be more than 24 (nautical) miles

Where bays are isolated from the main body of ocean water, they provide a degree of protection from coastal storms. Thus, over the centuries, people have chosen to settle around bay shores. The size of these settlements of from small waterfront communities to megacities with over ten million people, depends on a bay's aesthetic appeal and economic resources. The physical and hydrographic processes that occur in bays are generally less dynamic than those along open coastlines. Bays are usually characterized by restricted flushing, low current velocities, and high rates of sedimentation. They retain dissolved and suspended solids longer and are less suitable than coastal waters for receiving wastes.

Pritchard (56) defines an estuary as a semi-enclosed coastal body of water that has a free connection with the open sea and within which seawater is diluted with freshwater derived from land drainage. Estuaries are the tidal mouths of rivers where ports are located. Since they are typically areas of high biological productivity, estuaries usually sustain substantial fisheries.

Estuarine conditions vary seasonally, fortnightly, and diurnally (2) particularly with respect to salinity. Whereas salinities in the open ocean average about 35‰, salinities within most estuaries tend to decrease steadily as one moves inland, the highest salinity being about 30‰ at the estuary mouth. The upper limit, or head, of an estuary lies where salinity drops below about 0.1‰. This area is characterized by a short transition zone in which the ratios of the major dissolved constituents change rapidly and chlorinity (the most common measure of salinity) drops to about 0.01‰. Although sea salts no longer influence the freshwater flows beyond this point, rivers are affected by tides, most notably when river outflows, tidal ranges, and hydraulic geometry of the stream channel combine in the upstream movement of a steep, destructive wave or (tidal) bore (16).

Estuaries are classified according to their water budgets, their geomorphology (drowned river valley, fjord, tectonic, or bar-built), or their hydrography. Pritchard's (56) hydrographic classification into (1) salt wedge, (2) partially stratified, or (3) vertically mixed estuaries is as useful as any. Their vertical structures, averaged and idealized for purposes of numerical modeling (see Section 4.4), are shown in Figure 2.10. In real estuaries, where no two successive tidal cycles are alike, flows are modified by Coriolis effect and tidal pumping around islands (22), and current and salinity distributions are both seasonally and site-specific (see Appendix D).

Casual observations and interpretations of net downstream tidal·flows within estuaries have led many to recommend that wastes be released only during ebb tide. This doesn't work. Field and hydraulic model observations confirm the numerical models of advection and mixing in estuaries, where average concentrations are fixed by the tidal excursion and volume and by the total daily waste loadings. The effects of discharging twice the mass of wastewaters during half the time cannot be

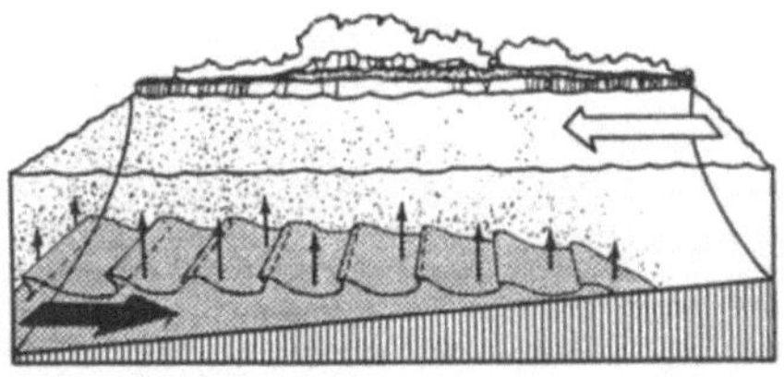

(1) Salt wedge estuary

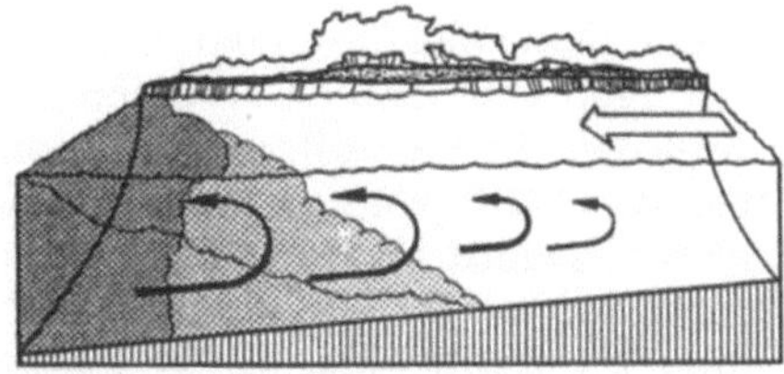

(3a) Vertically homogeneous estuary

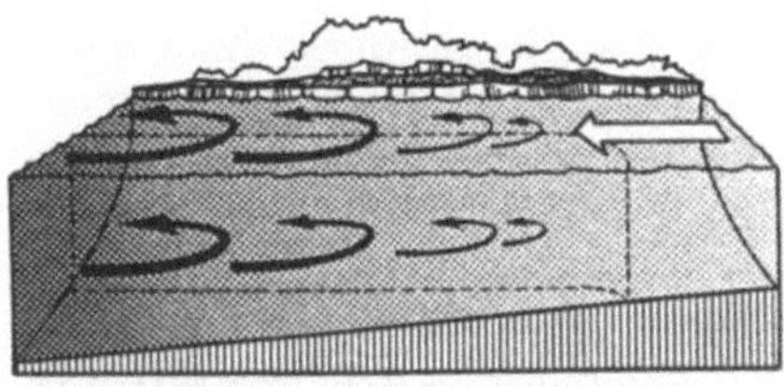

(2) Partially stratified estuary

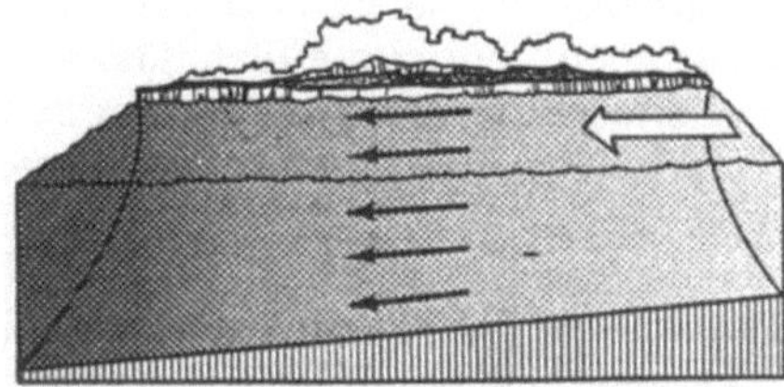

(3b) Sectionally homogeneous estuary

Figure 2.10. Hydrographic Classification of Estuaries. Source: Pritchard (56)

distinguished from the effects of a continuous discharge into an estuary. After all, the ocean also constitutes a discharge, this one at the mouth) (83)).

A salt wedge is formed in estuaries where the river flow dominates the circulation pattern. If friction were absent, the freshwater/saltwater interface would be horizontal and extend upriver to where the riverbed was at sea level. However, because there is a small amount of friction between the layers, the interface slopes slightly downward as the wedge points upstream. Where there is a steep density gradient at the interface, mixing between the layers is restricted (56).

If the velocity of the seaward-moving freshwater exceeds a certain value, internal waves form at the interface. As these waves break at their crests, saltwater is entrained into the upper (freshwater) layer. This is commonly known as a two-layer flow with entrainment, or a partially mixed estuary. In entrainment, saltwater mixes with the freshwater in the upper layer, but freshwater is not incorporated into the denser saltwater layer. In other words, the fluxes of both water and salinity are from the lower into the upper layer. Since the depth of the upper layer does not increase, its velocity is increased. A slow movement of the lower salt layer upstream compensates for the water lost by entrainment. The net flow of water is to the ocean, and most mixing of surface water with seawater occurs near the mouth of the estuary. Fjords and straits exhibit the same mixing characteristics.

In shallow estuaries, tidal currents often create vertical mixing that may extend throughout the water column. Even though salinity is relatively uniform vertically, there are still two layers separated by an area of no net motion between the seaward-flowing upper layer and the landward-flowing bottom layer. The marked salinity interface disappears and salinities continuously increase from the surface to the bottom, the greatest difference occurring near the level of no net motion. Stratification varies greatly in this type of estuary, depending on the magnitude of tidal currents relative to the strength of the river flow. If tidal currents are very strong compared with the river flow, an estuary will be vertically homogeneous with no measurable variation in salinity from top to bottom. The highest salinities will be at the mouth of the estuary, and decrease toward the head.

Salinities vary laterally in estuaries that are wide relative to their depth as a result of the Coriolis effect. The seaward flow is deflected to the right in the northern hemisphere. Thus average salinities on the right are lower and there is a compensatory flow of more saline waters on the left. Where the ratio of width to depth is relatively small, salinities may not vary greatly across the estuary; which then becomes laterally homogeneous. These are for time-averaged conditions; the fine structure can be complex with small eddies near the interface (32).

Another type of circulation may exist where a number of small estuaries are tributaries to a larger estuarine system. In Baltimore Harbor and Raritan Bay in the eastern United States, a three-layer pattern of flow occurs. Water moves inland both at the surface and in the bottom layers and moves seaward at mid-depth (32).

Biological attributes of estuaries. Estuaries are important for propagation, movement, and harvesting of marine and freshwater fish. Most estuarine organisms are euryhaline, tolerating a range of salinities. However, totally estuarine species cannot survive in either fresh or seawater (29, 58, 65).

Because of their limited mobility, the benthic communities in estuaries must be able to survive large diurnal and seasonal hydrographic changes. Generally, the number of species declines to a minimum as salinities decrease to approximately 5‰ and then rise with increasing freshwater. Nutrients in estuaries come from oceanic waters carried into the estuary; rainwater, soils in the drainage basin; and waste products. Nutrients are retained and ample food is available for the biota. Phytoplankton populations are maintained by nutrient availability and flow reversals that prevent their being washed out to sea. Zooplankton grazers in turn provide for larval and juvenile fish.

Eutrophication of natural waters is caused by excessive (overdose) enrichment by nitrogen and phosphorous compounds. This stimulates a sequence of (1) increases in algae and macrophytes, (2) oxygen supersaturation in the upper layer, (3) possible temporary enhancement followed by chronic deterioration of fisheries, and (4) deterioration of the physical water quality from overgrowth of plant life and reduced oxygen levels in the lower layers (43).

In estuaries with clear water and limited vertical mixing, most of the nutrients are in the sediments and used by the benthic communities. These estuaries may be considered biologically productive, but a lack of nutrients, especially carbon, in the water column often limits the productivity. In contrast, turbid estuaries tend to have greater vertical mixing and are more productive because excess nutrients are in solution or are suspended along with sediment, organic detritus, and plankton.

The standing stock of nutrients in estuarine waters depends on: (1) stream flow and quality, anthropogenic sources, rain, and influx of seawater; (2) metabolic processes of estuarine populations; and (3) the role of sediments as sources, reservoirs, and sinks of available nutrients (43).

The decomposition of plants whose growth was stimulated by eutrophication may give rise to oxygen deficiencies. More commonly, partial or total depletion of oxygen (anoxia) is caused by the biochemical and chemical oxygen demand of domestic sewage and industrial wastes Increased organic production may also be due to inputs of wastes containing inorganic nutrients. Although oxygen is supplied by reaeration and aquatic photosynthesis and consumed by respiration of pollution tolerant biota, respiration from consumers eating the primary biomass and from mineralizing bacteria living on decaying matter leads toward anoxia.

Straits and possibilities for environmental interventions. Straits connect adjacent seas with each other or the ocean. Surface salinities on either side of a strait reflect differences in precipitation, runoff, and evaporation, and hence density. Straits are similar to estuaries in that their salinities and flows are highly stratified, but they differ in that tides are less important than salinity differences in their stratification. Their flows are governed by the slopes of the surface and interface, modified by the Coriolis effect, which, Defant (16), describes as providing sufficient elevation to force the circulation,. This has been long known to submariners who could pass undetected within the Atlantic underflow along the southerly (right-hand) bank through the Strait of Gibraltar.

The Bosporus is the principal strait in which the two-layer flow has been considered important in waste disposal (see Chapter 15). Two-layer flows in the Strait of Gibraltar prompted Marchetti in 1979 to propose that stack gases from European power plants be discharged into the Mediterranean outflow which sinks and spreads in the Atlantic, in order to dispose of carbon dioxide (46) There would, of course, be problems of collecting and transporting the stack discharges and later of diffusing them into outflow. Similar schemes count on the already large holdings of CO_2 in surface waters where their residence time is 140 years.

These and other schemes in planetary engineering for ameliorating global warming were revived by the <u>Economist</u> (2/26/94). They would all be justified by economies of scale. Others included sprinkling iron particles over the southern ocean to enhance primary productivity, periodically using battleship guns to put the same amount of dust into the atmosphere as the 1991 Mount Pinatubo eruption to reduce insolation, and strip mines on the moon to supply aluminum foil in the stratosphere to reflect sunlight. Research would be useful. Other problems of limits to scale at the coastal city level are discussed in Chapter 16.

2.2   The Living Sea

The living sea shows the ability of living and non living matter to assume every possible form, and the capacity of life to adapt to food in any form that can provide for nutrition and reproduction (14).

Productivity depends upon temperature, salinity, and nutrients. These depend on latitudinal variations in climate (see references 5 and 38, and Annex Figures A3

32

through A-6). The influence of marine climate and of upwelling, Figure A-1, on the general geographic distribution of marine productivity is shown in Figure A-7.

Table 2.7 lists the seasonal control of the times of onset and duration of marine productivity rates and standing crops. Differences in the response of tropical and temperate ecosystems to these factors listed in Table 2.8 have led to increasing concern over the environmental effects of development on economically important tropical ecosystems whose specialization has limited their resilience.

Table 2.7. Duration of biological seasons. Shaded areas indicate periods of high plankton production. Source: Bogorov (5).

<table>
<tr><th>SEAS \ MONTHS</th><th>I</th><th>II</th><th>III</th><th>IV</th><th>V</th><th>VI</th><th>VII</th><th>VIII</th><th>IX</th><th>X</th><th>XI</th><th>XII</th></tr>
<tr><td>I. ARCTIC</td><td colspan="12"></td></tr>
<tr><td>A) ARCTIC BASIN</td><td colspan="7">WINTER</td><td colspan="1">SPR</td><td colspan="1">AUT</td><td colspan="3">WINTER</td></tr>
<tr><td>B) HIGH ARCTIC (SIBERIAN)</td><td colspan="6">WINTER</td><td colspan="1">SPR</td><td colspan="1">S</td><td colspan="1">AUT</td><td colspan="3">WINTER</td></tr>
<tr><td>C) LOW ARCTIC (MURMANSK)</td><td colspan="4">WINTER</td><td colspan="2">SPRING</td><td colspan="1">S</td><td colspan="2">AUTUMN</td><td colspan="3">WINTER</td></tr>
<tr><td>II. MODERATE CLIMATE</td><td colspan="12"></td></tr>
<tr><td>A) NORTHERN (NORWEGIAN WATERS)</td><td colspan="2">WINTER</td><td colspan="2">SPRING</td><td colspan="5">SUMMER</td><td colspan="1">AUT</td><td colspan="2">WINTER</td></tr>
<tr><td>B) SOUTHERN (ENGLISH WATERS)</td><td colspan="2">WINTER</td><td colspan="2">SPRING</td><td colspan="5">SUMMER</td><td colspan="2">AUTUMN</td><td colspan="1">WINTER</td></tr>
<tr><td>III. SUBTROPICAL</td><td colspan="1">W</td><td colspan="3">SPRING</td><td colspan="5">SUMMER</td><td colspan="2">AUTUMN</td><td colspan="1">W</td></tr>
<tr><td>IV. TROPICAL</td><td colspan="1">SPR</td><td colspan="8">SUMMER</td><td colspan="2">AUTUMN</td><td colspan="1">SPRING</td></tr>
</table>

Table 2.8. Comparison of shallow tropical and temperate marine ecosystems

Physical and Chemical			Biological Functions	
Temperature	mean	higher	Metabolic rates	higher
	range	much lower	Benthic productivity	higher
Light	mean	higher	Phytoplankton productivity	lower
	range	lower	Thermal tolerance	smaller range
Salinity		higher	maximum	nearer ambient
Oxygen		lower	Breeding seasons	longer
Carbon dioxide		lower		successional
Phosphorus		lower	Asexual reproduction	higher
Nitrogen		lower	Growth rates	higher
Transparency		higher		more variable
Tides		lower	Larval development	
			fish, zooplankton	faster
			benthic invertebrates	slower
Community Structure			Feeding habits	more specialized
			Niche width	smaller
Species diversity			Space sharing	greater
benthic invertebrates		higher	Minimum oxygen	closer to ambient
fish		higher	Algal invertebrate symbiosis	greater
zooplankton		higher	Poison defense	greater incidence
phytoplankton		higher	Cleaning symbiosis	greater incidence
macrophytes		lower	$CaCO_3$ precipitation	greater
mean size		smaller	Color polymorphism	conspicuous
Biomass		lower	Evolution	higher rates
Population density		lower		
Population size		smaller		
Predator population		higher		
Colonial life forms		more		
Zooplankton / phyto-plankton ratio		greater		
Eggs		smaller		
Planktonic larvae		more		
Meristic counts		lower		
Macrophyte taxa		more red and green algae		
Phytoplankton taxa		more flagellates		
Zooplankton taxa		more copepod		
Lipids in plankton		lower		

Source: Johannes and Betzer (38).

2.2.1 Ecological Relationships and Trophic Levels

The distribution and abundance of organisms in the sea depend on interactions between the organisms and their environment. Essentially all life in the sea, as on land, depends on the sun for energy. In the marine food chain, energy is transferred from plants through herbivores (grazers) to carnivores (animal eaters). Here, tiny floating plants (phytoplankton) are the primary producers and thus constitute the first trophic level; herbivorous grazers (zooplankton) are the second; and the larger predators, such as fishes, are the third and higher levels. Consumer organisms store some of this energy in order to carry out life functions; the rest is passed along to the next trophic level when the organisms eaten by another consumer. Ecological efficiency refers to the 5 to 50 percent of energy transferred by organisms between and within trophic levels. Thus organisms found at the bottom of a food chain are much more abundant than those at higher trophic levels. At each trophic level, energy is: (1) used to increase the size of the individuals in the population, (2) recycled by the decomposer chain, or (3) lost to the system in respiration (43).

Structure and organization within a community are determined by productivity, diversity, dominance, and stability. Dominance can be measured in relation to biomass (standing crop) which is the weight of living matter per unit area. Species diversity may be measured simply by counting the number of different species in a collection or by weighting each species by its relative abundance (43). Species diversity is greater in coastal regions than in the open ocean because physical and chemical conditions are more variable and food is more abundant near the coasts (29). In the open ocean, temperature and latitude have the largest influence on diversity; thus, the warmer, tropical waters generally support a larger variety of life forms. Diversity may also vary with depth in the water column and the abundance of available nutrients. In benthic communities, different substrates also affect the diversity of habitats. If a system is perturbed, community structure may be altered by certain fast-growing species that can force out previously dominant residents (29).

2.2.2 Marine Productivity

Besides light, plants require nutrients. Areas of high primary productivity have been grouped into three categories by the Food and Agricultural Organization of the United Nations (21): (1) upwelling areas found primarily off the western subtropical continental coasts (Peru, California, North and Southwestern Africa) and along the equator, where cold, nutrient-rich waters rise to the surface; (2) temperate and sub-Arctic waters off the Southern Ocean, North Atlantic, and North Pacific; and (3) shallow waters over those parts of the continental shelves that receive stream runoff. Marine life balance, succession, and distribution are governed by seasonal changes in temperature as shown in Table 2.6 (5,6) and by nutrient availability. In areas where nutrient supplies are low, productivity is low.

Although nutrients are essential, too much can reduce species diversity. This is desirable in agriculture and aquaculture (see Chapter 16) but not in natural systems. As noted above in the discussion of estuaries, excessive buildup of nitrogen and phosphorus compounds leads to eutrophication, with increased production of a few species of algae and phytoplankton and subsequent decline in other types of species. Because much of this plant material cannot be consumed by predators, it is instead decomposed by bacteria. This reduces the available oxygen in the water column and predators, including fish, disappear. Sewage effluents can place a high demand on the available oxygen if disposed of in areas with limited water exchange. In areas of extreme nutrient loading and poor flushing, anoxic conditions develop and biological life is limited to anaerobic bacteria.

2.2.3 Plankton

The plankton community consists mostly of tiny plants (phytoplankton) and a lesser number of small grazers ()zooplankton) that float and drift with the currents and tides. Most marine organisms are planktonic during at least the early stages of their life cycle. Plankton that live at the ocean surface are called neuston.

Phytoplankton in the sea are mostly one-celled diatoms, coccolithophores, chysophyceans, and dinoflagellates (see Figure 2.11). Other phytoplankton include the blue-green algae, which are sometimes classified with bacteria.

Zooplankton consist of protozoans (foraminifera, radiolarians, and ciliates), and non gelatinous and gelatinous metazoans. Non gelatinous forms include larval stages of crustaceans, copepods, euphausids, mysids, chaetognaths, molluscs, annelids, and rotifers. Gelatinous forms include the coelenterates, ctenophores, and tunicates. The smaller zooplankton include the flagellated protozoa and amoebae that feed on bacteria. Selective grazing by zooplankton may govern the distribution of certain phytoplankton species.

Although many pollutants have the potential to control plankton growth in the laboratory, field observations demonstrate that phytoplankton growth and reproduction tend to be less affected by domestic sewage discharge than by scientifically balanced fertilizers in agricultural runoff.   These effects are difficult to distinguish from natural phytoplankton blooms that deplete oxygen and produce toxic chemicals that kill fish and shellfish or possibly cause paralytic shellfish poisoning (7). Seasonal blooms in coastal waters can be remarkably enhanced by river basin development projects that reduce seasonal high flows and increase summer high flows (see Section 3.3.1)

2.2.4   Nekton

The nekton include free-swimming fish, squid (molluscs), and cetaceans (mammals). An enormous diversity of forms, sizes, and shapes have evolved as a result of the competition between these various animals for food and habitat.

The larger nekton require greater food supplies for energy but are usually fewer in number. Since man is a competitor in the marine food chain, the natural balance in fishery populations is upset by overfishing certain commercially important species. Ocean disposal of wastes may also affect the production and growth of fish. Some waste inputs, particularly toxic chemicals, could affect local fish distribution and productivity, but the mobility of the nekton allows them to avoid affected areas. Many other fish prefer to feed on food associated with disposal areas, and productivity and growth may actually be enhanced by nutrient and organic carbon inputs associated with domestic waste discharges.

2.2.5 Bacteria

Although terrestrial bacteria are a principal design criterion in ocean disposal (Section 3.2), marine bacteria are of greater ecological importance because they are the primary decomposers of organic matter in the oceans (76). Marine bacteria utilize both dissolved and particulate organic matter found in the remains of plants and animals which originates during photosynthesis and excretion. With the breakdown (mineralization) of organics by bacteria, carbon dioxide and nutrients are released in the form of simple, soluble inorganic ions that can be utilized by plants and phytoplankton. The rates of mineralization of sewage constituents can be quite rapid, exceeding those for marine nutrients by several fold (31).

Two major types of bacteria break down of compounds in the oceans. Organotrophic bacteria use organic compounds as an energy source, while chemolithotrophic bacteria oxidize reduced inorganic compounds such as ammonia and hydrogen sulfide into nitrate and sulfate for energy.

The numbers of bacteria present in the oceans are directly proportional to the amount of food (organic matter) in their environment. Most bacteria are found in the upper waters, or photic zone, where photosynthesis occurs and organic matter is produced, and at the bottom sediment surface, where organic materials accumulate. In general, the average concentrations of bacteria in sediments are greater than those in the photic coastal waters (approximately 10^8 to 10^9/mL in interstitial waters of nearshore sediments versus 10^6/mL in coastal waters). Seasonal variations of bacteria (and zooplankton) in seawater tend to lag slightly behind phytoplankton concentrations.

Sewage and other natural organic materials discharged to the ocean are mostly dilute organic matter subject to degradation by bacteria. As particulates from sewage accumulate on the seafloor surface, they are mineralized at and below the surface of the sediments. Below the surface of the sediments, rapid bacterial metabolism of organic matter results in depletion of the available oxygen. In such anaerobic conditions, organisms with greater oxygen requirements will be eliminated in favor of anaerobic bacteria. These anaerobes produce hydrogen sulfide, ammonia, and methane as by-products of anaerobic degradation of organic matter. In most coastal sediments, the number of aerobic bacteria declines sharply downward from the surface, and the proportion of anaerobic bacteria increases.

Eventually bacteria break down petroleum hydrocarbons, chlorinated hydrocarbons, and other complex organic compounds into products some of which are more toxic than the original compound. Bacteria have been implicated in the methylation of mercury, whereby the inorganic form of this element is converted into an organic form . Mercury enters the marine environment sublimation and air transport from the earth's crust ($\sim 2 \times 10^4$ to 2×10^5 t/y) and river discharges ($\sim 4 \times 10^3$ t/y) that equal about half the world's industrial production (86). Microbiological methylation produces mercury's most toxic form (see Chapter 15).

2.2.6 Benthos

Sessile (attached), creeping, or burrowing organisms living on or in the sea bottom are known as the benthos. The numbers of benthic organisms generally

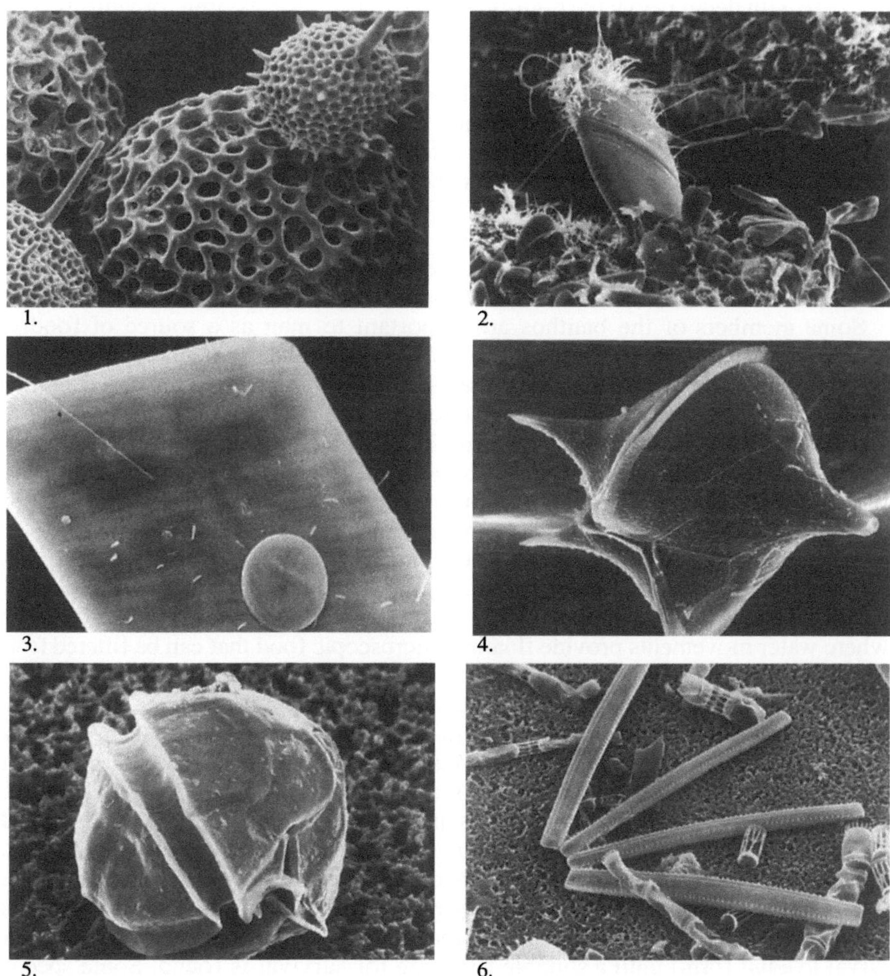

Figure 2.11. Marine phytoplankton. (1) Radiolarian skeletons (210x). (2) Stalked diatoms, *Liconophora sp.*, on whelk egg (265x). (3) Rare bacterium, *Flexobacterium sp.*, attachment to pennate diatoms, the rod-shaped *Rhabdonema adriaticum* and the disc-shaped *Cocconeis sp.* (950x). (4) The dinoflagellate, *Peridinium sp.* (485x). (5) The red tide dinoflagellate, *Gonyaulax temarensis* (2100x). (6) Mixed plankton, Naraganset Bay, RI. (700x). Source: Sieburth (92).

decrease with depth below the high-tide level. Since most benthic organisms have limited mobility, they are particularly susceptible to and hence reliable indicators of pollution changes (see Chapter 14).

Sessile members of the benthos include sponges, barnacles, mussels, oysters, crinoids (feather stars and sea lilies), and some worms. Attached plants such as kelp, eel grasses, and some diatoms are often included in this group of organisms. Benthic organisms that move at or over the sediment surface include echinoderms (starfish and urchins), molluscs (snails, clams, and scallops), and crustaceans (lobsters, shrimp, some copepods, and amphipods). Some fish with limited mobility are also considered to be benthic animals. Burrowing forms of benthic life include most worms and clams, and some crustaceans.

Some members of the benthos are important to man as a source of food or income, but most benthic forms are of little direct value. Nonetheless, they play an important role in marine food chains. The benthos constitute a substantial proportion of the marine biomass of the lower trophic levels and therefore provide a large amount of food for the higher trophic levels in most marine ecosystems.

The distribution of benthic forms is largely influenced by the substrates (bottom types) where they occur. These substrates vary from clean, firm rocks to shifting sands to soft muds, and they support an enormous diversity of organisms.

Benthic survival strategies include attachment to a firm surface, free movement on the bottom, or concealment below the surface. These strategies correspond to the organism's method of obtaining food. A sessile life style is possible only where water movements provide floating microscopic food that can be filtered from the water and for dispersal of reproductive and larval life stages and metabolic products. Turbidity is a problem for filter-feeders, since high concentrations of fine particles suspended in near-bottom waters can clog the filtration devices that these animals use to collect food particles. Most free-moving benthic organisms obtain their food by catching it, but others are scavengers of bottom detritus.

Shallow-water marine communities usually have abundant sources of food. These communities are generally dominated by sessile filter feeders and substrate grazers. Sessile organisms commonly shed their gametes (eggs and sperm) into the water and thus ensure widespread dispersal. Many free swimming larvae of the benthos have the ability to postpone metamorphosis (that is, the change from larval to adult form) until a suitable substrate for survival is found. Some species prefer to colonize a substrate similar to themselves (e.g., oysters settle on oyster shells), but this strategy can result in overcrowding and mortalities. Reef communities may develop when one organism predominates and creates its own environment. Colonial coelenterates living in shallow tropical marine waters at temperatures greater than 20° C, for example, form coral reefs with their skeletons. The reefs support a large diversity of life since they provide: (1) a firm substrate for attachment, (2) hiding places, (3) calcareous rock that can be readily penetrated by boring or corrosive organisms, and (4) abundant calcareous sediment that burrowing animals can inhabit. A variety of predatory animals from many trophic levels may use the coral reefs as hunting grounds. All coral reefs support symbiotic relationships in which one organism benefits from, or in some cases may even demand, the presence of another. Coral itself participates in such a relationship, since internal symbiotic photosynthetic organisms (zooxanthellae) provide coral with its needed oxygen and metabolize the coral's waste products so that they do not reach toxic levels. Reef communities occur where many elegantly balanced conditions exist. The disruption of any of these conditions can upset this delicate balance, destroy the reef assemblages and, ultimately, kill the coral itself.

Since benthic organisms are unable to escape the contaminants entering their environment, these organisms are often used to monitor the effects of waste discharges. Adverse effects occur if a completely different substrate is imposed upon a benthic community, or if anoxic conditions resulting from the accumulation of excessive amounts of organic material deprive the community of needed oxygen. Even so, some benthic life forms still manage to exist in highly contaminated bottom sediments. Although their diversity may be diminished, their biomass is often comparable to or greater than that found in uncontaminated areas with more diverse assemblages.

A number of approaches have been developed to measure the effects of waste inputs on benthic populations. These include the sentinel organism approach, which measures changes in contaminant concentrations in a stationary species over time. Ecological approaches include purely descriptive models usually based on one or more cluster analyses that are not intended for prediction but that support zero-discharge goals   The ecological model that promises to have the best predictive capability is the infaunal trophic index (ITI) which examines the feeding strategies of benthic organisms, (see Sections 14.4 and 15.4).

2.2.7   Fisheries

Most harvesting of fish and shellfish occurs along the ocean margins at depths of less than 100 m. Although fishery harvests provide only a small percentage of the total calorie requirements of the world's population, they are an excellent source of animal protein. It has been determined that about 10 grams of animal protein per person daily can prevent protein deficiency, while 36 grams are suggested for minimum good nutrition (43).   The 1989 fishery harvest of about 87 million tons contained 13 million tons of pure protein and constituted principal sources of protein and income for coastal peoples around the world

Commercial fishing investments require optimism about future catches. Following a decade of unprecedented growth during which the catch reached about 70 million metric tons per year, a fishing industry spokesmen predicted in 1980 that 100m tons per year of fish could be harvested for human consumption. This would be realized by increased fishing effort, improved fisheries stocks management, presumably by the fishermen themselves who increased their fleets by half. Their ambitious goals were not to be met. Hoped for consumer acceptance of dogfish shark and other less desirable fish that were realized only by developing an agricultural market for "trash" fish. It was a case of market failure chronicled by <u>The Economist</u> (3/19/94). Fish harvests depend not upon the numbers of fishing boats but upon the reproductive success of the fish. The harvest began to decline from its 87 million ton peak although both the level of effort and the portion marketed for fertilizer increased. Even then owners of the world's fleets claimed to have lost $22 billion. The editors took the extraordinary position of recommending government regulatory intervention in the market. The most favorable current estimate for a sustainable global marine fishery is less than

90 million tons per year, including trash fish, and even this will require mandates on quotas by national authorities. These have so far been unacceptable to fishing industry constituents who, in the United States, in 1982 successfully promoted indirect controls such as mesh and fish size as alternatives to quotas (89).

International controls for fisheries sustainability inherently conflict with basic principles of such other international instruments such as the General Agreement on Trade and Tariffs and the original formulation of the Law of the Sea. For example, a 1991 international trade panel ruling negated a unilateral ban on tuna imports imposed because Mexican fishing methods were more destructive to dolphins than U.S. methods. Bilateral controls being sought since 1975 to end the "Cod War" between Great Britain and Iceland are similarly vulnerable to market forces and to unilateral "exclusive economic zones." (89).

Aquaculture systems are meeting an increasing portion of the demand (94, 95). 1989-1991 production was 12 mt, mostly fresh water varieties. Waste-fed aquaculture systems can provide additional protein for animals or people (20). Wastewater-fed ponds are one of options in tropical climates. (see Sec. 2.4.1).

Effects of Waste Discharges. The effects of a waste discharge upon a fisheries depend on the nature of the wastes, the transport of the wastes in the marine environment, the types of fisheries involved, the ecological and commercial importance of the fishery, and the residence time of the stocks in the area of discharge. Marine organisms deal with stress by (1) escape--benthic species can escape by burrowing into the sediments, while finfish escape by swimming; (2) reduction of contact--some fish species secrete a mucous layer, while shellfish species may close their valves and stop feeding to avoid changes in their environment; (3) regulation--many fish are able to control their environment by regulating chemical concentrations, volumes of body fluids or fluid pressures within their bodies; (4) acclimation--a number of species have adapted to man-induced changes in their environment with no apparent harm because they are used to large changes in their coastal-zone environment.

In general, fish are attracted to areas around municipal outfalls because there they find food to their liking, the food consisting of grazers of phytoplankton living in enriched waters scavengers living on whatever they and grazers sewage outfall will attract large numbers of free-swimming fish, although some species may choose to avoid the immediate area of the discharge. Since the discharge provides an ample source of food, energy that would otherwise be spent in foraging for food is used for enhanced growth and recruitment of the species. If a discharge is discontinued, the enhanced species will return to normal size and abundance. In contrast, local populations of shellfish are likely to be adversely affected by pollutant inputs from sewage outfalls. The effects on mobile species, particularly contaminant concentrations within the fish themselves, depend on their residence time in the discharge area. As fish move in and out of the discharge area, effects may be short-lived ant transitory. However, the potential for bioaccumulation of organic chemicals in finfish is increased with prolonged residence in an affected area. The effects of discharges are not always local or easily identified. For example, the release of a persistent insecticide into a coastal river in the eastern United States resulted in the contamination of fish hundreds of kilometers from the discharge site

(64). In southern California, seabird mortalities were attributed to consumption of fish contaminated by a pesticide discharged from an outfall (75).

Tastes and odors can often be traced to the bioaccumulation of pollutants. Crustaceans, fish, and molluscs exposed to oil can acquire an objectionable, oily taste. With increased inputs of oil and petroleum by-products into the marine environment, the likelihood of fish tainting increases. Susceptibility to tainting varies among species and depends on the condition of the fish, the nature of the oily compound, and exposure time.

2.3   Special Topics

2.3.1   Public Health Aspects

Both natural and man-made conditions in the sea give rise to certain public health concerns (73). Some marine plankton, nekton, benthos, and bacteria are naturally toxic or infectious. In some areas of the world ocean, seasonal occurrences and occasional dense blooms or "red tides" of dinoflagellates, notably *Gonyaulax sp*, contain neurotoxins that can be concentrated by edible shellfish and cause human paralysis or death It has been speculated that some dinoflagellate blooms may be induced by nutrient enrichment of open coastal waters from sewage. No evidence for this has been found by several intensive investigations (7, 48). However, some writers arguing from first principles and using selected data from embayments with limited circulation, have assembled putative evidence that the problem should be real and worthy of further investigation (17, 68). And , as many fishermen and swimmers know, some bottom fish, sea urchins, and jellyfish carry natural toxins.

A few marine bacteria , particularly *Vibrio parahaemolyticus*, are pathogenic to humans. often causing minor cases of food poisoning and sometimes even death. Consumption of raw fish and shellfish is one of the most common causes of *Vibrio*-associated illnesses. Another microorganism, *Mycobacterium marinum*, causes skin lesions. These pathogens are opportunistic microbes that can adapt to a tissue or host other than the one in which they are normally found. They are killed by cooking.

Chemical contaminants of interest to health officials include certain inorganic and organic chemicals and radionuclides. Acids, alkalis, and cyanides discharged to the ocean are quickly neutralized and are only locally significant. Except for artificial radionuclides, the metals discharged into the ocean are also normal constituents of the sea. . Earlier concerns about the toxicity of heavy metals in the marine environment have been reduced. Recent evidence shows that, in general, concentrations of these metals do not increase upward in the food chain. Ordinarily, they are not known to pose a threat to people eating seafood (2), although they may exceed the precautionary levels established by various standards or guidelines. However, extraordinary conditions and events such as environmental methyl mercury and cadmium poisoning have occurred. (83).. In contrast, toxic synthetic organic chemicals are persistent in the sea, are known to accumulate in food chains, and are found in increasing amounts in the marine environment. The same chemicals have been linked to sublethal effects on marine life and to mortalities in birds that feed on fish with high levels of these chemicals (74, 75). There is evidence of exposure with presumptions of effects. Careful research and monitoring have revealed no evidence of human morbidity or mortality from eating fish and shellfish with high levels of a known carcinogen, kepone, in an area of the James River, North Carolina where sediments and their interstitial waters were their sources. Meanwhile, economic losses have been substantial in cases where fishery areas have been closed because organic contaminant levels exceeded limits (36).

Radionuclides hare introduced into the marine environment by both nature and man. Natural sources include the interaction of cosmic rays with atoms in the atmosphere and the weathering of rocks. Artificial radionuclides come from nuclear power production, nuclear detonations, and pharmaceutical and industrial inputs. While there has been no measurable harm to marine organisms (52), the Chernobyl and similar incidents in and north of the Russian Republic and elsewhere, are reminders of the need for conservative approaches to radioactive risk.

It is essential to know the transmission routes by which pathogens reach human populations in order to predict their effects on human health (for a similar model, see Figure 3.1). By far the most important route of transmission is the consumption of contaminated shellfish. Recreational use of contaminated waters raises an important aesthetic issue but is not a demonstrable public health problem.. Well-managed harvesting, depuration, and sanitary handling of shellfish can prevent disease outbreaks caused by eating raw shellfish. In any event, adequate cooking destroys all accumulated pathogens. Although increasing numbers of antibiotic-resistant microbes are being found particularly in hospitals where they have caused staphylococcal disease outbreaks. They have also been found in the sea (28, 42, 67), but are not suspected to have caused similar problems.

2.3.2 Environmental Toxicity

The toxicity of a substance depends on its concentration in the environment, its availability to an organism, and the susceptibility of the organism. Any substance can become toxic when its concentration exceeds some threshold. Persistent substances are most likely to accumulate to toxic levels, although others also accumulate if they settle quickly to the ocean bottom and thereby avoid the great ability of the ocean to absorb or disperse them. These pollutants can be resuspended by currents or waves during storms,

No environmental protection entity wants to see marine organisms killed with toxicants; neither does it want to see them sick. The problem is that sublethal or chronic effects are not revealed by the bioassay tests in which organisms are subjected to a range of concentrations of a known or suspected pollutant. Results are reported as an LD50 or LC50, which is the lethal dose or the concentration at which half of the laboratory test specimens are killed within a specified time limit. Such tests are of dubious value. Environmental changes in salinity, pH, turbidity, temperature, concentrations of other substances, and the adaptation of the organisms are not revealed by the laboratory LD50 value.. For example, chlorinated hydrocarbon pesticides are more toxic in summer; detergents are more toxic at higher salinities; and some organisms are more sensitive during their reproductive or early developmental stages. In addition, organisms may be able to adapt to gradual changes in a toxicant level in their environment, but exposure concentrations are increased abruptly during bioassays.

Some toxicants accumulate in sediments or biological systems, where they may be toxic both to marine organisms and man. Bioaccumulation is assessed by comparing concentrations of the substance in the water column, the sediments, and certain organisms. The bioconcentration factor, BCF, is the ratio of the concentration in the organism that in the surrounding water indicates pathways of a toxicant in the marine environment. Flocculation and sedimentation of particles with adsorbed toxicants may cause pollutants to become entrapped in the sediments. Some of these pollutants reenter the water column by physical resuspension, by metabolic activity (excretion of the benthos), or by accumulation within benthic organisms that serve as food for pelagic species.

Wastes discharged into the marine environment undergo various biochemical transformations, such as decomposition of organic matter, alteration of the physical or chemical form of the constituents, adsorption onto particles and sediments, and incorporation into living matter. Some of these transformations and interactions increase toxicity, whereas others decrease it. Where mixing and circulation are limited, and where added nutrients and organics increase the biological activity, the oxygen content of the water is decreased.

Toxic inorganic chemicals include acids, alkalis, cyanides, and metals. The first three are quickly neutralized when dispersed and diluted in seawater, but the effect of metals is difficult to demonstrate. Metal compounds form a major portion of the earth's crust and are transported to the ocean by rivers, glaciers, and wind. These metals are naturally occurring constituents of the sea, and difficult to distinguish from metals in wastes. Man-made sources of metals include industrial and sewage effluents, combustion of fossil fuels, and resuspension of previously dumped materials. When the normal background concentrations of certain in are substantially increased, the potential for toxicity exists. Several factors influence the toxicity of metals to marine organisms:

1. The physiochemical form of the metal
2. The synergistic or ameliorative action of other metals or compounds
3. The physical properties of the surrounding seawater
4. The physiological condition of the marine organisms.

Acute toxic effects have been demonstrated by laboratory experiments, primarily through the interaction of metals with animal tissue enzymes and resulting alteration of enzyme activity. These effects are most likely to occur near large sources, where the highest pollutant concentrations are found (53). Opinions differ as to which metals are more toxic, but there is some agreement that, in decreasing order, mercury, cadmium, lead, silver, copper, zinc, chromium, and selenium are of primary concern (52, 53):. Antimony, arsenic, nickel, and vanadium secondary.

Other trace metals such as thallium may be damaging to marine ecosystems, but they are not yet used to any great extent by man and inputs are low. Elevated metal concentrations have been detected frequently in organisms from contaminated areas, but few acute cases of trace metal toxicity in the have been documented. One case involved a substantial abalone kill attributed to copper pollution from the discharge of seawater that had been held for some time in copper condensers at a coastal power plant (45). Laboratory experiments indicating that elevated concentrations of certain trace metals in seawater are toxic to marine organisms are difficult to interpret since toxic effects have been reported at concentrations well within natural background variations. This discrepancy is due either to the inability to reproduce field conditions in the laboratory, or to the natural detoxification processes of organisms that have had time to adjust to these levels in the environment, but not in the laboratory. Studies have now shown that high concentrations of toxic metals in tissues are not necessarily the cause of metabolic disorders in marine life (12) In general, concentrations of metals do not increase with higher trophic levels in the food chain (75). Instead, the animals detoxify themselves by sequestering (chemically binding and storing) the toxic metal with other substances found within their bodies.

Organic chemicals ultimately decompose into carbon dioxide, water, and other simple inorganic compounds. However, many synthetic organic chemicals, whose use depends upon their toxicity such as in anti-fouling paints, are extremely persistent in the environment and often accumulate in the food chain.

The chemical manufacturing industry has grown phenomenally during the past half century and has produced hundreds of thousands of new synthetic organic products and pharmaceuticals whose short term benefits are unquestioned but whose long term costs are rarely known. The polynuclear aromatic hydrocarbons, PAIIs, are innovations with lower molecular weights, lower bioaccumulation factors, and higher carcinogenicity than most other halogenated hydrocarbons; they are producing new benefits with unknown costs throughout the world. Their stability ensures that many of them will join DDT, kepone, and PCBs that have previously received notoriety in coastal or estuarine environments that contributed to their being banned in the United States (39, 45, 47, 60, 64, 66). Chlorinated hydrocarbons are only slightly soluble in water, are found in the food of living organisms, and tend to remain in the fatty tissues for a long time. Laboratory tests for lethal toxicity do not provide for predicting toxicity in the field. Many sublethal effects have been documented in marine organisms exposed to these chlorinated hydrocarbons (60). For example, DDT has been correlated with a decrease in the reproductive success of spotted sea trout in the southeastern United States (9) and of brown pelicans in California (39).

Although DDT can be replaced with malathion, parathion, and other biodegradable organophosphate pesticides that affect the nervous systems of marine animals and humans (53), its lower price and still high though decreasing effectiveness assures its continued use. Once DDT and its congeners get into the environment, they remain for years. Large residuals of DDT manufacturing wastes originating in the City of Los Angeles have persisted in the sediments near a Los Angeles County Sanitation Districts' outfall for over 30 years after the manufacturer ceased making it. Meanwhile, although domestic use of DDT has long since been banned by the USEPA, at least 1.5 tons per day are revealed by U.S. Customs documents to be manufactured and shipped to developing countries where it is applied for mosquito control and to food crops destined for export (66,

93). Some of it remains in the soil, some is washed into the coastal zone, and some is returned as residuals to the United States.

DDT also demonstrates the institutional difficulties in international information and technology transfer where short term marketing benefits compete with the long-term environmental and economic development benefits (see Chapter 16).

Other chlorinated hydrocarbons are created when sewage effluents are chlorinated to meet end-of-pipe bacteriological standards and then sometimes dechlorinated to make the effluent less toxic. Chlorine interacts with the organic materials in seawater to produce chlorinated organic compounds such as chloroform, a known carcinogen and mutagen.

Oil spills contribute about 12 percent to the total petroleum in the oceans, the remainder is from river runoff (35%), normal shipping operations (34%), natural seeps (8%), atmospheric fallout (9%), and offshore oil production (2%) (19). Small amounts derive from waste discharges and urban runoff (14). Locally high concentrations of petroleum hydrocarbons in the parts-per-million range can kill most marine species, but sublethal effects of petroleum residues are largely unknown.

Measurements of Toxicity.   Laboratory determinations of the LC50 (or LD50) are the most readily available, although inconsistent, source of data on the relative toxicity of various chemicals. The great variability in these test results and their limited applicability to the marine environment are due to several factors:

1. Organisms used in the tests are physiologically different from those found in the discharge area.
2. Test conditions--including time, temperature, pH, and salinity regimes--often vary and make it difficult to compare the data. Some tests use the same water (static system) throughout the time period and some tests use flowing water. Some are short-term acute tests, whereas others are chronic tests, which are more sensitive and more closely approach environmental conditions. The latter tests do not attempt to destroy the organism, but rather are aimed at examining the sublethal effects over a longer period of time.
3. The developmental stage of the test species often varies in separate experiments. Fry and larval stages tend to be more sensitive than adults to toxic substances.
4. Different forms of the toxin are used for different experiments. The ionic form of the inorganic metals used in laboratory toxicity tests are often substantially different from the form of the metals found in the sea and from those metals found in effluents, which are usually attached to particulates.

Thus, the best that can be expected from laboratory studies of toxins are suggestions of relative toxicity of materials under the laboratory conditions in which they were measured. They cannot be used to predict their effects in the marine environment.

Bioaccumulation.   Bioaccumulation is the ability of an organism to concentrate, often by many orders of magnitude, the chemical components of its environment in its tissues. The accumulation of contaminants by an organism is dependent upon exposure time, concentration, and chemical form of the contaminant. Uptake of contaminants from all available sources (water, food, and sediments) determines the bioconcentration factor (BCF). Bascom (2, 3) argues that, contrary to conventional wisdom, studies now show that metals in their usual inorganic forms are not concentrated through the marine food web. In areas near sewage discharge, bioaccumulation of trace metals or organic compounds, such as PCBs or aromatic hydrocarbons, in marine biota may be observed. However, elevated tissue concentrations of these substances in a discharge area are not a signal of significant adverse effects on the marine biota. For example, caged mussels placed in highly contaminated areas of the New York Bight apex were shown, through a variety of biochemical and growth rate measurements, to suffer no negative biochemical effects, even though these mussels did accumulate significantly higher concentrations of several metals (Zn, Pb, and Cd), PCB, and polyaromatic hydrocarbons than did mussels grown at cleaner reference sites. In fact, the biological measurements indicated that the mussels at the contaminated (and enriched) sites were probably growing more rapidly than mussels at the cleaner sites (2). Bioaccumulation of trace metals and toxic organics is therefore an insufficient predictor of adverse effects of sewage disposal unless concentrations in seafood species approach maximum safe levels for human intake.

The accumulation of toxic substances in people depends on the amounts of a particular seafood consumed, concentrations in the seafood (which depend upon the pollutant concentration and the proximity and residence time of the species in the discharge area), and upon how the seafood is prepared. Although sedentary shellfish might appear to be of particular concern, both laboratory and field studies of kepone contamination have shown that oysters do not concentrate organic toxins. Furthermore, when relocated to a cleaner environment, organisms are able to purge organic toxins from their systems within a relatively short time (26).

Ameliorative Effects. Marine animals have evolved natural protective mechanisms to survive gradual variations in the amounts of metals present in seawater. One is a sequestering protein, metallothionein, which is produced within organisms to maintain a reserve supply of essential metals (e.g., zinc and copper) required for enzymatic activity (2, 3). The metallothionein also holds excess or otherwise nonessential metals and thus prevents them from reaching the enzymatic and genetic systems, where toxic effects can occur.

A number of other interactions of a protective nature are known to occur in the marine environment. Mercury, certainly the most studied and possibly the most toxic of all metals, is a normal constituent of seawater and is accumulated in fish. Selenium, another potentially toxic metal, is required by most animals and is also a trace element in the sea. Experiments with a number of animals have confirmed an antagonistic interaction between these two metals that limits the toxicity of both the mercury and the selenium (25). In one study, a group of quail were fed tuna contaminated with methyl-mercury with no apparent detrimental effects, whereas another group was poisoned by an equivalent amount of methyl-mercury in a corn/soya diet. These results support the contention that selenium, which occurs naturally in high levels in tuna, suppresses the toxicity of mercury.

Although the research is still in an early stage, it appears that selenium may have similar ameliorative effects on other heavy metals, such as silver. A number of other compounds also have been found to suppress the toxicity of metals. Vitamin E, which contains selenium, has been shown to be effective against silver toxicity; and sulfur amino acids, in combination with forms of selenium and arsenic, are especially effective against methyl-mercury toxicity. Moreover, recent research indicates that there are other ameliorative interactions within the marine environment.

Red Tides. Dense, short-term blooms of various plankton, long known as "red tides," occur naturally, particularly in enriched coastal waters (7). There are two principal problems associated with red tide blooms. Particularly in southern California, shellfish may become contaminated during blooms of the dinoflagellate, *Gonyaulax polyedra*, and rendered unsuitable for human consumption as a result of accumulation of the neurotoxin responsible for paralytic shellfish poisoning. The usual effect of dense blooms, however, is oxygen consumption during the decomposition of the dead plankton cells, which can create hypoxic or anoxic conditions inimical to marine life in the water column and near the sea bottom.

2.3.3. Wave and Current Measurements.

Liquid-level recorders within stilling chambers connected to the sea through a small hole to damp out gravity waves are used for measuring tides. Wave lengths, periods, and directions of gravity waves are measured along perpendicular rows (Mills cross) of sensors usually spaced at equal intervals (L). More efficient arrangements are possible. Kinsman (41) shows that four sensors can do the work of seven equally spaced ones by placing them at intervals of L/2, L, and 3L/2 for estimates of covariance at 0, ±L/2, ±L, ±3L/2, ±2L, ±5L/2 , and ±3L.

This example shows how the design for sampling in space can provide maximum efficiencies in data collection and analysis. The same can be done for sampling in time to determine mean values and variance in currents and water quality. Time series analysis is discussed in connection with monitoring in Chapter 10.

2.3.4   A Historical Note on Biological Oceanography

Applied oceanography began with ancient mariners, navies, and fisheries and continued with Ptolemy, Galileo, Columbus, and Franklin. Scientific oceanography began with the British *Challenger* expedition of 1873-1876 that, among other things, provided the epitaph of a much- cherished theory published in 1868 by T.H. Huxley on primordial ooze, a "living slime" called *Bathyblus haekelii,* All this was during the period of scientific and religious shock that followed Charles Darwiin's publication in 1849 of *On the Origin of Species by Means of Natural Selection.*

Having been forced to consider "missing links" among primates, including man, evolutionist anthropologists looked for "fossil cultures" among bushmen in the Kalahari and Australian deserts and the American west. This would legitimize both the evolutionary pinnacle of northern European Anglo-Saxon Protestants and the continuing of colonial expansion. Biologists sought "living fossils" in the depths of the sea. Occasional strandings or net catches of rare fish that required taxonomic clarification kept imaginations fired and funded by Calvinist and mercantilist zeal. 1857 soundings by H.M.S. *Cyclops* for transatlantic cable crossings recovered muds now known as calcareous oozes that were given to Huxley. The oozes could not be distinguished from chalk from British cliffs that had been pulverized in water The ooze was concluded to be the progenitor of future chalks and so had to be living. It would explain much.

In 1868, Huxley reexamined the 1857 samples and found them covered by a sort of mucus which enclosed bits and clusters of granules. Under the microscope, the granules seemed to sink and shift, leading to the idea "that the granule heaps and the transparent gelatinous matter in which they are imbedded represent masses of protoplasm." Huxley named the stuff *Bathybius haeckelii* after the German naturalist Ernst Haeckel who was delighted with the honor and called it primordial slime (*Urschbleim*) The abyss was elevated to the position of being the source of all life. When the *Challenger* brought up samples that contained tube-worms from the mid-Atlantic ridge, *Bathybius* became even more famous: here was proof. In contrast, much effort devoted to microscopic examinations of muds revealed nothing more than the calcareous litter that had been there from the first; there were no intermediate forms. Finally, on the *Challenger* itself between Hong Kong and Japan in 1875, the expedition chemist, Buchanan, observed some fresh gelatinous substance in the water overlying the muds which had been preserved with alcohol, but none on the muds Buchanan tested the substance and found that it was a colloidal precipitate of alcohol and the calcium sulfate in sea water. It was the extinction of *Bathybius*.

Improving our understanding of coastal cities' dependence upon marine life and processes is still interdisciplinary (see Chapters 3, 4, 10, and 11), still imaginative with new species and extinctions, and is reflected in the recommendations of Chapter 12. Oceanography, engineering, people, and planning for sustainable ecological systems can come together in curious ways.

<u>Sources</u>: Schlee, S. 1973 The Edge of an Unfamiliar World. E.P. Dutton, New York; Herdman, W.A. 1929 Founders of Oceanography and Their Work. Edward Arnold & Co, London; Thompson, C. Wyville (1872) The continuity of chalk, Nature 3 (64)226-227, and (1874) The Depths of the Sea, London, Macmillan & Co. Langness, L.L. (1987) The Study of Culture, Novato, CA, Chandler & Sharp.

2.4   References

1. APHA. 1980. Standard Methods for the Examination of Water and Wastewater, 18th ed. American Public Health Association, Washington, DC
2. Bascom, W. 1981. The non-toxicity of metals in the sea. Specialty Conference on Disposal of Sludge in the Sea. International Association for Water Pollution Research and Control, London.
3. Bascom, W. 1982. The effects of waste disposal on the coastal waters of Southern California. Environmental Science & Technology., 16(4):232.

4. Bigelow, H.B., and Edmondson, W.T. 1947. Wind Waves at Sea, Breakers and Surf. Pub. 602, U.S. Navy Hydrographic Office, Washington, DC.

5. Bogorov, B. G. 1958. Perspectives in the study of seasonal changes of plankton and of the number of generations at different latitudes, in A.A. Buzatti-Traverso, ed., Perspectives in Marine Biology. Univ. Calif. Press, Berkeley, 14-151.

6. Brett, J. R. 1970. (Effects of) temperature on fishes, Chap. 3.32 in 0. Kinne, ref. (40).

7. Brongersma-Sanders, M. 1957. Mass mortality in the sea. Geological Society of America, Memoir 67, vol. 1, 413-428.

8. Brooks, N. H. 1972. Dispersion in hydrologic and coastal environments. Report. No. KH-R-29. California Institute of Technology, Pasadena, CA. Also abstracted in Sec. 8, University of California Water Resources Series, Program VII, Pollution of Coastal and Estuarine Waters, 1970, and in (74).

9. Butler, P. A., Childress, R., and Wilson, A. J. 1972. The association of DDT with losses in marine productivity. In M. Ruivo, ed., Marine Pollution and Sea Life. Fishing News (Books), Ltd., London.

10. Cabelli, V. J. 1983. Health Effects Criteria for Marine Recreational Waters. Pub. EPA-600/1-80-031.   United States Environmental Protection Agency, Washington, DC.

11. Cabelli, V. J., Levin, M. A., and Dufour, A. P. 1983. Public health consequences of estuarine and marine pollution . In Myers, ref. (49), 519-576.

12. Calabrese, A., Gould, E., and Thurberg, E. P. 1982. Effects of toxic metals in marine animals of the New York Bight. In Mayer, ref. (48), pp. 293-298.

13. CERC.   1984.   Shore Protection Manual, 2 vols.   Coastal Engineering Research Center, U.S. Army Corps of Engineers, Vicksburg, MS.

14. Collier, A. W. 1950. Oceans and coastal waters and life-supporting environments, Chap. 1 in Kinne, ref. (40).

15. DAMOC. 1971. Master Plan and Feasibility Report for Water Supply and Sewerage for the Istanbul Region. Vol. 3 of 4. Daniel, Mann, Johnson, and Mendenhall, Los Angeles, CA

16. Defant, A. 1961. Physical Oceanography, 2 vols.   Pergamon Press, Oxford.

17. Degolis, D. F., Smodlaka, I., Skrivanic, A., and Precali, R. 1979. Increased eutrophication of the northern Adriatic Sea. Marine Pollution Bulletin, 10:298-301.

18. Eckart, C. A. 1948. An analysis of stirring and mixing processes in incompressible Fluids. Jour. Marine Research, 7(3):265-275.

19. Exxon Corporation. 1985. Fate and effects of oil in the sea. New York, NY.

20. Edwards, P. 1985. Aquaculture: a component of low cost sanitation technology. Technical Paper no. 36. World Bank, Washington, DC.

21. FAO. 1972. Atlas of the Living Resources of the Sea. Food and Agriculture Organization of the United Nations, Rome.

22. Fischer, H. B., List, E. J., Koh, R. C. Y., Imberger, J., and Brooks, N. H. Mixing in Inland and Coastal Waters. Academic Press, New York, NY

23. Fleming, R. H. 1957. General features of the ocean. Geol. Soc. Amer. Mem. 67, vol. 1, 95.

24. Fleming, N. C. 1978. Holocene eustatic changes and coastal tectonics in the northeast Mediterranean. Philosophical Transactions of the Royal Society of London, Part A, vol. 289, no. 1362:405-458.

25. Ganther, H. E. 1980. Interactions of vitamin E and selenium with mercury and silver. Annals of the New York Academy of Sciences, 355:212-266.

26. Ganther, H. E., and Sunde, M. L. 1974. Effect on tuna fish and selenium on toxicity of methylmercury: a progress report. Jour. Food Sciences, 39.

27. Garrett, C. Coastal dynamics, mixing, and fronts. 1983. In P. G. Brewer, ed. Oceanography: Present and Future. Springer-Verlag, New York, 69-86.

28. Grabow, W. O. K., Prozesky, O. W., and Smith, L. S. 1974. Drug resistant coliforms call for review of water quality standards. Water Research, 1-9.

29. Gross, G. M. 1971. Oceanography, 2d ed. Charles E. Merrill Publishing Co., Columbus, OH.

30. Gunnerson, C. G. 1961. Marine disposal of wastes. Jour. San. Engr. Div. Am. Soc. Civ. Engrs., 87(SAl):23-56.

31. Gunnerson, C. G. 1963. Mineralization of organic matter in Santa Monica Bay. In C. H. Oppenheimer, ed., Marine Microbiology. C. C. Thomas, Publishers, Springfield, Il. 641-653.

32. Gunnerson, C. G. 1967. Hydrologic data collection in tidal estuaries. Water Resources Research, 3(2):491-504.

33. Hansen, D. V. 1977. Circulation. Mesa New York Bight Atlas Monograph 3. New York Sea Grant Institute, Albany.

34. Harvey, H. W. 1955. The Chemistry and Fertility of Sea Waters. Cambridge Univ. Press, Cambridge, England.

35. Hicks, S. D. 1983. Sea Level Variations for the United States, 1855-1980. National Oceanic and Atmospheric Administration, Washington, DC.

36. Huggett, R. J., and Bender, M. E. 1980. Kepone in the James River. Environ. Sci. Technol., 14: 919.

37. Hume, N. B., Gunnerson, C. G., and Imel, C. E. 1962. Characteristics and effects of Hyperion effluents in Santa Monica Bay. Jour. Water Poll. Cont. Fed., 34(1):15-35.

38. Johannes, R. E., and Betzer, S. B. 1975. Marine communities respond differently to pollution in the Tropics than at high latitudes. In E. J. Ferguson Wood and R. E. Johannes, Tropical Marine Pollution. Elsevier, Amsterdam.

39. Keith, J. O., Woods, L. A., Jr., and Hunt, E. G. 1970. Reproductive failures in brown pelicans on the Pacific Coast. Transactions 35th North American Wildlife and Natural Resources Conference. Wildlife Management Institute, Washington, D.C., 56-63.

40. Kinne, O. 1980. Marine Ecology, 2 vols. Wiley-Interscience, London.

41. Kinsman, B. 1965. Wind Waves. Prentice-Hall, Englewood Cliffs, NJ.

42. Koditschek, L. K., and Guyre, P. 1974. Antibiotic resistant coliforms in New York Bight. Marine Pollution Bulletin, 5:71-74.

43. Krebs, C. J. 1972. Ecology-: the Experimental Analysis of Distribution and Abundance. Harper and Row, New York, NY.

44. Luce, J. V. 1971. Lost Atlantis: New Light on an Old Legend. McGraw-Hill, New York, NY.

45. Martin, M. 1977. Copper toxicity experiments in relation to abalone deaths in a power plant's cooling waters. Calif. Fish and Game, 63. 95-100.

46. Marchetti, C. 1979. Constructive solutions to the CO_2 problem. Internal report, International Institute for Applied Systems Analysis, Laxenburg, Austria.
47. Massachusetts Coastal Zone Management Office. 1982. PCB pollution in the New Bedford, Massachusetts area: a status report. Boston.
48. Mayer, G. F., ed. 1982. Ecological Stress and the New York Bight. Estuarine Research Foundation. Columbia, SC.
49. Myers, E. P., ed. 1983. Ocean Disposal of Municipal Wastewater, 2 vols. Sea Grant Program, Massachusetts Institute of Technology, Cambridge, MA.
50. NACOA. 1981. The Role of the Ocean in a Waste Management Strategy. National Advisory Committee on Oceans and Atmosphere, Washington ,DC
51. Ninkovich, D., and Heezen, B. C. 1965. Santorini Tephra, in Submarine Geology and Geophysics, Colston Papers, no. 17, Bristol, U.K.
52. NOAA. 1979. Proceedings of a Workshop on Scientific Problems Relating to Ocean Pollution Environmental Research Laboratories. National Oceanic and Atmospheric Administration, Boulder, CO..
53. NOAA. 1979. Federal Plan for Ocean Pollution Research Development, and Monitoring Fiscal Years 1979-1983. National Oceanic and Atmospheric Administration, Washington, DC.
54. NOAA. 1985. Tide Tables. (1) West Coast of North and South America, (2) East Coast of North and South America, (3) Europe and West Coast of Africa, and (4) Central and Western Pacific Ocean and Indian Ocean. National Oceanic and Atmospheric Administration, Washington, DC.
55. Pilkey, O. H. 1980. Shoreline research. In P. G. Brewer, ed., Oceanography: the Present and the Future. Springer-Verlag, New York, NY, 87-100.
56. Pritchard, D. W. 1970. Estuarine analysis. Sec. 6. Pollution of Coastal and Estuarine Waters Program, VII, Water Research Education Series, University of California, Berkeley, CA..
57. Robinson, M. A. 1980. World Fisheries to 2000. Marine Policy, IPC Business Press, 20.
58. Ross, D. A. 1970. Introduction to Oceanography. Appleton-Century, Crofts. New York, NY.
59. Shepard, F. P. 1973. Submarine Geology, 3d ed. Harper and Row, New York, NY.
60. Sherwood, M. J. 1982. Fin erosion, liver condition, and trace contaminant exposure in fishes from three coastal regions. In Mayer, ref. (48), 359.
61. Simpson, J. H., and Hunter, J. R. 1974. Fronts in the Irish Sea. Nature, 250:404 406.
62. Smith, D. D., and Brown, R. P. 1971. Ocean Disposal of Barge-Delivered Liquid and Solid Wastes from U.S. Coastal Cities. U.S. Environmental Protection Agency, Solid Waste Program, Cincinnati, OH.
63. Sternberg, R. W. 1972. Predicting initial motion and bed-load transport of sediment particles in the shallow marine environment. In D. J. P. Swift, D. B., and O. H. Pilkey, eds., Shelf Sediment Transport Processes and Patterns. Dowden Hutchinson and Ross, Inc., Stroudsburg, PA, pp. 61-82.
64. Sterrett, F. S., and Boss, C. A. 1977. Careless kepone. Environment, 19(2):14.

52

65. Sverdrup, H. U., Johnson, M. W., and Fleming, R. H. 1942. The Oceans. Prentice-Hall, Englewood Cliffs, NJ.

66. Taubenfield, R. 1971. DDT: The United States and the developing countries. In W. H. Matthews, F. E. Smith, and E. D. Goldberg, eds., Man's Impact on Terrestrial and Oceanic Ecosystems, MIT Press, Cambridge, MA., 503-506.

67. Timoney, J. F., and Port, J. C. 1982. Heavy metal and antibiotic resistance in *Bacillus* and *Vibrio* from sediments in the New York Bight. Mayer, ref. (48), 235-248.

68. Tsuji, T., Seki, H., and Matori, A. 1974. Results of red tide formation in Tokyo Bay. Jour. Wat. Poll. Cont. Fed., 46(1):165.

69. Trask, P. D., ed. 1939. Recent Marine Sediments. Amer. Assoc. Petroleum Geologists, Tulsa, OK. (See 5-21, F. Hjulstrom, Transportation of Detritus in Water.)

70. U.S. Department of State. 1981. Draft convention on the law of the sea.. Washington, DC.

71. Weyl, P. K. 1970. Oceanography: An Introduction to the Marine Environment. Wiley and Sons, New York, NY.

72. Whitehouse, V., Jeffrey, G. L. M., and Debbrecht, J. D. 1960. Differential settling tendencies of clay minerals in saline waters. In Seventh National Congress on Clay and Clay Minerals. Pergamon Press, London.

73. WHO. 1979. Principles and guidelines for the discharge of wastes into the marine Environment. World Health Organization, Regional Office for Europe, Copenhagen.

74. Young, D. R., and Mearns, A. J. 1979. Pollutant flow through food webs. In W. Bascom, ed., Annual Report for the Year 1979. Southern California Coastal Water Research Project, Long Beach, CA. .

75. Young, D. R., Heesen, T. C., Esra, G. M., and Howard, E. B. 1979. DDE - contaminated fish off Los Angeles are suspected cause in deaths of captive marine birds. Bull. Environmental Contaminant Toxicology, 21:584-590.

76. Zobell, C. E. 1946. Marine Microbiology. Chronica Botanica Co., Waltham, Mass.

78. Emery, K.O., and Aubrey D.G. 1991. Sea Levels, Land Levels, and Tide Gauges. Springer Verlag, New York, NY.

79. Emery, K.O., and Gunnerson, C.G. 1973. Internal swash and surf. Proc. National Academy of Sciences, 70(8) 2379-2380.

80. Glantz, M.H. 1984. Floods, fire, and famine; Is El Niño to Blame?. Oceanus, 27((2)14-19.

81. Gunnerson, C.G. 1988. Report to Government of Indonesia. Motor Columbus Engineers, Baden/Jakarta.

82 Gunnerson, C.G., McCullough, C.A., and Bailey, T.E. 1966., Sanitary Engineering Div., Am Soc Civil Engrs, New York, 92(SA5):23-45

83. Iijima, N., Editor 1979. Pollution Japan: a historical chronology of environmental and occupational diseases, 1469-1975. Pergamon, for Asahi Evening News, Tokyo.

84. Open University. 1989a. The Ocean Basins: their Structure and Evolution. Pergamon, Oxford, U.K.

85. Open University. 1989b. Seawater: Its Composition, Properties, and Behavior. Pergamon, Oxford, U.K.

86. Open University. 1989c. Ocean Circulation. Pergamon, Oxford, U.K.

87. Open University. 1989d. Waves, Tides, and Shallow Water Processes. Pergamon, Oxford, U.K.

88. Open University. 1989e. Ocean Chemistry and Deep Sea Sediments. Pergamon, Oxford, U.K.

89. Open University. 1991. Case Studies in Oceanography and Marine Affairs. Pergamon, Oxford, U.K.

90. Roberts, P.J.W. 1993. Near field modeling of the Mamala Bay outfalls. Ozturk, I., Editor, Proceedings, International Specialized Conference on Marine Disposal Systems, Iller Bankasi, Istanbul.

91. Schneider, S.H. 1994. Detecting climate change signals: are there any signals? Science 261 (21 Jan 1994) 341-347.

92. Sieburth, J.M. 1975. Microbial Seascapes. University Park Press, Baltimore, MD.

93 Smith, C. 1993. Exporting risk: the unethical trade in hazardous pesticides. Special report, Foundation for Advancements in Science and Education, Los Angeles, CA

94. World Resources Institute. 1993. World Resources 1992-93. Oxford, New York

95 World Resources Institute. 1994(. World Resources 1994-95. Oxford, New York.

96. Rasmussen, E.M. 1984.. El Niño: the ocean / atmosphere connection. Oceanus, 27(2)4-12.

97. Pineda, J., 1994. Internal tidal bores in the nearshore: warm water fronts, seaward gravity currents, and the onshore transport of Neustonic larvae. Jour. Marine Res. 52(3) 427-458.

3 Ecological Design

People live in cities because there they can secure what they need and want at less economic cost than they can in towns and villages. Their immediate needs include ready access to water, food, and shelter. These are met in the near term at increasing resource and environmental costs in the surrounding and supporting area. In time, public health becomes an issue because crowding and enhanced disease transmission threaten it. After the basic needs of nutrition and shelter are met, not always easily, the demands for individual and community health can be addressed. It is at this point that DeAnne Julius' (46) argument that the objective of achieving public health benefits is more important than measuring them becomes paramount.

This chapter provides information on the rationale and selection of design criteria for the protection of public health and of the natural ecological systems of which man is a part, with due regard for the interwoven aspects of economics, ecology, and ethics recently reviewed by Daly and Townsend (41).

3.1 Public Health

The risk of communicable disease transmission through drinking water and the benefits of water treatment systems that range from simple boiling to conventional physical and chemical works are well known. In contrast, information on the marginal health or ecological benefits of alternative levels of wastewater treatment for disposal or reuse is not. This has led to inferences being drawn from the limited amount of credible epidemiological data on swimming in marginally polluted ocean waters reported by Cabelli (3), and on wastewater pond systems for agriculture and aquaculture in developing countries (35, 54).

What is known is that public health and other human rights are matters of increasing concern in water and sanitation, particularly in urban areas Affordable appropriate technologies that adapt modern practices to traditional ones that can be upgraded as more resources become available have been developed for low-income country environments (46) State-of-the-art technologies serving zero-discharge policies are evolving although expensive symbols of progress and modernity. A more serious analytical problem in technology-driven policies lies in programs becoming impervious to rigorous priority setting because, as their scale increases, post-audits of their marginal health and ecological benefits become less interesting than building new projects.

Table 3.1. Pathways and interventions in waterborne disease transmission

(1) ◊1	(2)	(3)	(4)	(5) ◊2	(6)	(7)	(8)
Excreted---->Persistence------>Latency------>Resistance------>Infective-------->Immunity------>Infection[3]---->Sickness[4]							
Load (host #1)[5]	in the Environment	Multipli-cation		Dose	(latency)		(host #2)
Alternative source, pathway, or reinfection by Host #2	From minutes for some viruses to 1 year for helminths (e.g. *Ascaris*) Environmental factors[7]	Intermediate host (snail. mosquito, rodent, dog, monkey, flies, ticks)	Subcliniccal effect, carrier	From 1 for helminths to 10^6 to 10^9 for *Salmonella* . to10^{11} for *V. cholerae*[6]	None for helminths. Variable for bacteria . From none to lifetime for some viruses	Death (mortality)	Recovery (morbidity)

TECHNOLOGICAL INTERVENTIONS

SANITARY / ENVIRONMENTAL / WATER ENGINEERING [8]

Separation	Drainage	Irrigation	Shelter
Removal	Larvicide	Drainage	Shelter
Treatment	Pesticides	Quarantine	Transportation
Reclamation	Weedicide		Irrigation

TRADITIONAL MEDICINE
Wellness model (man with nature)
Exposure, Quarantine, Pain control

MODERN MEDICINE
Sickness model (man over nature)
Vaccination, Inoculation,
Invasive diagnosis and treatment, Pain control

EDUCATIONAL AND CULTURAL INTERVENTIONS
Taboos, Hygiene Training, Education, Public Participation, Technology Transfer (information, skills, and practice)

Notes to Table 3.1: (1) Quantifying potential risk of environmental exposure begins at this point with environmental studies (2) Quantifying risk of exposure ends here as an essential but insufficient element in risk assessment; it is often used to support zero-discharge propositions. Credible risk assessment continues with epidemiological surveys to determine risk of infection also necessary for cost-benefit and priorities analyses. (3) Measured as incidence (new cases per thousand per year). (4) Measured as prevalence (total cases per thousand). (5) Can be reinfected by host #2, particularly under crowded conditions in homes and communities, and for some pathogens in swimming pools and bathing beaches. (6) Manson-Bahr, 1972: "Pure cultures have been swallowed many times and in only one case has true cholera been produced." (7) Adaptation, crowding, competition, parasitism, predation, seasonality, pollution (e.g., Manson-Bahr, 1972: "Vibrios can live for 2 weeks in reservoir water but contaminated water is unfavourable for their survival."). temperature, toxicity, etc. (8) Technology selection issues include appropriate technology, comparative costing, complexity, construction employment benefits, cost recovery, cultural norms, discharge standards, differences in borrowing and disbursement targets, information and technology transfer, institutional framework, market forces, matching capacity to demand, modernity, population density, public participation, resource recovery, sustainability, water service levels, scale, and willingness to pay. See. Chapter 12. Sources: Feachem, et al (6), Kalbermatten, et al (46), Shuval, et al (35), Wilcox and Manson Bahr (61).

Engineering and medical interventions along the fecal-oral route of infection are identified on Table 3.1. Although the emphasis in this manual is on the engineering interventions, these are usually made on the basis of anticipated health benefits. Both the incidence and prevalence of disease are variable and site-specific, so that the best practice in setting priorities for improvements in water supply, sanitation, and drainage requires credible epidemiological evidence as well as environmental, economic, and institutional analyses. The diamonds on Table 3.1 indicate the decision-making areas where the environmental and epidemiological surveys are used in technology selection and in developing the educational, often cross-cultural, interventions that usually accompany the structural ones. Both prospective and retrospective epidemiological surveys are important, and the latter are an essential element of post-audits of environmental programs and projects.

Wastewater technology selection is increasingly driven by policy decisions of international, bilateral, national, state, community, and non-governmental organizations on issues ranging from environmental to effluent standards, and from there to trade policies and related market-driven forces. Programs and protocols published as AGENDA 21 by the 1992 United Nations Conference Environment and Development are being increasingly applied to waste management policies. The applicability of these factors and recent trends from receiving water- to technology- to risk- to zero discharge-based standards mandated by market forces in the United States, are discussed in Chapter 10.

Municipal sanitation and sewerage systems provide health benefits by separating people from their excreta at the household and community level. The myriad household problems thus solved are aggregated into a municipal problem of collection, treatment and disposal where the whole is never less than the sum of its parts. Conventional primary and secondary sewage treatments were developed to protect ecological systems by reducing suspended solids and oxygen demand of organic wastes that would otherwise create fish kills and nuisances in fresh waters. They were not designed for removing pathogens, so that chlorination was added to destroy bacteria and protozoa in trickling filter, activated sludge, and physical-chemical effluents containing, say, 15 mg/l suspended solids.

Meanwhile, astronomical numbers of every water-borne pathogen known to man have long been discharged to marine, estuarine, and other surface waters. This observation lead HI. Shuval to ask why aren't more people sick? (59). The answer is that intervening factors reduce the incidence of infection. Following Feachem, (6), Shuval, *et al,* (35) demonstrated defined the intervening factors listed on Table 3.1 as pathogen persistence in the environment, latency that provides for increasing concentrations of pathogens in intermediate hosts, the minimum infective dose, host immunity, and the relative importance of alternative pathways of infection. High developing country occupational exposures resulted in a low incidence of excess infection due to viruses, medium excess infection due to bacteria, and high excess incidence due to helminths. Much lower excess incidence is found in industrial countries where, because of better nutrition, hygiene, and shelter, bacteria and protozoa are demonstrable problems, followed by viruses and finally helminths. These findings make it possible to set site-specific priorities on investments in wastewater treatment and receiving water monitoring based on their marginal health benefits in both industrial and developing countries.

3.1.1 Marine Recreational Water

The same intervening factors operate in marine recreational waters where any excess incidence of infection due to swimming, salt spray, or other contact activity lies at or below the limits of finding it. This has lead some U.S. governmental and nongovernmental entities to switch from bacteriological standards for receiving waters to technological standards that are operationally defined as a minimum of secondary (biological) treatment (49). Figure 3.1 summarizes the only definitive epidemiological survey to date made by the U.S. Environmental Protection Agency of the effects of swimming in polluted marine waters (3). Emphasis was upon substituting enterococci for coliforms as an indicator of pollution. The survey was designed to identify excess incidence of gastro-intestinal upsets among New York City residents, Alexandria, Egypt, residents, and visitors to Alexandria, beaches, mostly from Cairo. Predictably, this work confirmed that New York City marine recreational waters are less polluted than those of Alexandria; that New York City residents are more susceptible to infection from a given level of exposure than are Egyptians with a higher level of acquired immunity; and that immunities respond to locally dominant strains of pathogens. Cabelli (3) also reported that incidence among children (aged < 10 years) swimming in polluted water was about 1.5 times that for adults and concluded that much of peoples' immunity was acquired during childhood. This is consistent with the findings of Shuval et al. (35) on the roles of immunity and minimum infective doses summarized on Table 3.1

Coliform Bacteria in Sea Water. Concentrations of coliform bacteria are principal indicators of possible risk of infection from food, and from water used for drinking, cooking, bathing, irrigation, aquaculture, fisheries, and recreation. The same intervening factors that limit infection from high risk occupational and consumer exposures from wastewater irrigation apply, although with less intensity, to swimming in seawater. The combined effects of the environmental factors can be measured and used for system planning and design. Some investigators of water quality have advocated using other microorganisms in water quality decisions (3, 6, 35). Others advocate risk assessment models that lead to the institutionally attractive but thermodynamically flawed technological goal of zero pollutant discharge (see Chapter 16). Credible priorities based on these preferences require much more epidemiological evidence than is available even in high-income industrial countries. So, the use of coliform indices continues.

In situ studies of coliform persistence in coastal waters have been used since 1955 to develop functional criteria for ocean disposal works (10). Samples are taken periodically from a portion of a continuous discharge from a channel or outfall The water to be sampled is tagged by dye, drogue (floats with large submerged surfaces), fluorescent or radioactive material, and followed from the point of discharge for as long as is feasible. This measures the combined effects of dilution, sedimentation, and mortality, all of which are site specific.

Location, time, depth, temperature, salinity (conductivity), sea state and weather are recorded. Most field work is done during daylight hours that correspond to most recreational water use, and include afternoon sea breezes that push the upper layer shoreward. Vertical mixing in the surf zone and resuspension of sediments attenuate bactericidal effects of sunlight and enhance effects of solar heating Sedimentation of effluent solids is no longer a factor and dilution measured by salinity or other tracer is reduced to one half of its offshore value. Samples should be assayed on board or, after being iced, as quickly as possible, on shore. Table 3.2 lists representative results.

Meanwhile, there is an indigenous research interest in separating coliform decay rates into as many factors as possible (49, 53). Some "field studies"use floating plasticbags inoculated with serial dilutions of effluents in sea water so that the overriding effects of sedimentation are removed. Most are in laboratories seeking descriptive models of neat disaggregated effects on coliform mortality in different kinds of water inoculated with raw sewage, of environmental variables including temperature, sunlight, phage, speciation in ocean chemistry, biological synergism and antagonism, etc. Assesssing increased bacterial resistance induced by massive use of antibiotics. requires due regard for the ecological reality that host-parasite relationships are dynamic; if all the hosts die, so do the parasites, so that at least some of the hosts must survive (see Sec. 12.5.2 for the analog from economics known as the prisoners dilemma). Because they are cheaper and useful for academic research, than *in situ* studies, laboratory studies are more abundant (5, 49) but they cannot be used for design or monitoring Meanwhile, simplicity in enforcement supports end-of-pipe or other geograhically defined standards that promote moving sedimentation onto the shore, a measure that also serves the expensive American zero-discharge persuasion. In sum, where environmental and economic costs and benefits of receiving water standards are more important, *in situ* determinations of T$_{90}$s continue to be more appropriate for system design.

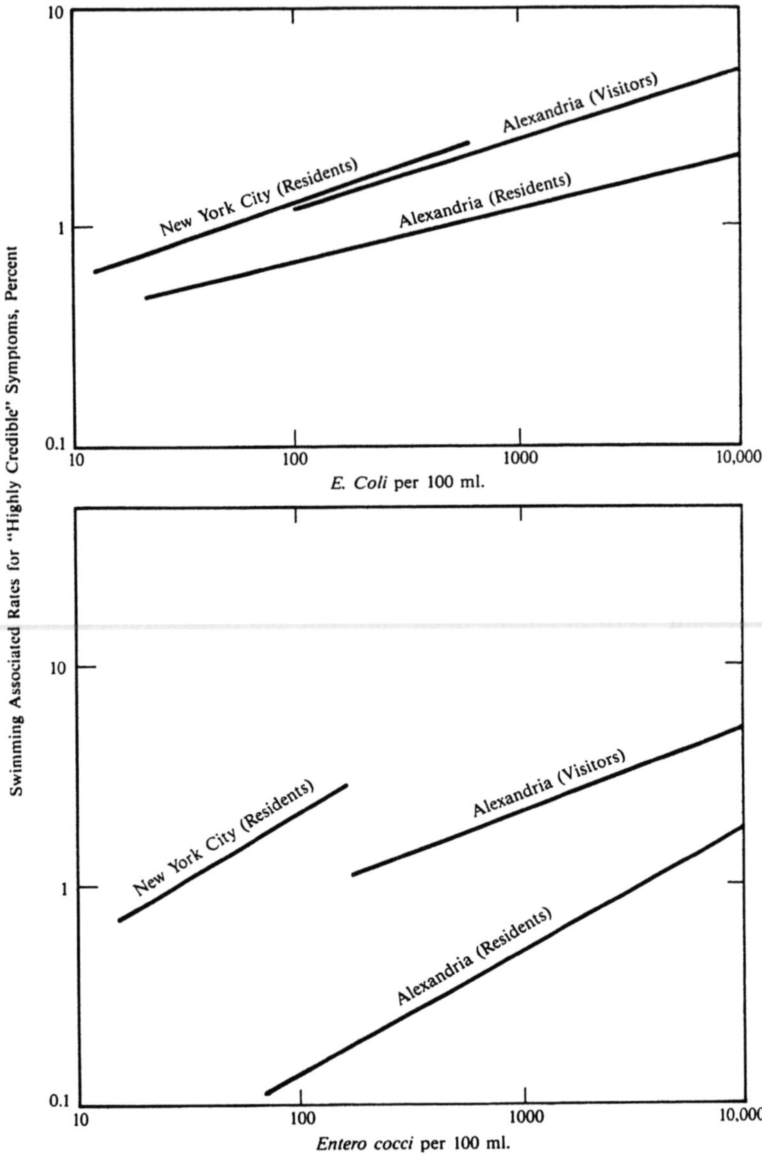

Figure 3.1. Reported gastro-intestinal effects of swimming in polluted waters of New York City and Alexandria, Egypt (after Cabelli, ref. 3)

Table 3.2. Reported *in-situ* persistence of coliform bacteria due to mortality and sedimentation in ocean waters in hours for 90 percent reductions [1] and as $k_e =$ ln $10/T_{90}$. Persistence is least for raw sewage and greatest for secondary effluent

Location	Date	T_{90}(h)	k_e(1/h)	Remarks
Raw or screened sewage [2]				
Honolulu	1970	< 0.75	> 3.1	T_{90}s varied from 0.2 h during trade winds to 0.75h other times
TItahi Bay, NZ	1959-60	0.65	3.5	
Rio de Janeiro	1963	1.0-1.2	1.9-2.3	
Israel	1975	<1.0	>2.3	Polio virus T_{90} > 24 h
Istanbul	1968	0.8-1.7	1.4-2.9	Mean value 1.1 h
Genofte, Denmark	1970	1.2	2.0	
Tema, Ghana	1964	1.3	1.7	
Nice, France	1966	1.1	2.1	
England	1965	0.78-3.5	0.66-2.9	
Manila Bay, PI	1968-69	1.8-3.4	0.67-1.3	High T_{90} w/partly old effluent
England	1969-73	1.4-5.3	0.43-1.6	Median of 11 results 3.2 h.
Mayaguez Bay, PR	c1982	0.7	3.3	
Montevideo	c1982	1.5	1.5	
Santos, Brazil	1986	0.8-1.7	1.4-2.9	
Porlaleza, Brazil	1986	1.1-1.5	1.5-2.1	
Maceio, Brazil	1986	1.2-1.5	1.5-1.9	
Alexandria, Egypt	1978	0.4-0.5	4.6-5.8	1.0h T_{90} used adopted for institutional reasons
Primary effluent				
Ventura, California	1966	1.7	1.4	
Seaside, New Jersey	1966	1.8	1.3	
Orange County CA	1954-56	1.8-2.1	1.1-1.3	
Santa Barbara, CA	1967	2.4	0.96	
Los Angeles, CA	1954-56	4.1[3]	0.56	
Secondary effluent				
Los Angeles, CA	1954-56	9.6[4]	0.24	
Combined				
The Hague	1968-69	5-175	0.01-0.5	Higher T_{90} value may represent mixing with old effluent or resuspended sediments

[1] The overall T_{90} is the harmonic mean of mortality (competition, predation, sunlight, etc.), sedimentation, and dilution: $1/T_{90o/a} = 1/T_{90m} = 1/T_{90s} + 1/T_{90d}$.

[2] Includes effluents from grit removal, flotation, comminution, and/or screening

[3] T_{90} due to mortality (T_{90m}) 17.8h + sedimentation (T_{90s}) 5.3h. Chlorinated primary effluent T_{90m} 42h. Both are within the range of laboratory determinations but are operationally negligible term because of dominance of sedimentation.

[4] T_{90m} = 17.8h. T_{90d} = 21h.

Sources: Feachem, et al (6), French (44), Gameson (7), Gunnerson (12), Ludwig, R.G. (personal communication, 1986), and Pearson (29,30).

Reported times for 90 percent reduction are usually those for mortality plus sedimentation. They range from 0.2 to 2 hours for raw sewage, from about 2 to 4h for primary effluent, and increase to 9.6h for secondary effluent. Chlorination reduces initial concentrations of coliforms, but those remaining are more persistent. The City of Los Angeles' Santa Monica Bay studies of Hyperion effluent are the only ones reported to date in which environmental conditions made it possible to separate the *in situ* effects of mortality from flocculation and sedimentation. Mortality is defined as the combined effect of competition, predation, sunlight, and other environmental factors Chlorination of primary effluent increased the T_{90} due to mortality alone from 17.8 to 39.5 hours, values that are within the ranges of most laboratory findings.

The site- and effluent-specific nature of sedimentation factors has been confirmed in the laboratory by Hering and Abati (13). Figure 3.2 shows differences both in initial flocculation times and in subsequent sedimentation for six different Southern California effluents. Their results are qualitatively consistent with earlier empirical work on effluents (11, 22, 36) and more recent work on sludge and artificial seawater mixtures by Koh (20, 40), Hunt (16), and their associates.

Appropriating site-specific decay coefficients due to mortality and/or sedimentation from one location to another opens the door to error, particularly when adopting high values for T_{90} or low values for k_e. For clarity in explaining field results and linking them to design criteria, both T_{90}s and their k_e equivalents should be presented since very few engineers and even fewer others think to the base e (see Tschobanoglous and Schroeder (55), pp 363-373). Although design travel times and outfall lengths are directly proportional to design T_{90}s, outfall costs are not and "conservative" decay factors can be expensive. Costs increase linearly with pipe diameter, plus step functions imposed by changes in constructions practices (see Chapter 12). It is more conservative to determine site-specific T_{90}s. More important, operational experience demonstrates that, when bacteriological criteria are met as around Los Angeles' Hyperion outfall or Rio de Janiero's Ipenema outfall, cultural, aesthetic, and other requirements are also met (7, 17, 24, 25, 29, 30, 33, 60).

Microbiological Standards for Bathing Waters. No distinction is usually made between fresh and salt water bathing when setting microbiological standards, criteria, or guidelines except where reclaimed waters are involved. Table 3.3 lists numerical limits on enteric microorganisms in the environment set by international, national, and California authorities. The scarcity of credible epidemiological data showing that swimming in polluted water is hazardous is revealed by the differences in both concentrations and classifications of pathogens or indicators to be monitored. It follows that risks of infection by pathogens in sea water have not been considered sufficient to warrant the costs of more definitive epidemiological surveys Where the designs are based on environmental rather than technological standards, the criteria listed in Table 3.3 can be used for designing receiving water studies and the disposal works to meet those values. Microbiological samples should be assayed as quickly as possible by inoculating the growth media on board the research vessel or, after being iced, as quickly as possible on shore.

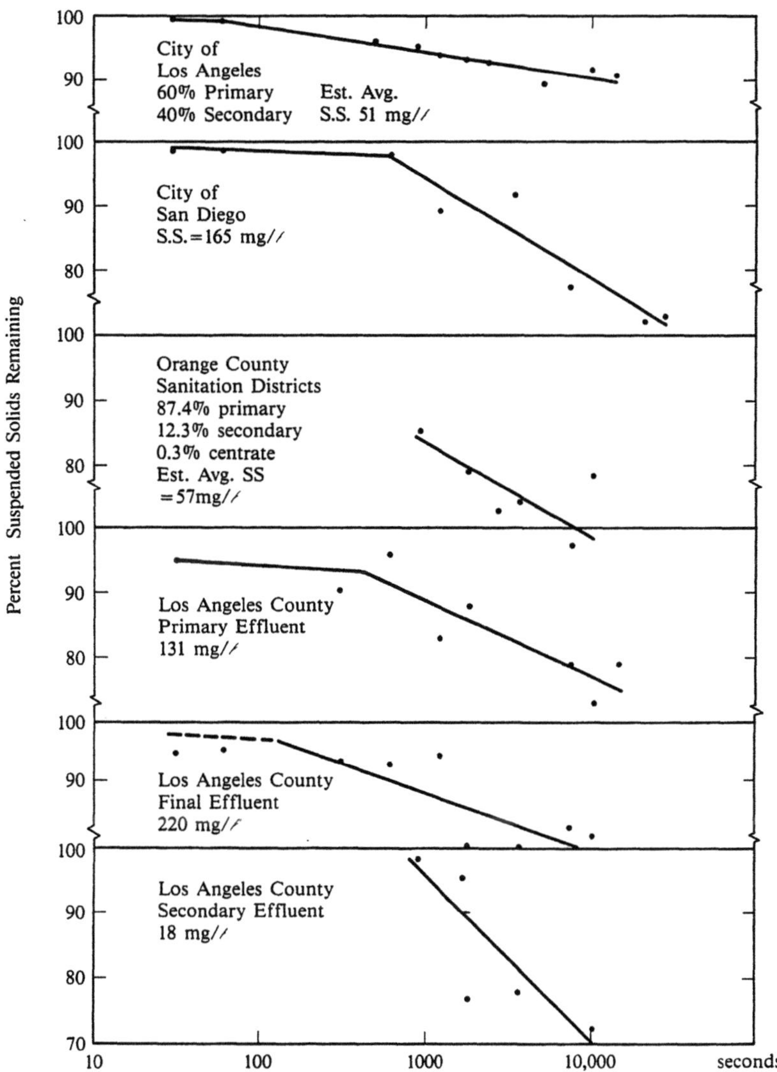

Figure 3.2. Laboratory settling velocities for Southern California effluent solids in seawater at 80:1 dilution (after Hering and Abati, 13).

64

Table 3.3. International and national requirements for ocean bathing water

	Total coli per 100 ml	E. coli per 100 ml	Fecal coli per 100 ml	Fecal strep per 100 ml	Other
WHO					
1977 guideline	1000 (g.m.)	–	–	–	–
EC					
1976 guideline	500	–	100	1[1]	salmonella[1] enterovirus[1]
limit	10000	–	200	--	0 salmonella/L
1981 guideline	95%<100	95%<2000	90%<100	96%=0	90%=0 enterococci
FRANCE standard					
1969 very good	<50	<20	–	–	–
good	50 to 500	20 to 200	–	–	--
moderate	500 to 5000	200 to 2000	–	–	--
BRAZIL					
1976 satisfactory	80%<5000	–	90%<1000	–	–
CUBA					
1986	1000 (geom mean of	=	200 (g m)	=	=
	5 samples aver 30days)	–	90%<4 00	–	–
MEXICO					
–	80%<1000,100%<10000	--	–	–	–
JAPAN					
1981	1000	–	–	–	--
USSR					
1977	100	–	–	–	–
USEPA					
1978 criteria	–	–	90%%<4 00	–	–
1986 criteria	–	–	200 (g.m.)	35 (g.m.)	–
CALIFORNIA					
1943 standard	1000 (80 pctl)[2]	–	–	–	–

[1] - to be checked if quality of water deteriorates

[2] - Equivalent to an empirically verified geometric mean (g.m.) of 2.3 B.Coli/cc (Gunnerson 1955). Earliest reference to the 10/cc B. coli index (indicated number) is Rawn and Palmer (1929), later promulgated by the California Board of Public Health as corresponding to edge of visible sewage field in sea water and to prevailing USPHS drinking water standards along with amount possibly swallowed by a swimmer. The 1978 technology-based standard for reclaimed domestic wastewaters tin recreational fresh water impoundments was set two orders of magnitude lower with a median value of 2.2 coliforms/100 ml with not more than one sample in excess of 23/100 ml during a 30-day sampling period The same limit is imposed for spray irrigation of food crops where spray imposes putative occupational or innocent passage contact (California Administrative Code, Title 22, Chapter 3, Register 77, No. 42, 10-15-77)

Additional sources - Degrémont (42), Fresinius (43), Salas (53)

3.1.2   Shellfish and Finfish

As indicated on Table 3.1, the minimum Infective dose for pathogens varies from a single egg for helminths to one million or more for salmonella. Such large numbers are extremely difficult to remove from the water by technological methods, but a shellfish can ingest a sufficient portion of them to become contaminated. Oysters, clams, and mussels are particularly susceptible since they are attached or sedentary during their adult life and concentrate microbial and other constituents, nutrients and contaminants, from their surrounding environment. Since sedimentation creates a sink for infectious agents attached to large particulates, it makes these pathogens more available to the benthic shellfish located near the contaminant source (9). The shellfish themselves are not harmed by sewage microorganisms, but humans are exposed to a number of diseases--for example, typhoid fever, dysentery, and other gastrointestinal illnesses. Cooking would make them safe, but many people prefer to eat raw fish and shellfish, particularly bivalve molluscs.

Most shellfish are filter-feeding animals that concentrate and incorporate pathogens and toxic elements or compounds from the environment so that coli 90%<4.3, 30d g.m.14), California (total coli 90%<70), Mexico and Venezuela (same as U.S. for total coli), Peru (total coli 80%<1000 and fecal coli 80%<200), and Japan (total coli 70). and Japan (53). They are not 100 percent effective, and even "acceptable" areas may contain contaminated shellfish. Where pathogens and other contaminants in the edible portions of shellfish exceed standards for human consumption, these levels may be brought into compliance by removing the shellfish to clean waters for a period of time. The two methods commonly used to rid shellfish of contaminants are relaying and depuration. Relaying involves placing contaminated shellfish in clean areas and then reharvesting them, usually several months later. This method was used in the mid 1970s, for example, to treat oyster beds in a small portion of the eastern United States that had been closed because they were contaminated by a synthetic organic pesticide. The young seed oysters were relaid in uncontaminated waters, where they matured and were later harvested with no trace of chemical contamination (15). Depuration involves placing the contaminated shellfish in tanks with clean water. The time required for purification is determined by water temperature, salinity, initial bacteriological quality (or contaminant level), and species of shellfish (14). For sewage-contaminated shellfish, 24 to 72 hours may be required.

The Minamata incident has given rise to concern, sometimes exaggerated (see Chapter 15), over the potential effects of toxic elements and substances in seafood. Table 3.4 shows that even extremely small amounts of some toxic materials are considered hazardous. However, daily exposures to lower levels in food are considered normal for some or all of these toxic elements and do not result in any known health effects (19). These safety limits are almost always met in seafood, although fish and shellfish in certain limited urban coastal areas occasionally surpass these levels for some chemicals and, as a result, fishing is restricted.

People with a high seafood intake are obviously the ones most likely to be affected by any toxic materials present in edible fish and shellfish tissues. One study (34) investigated the possible health effects of the consumption of fish from an area of the United Kingdom with known sewage sludge inputs containing

Table 3.4. Guideline Values for Toxic Substances in Foods

Element or Compound	WHO/FAO Provisional Tolerable Weekly Intake (mg/person)	USFDA Action levels, ppm
Cadmium	0.4-0.5	–
Lead	3.0	–
Mercury (total)	0.3 (seafood, edible portion)	1.0
Methyl mercury	0.2	—
Chlorinated pesticides	0.03-10 (varies with type of compound and food)	
PCB's	–	5.0 (in fish)

Note: Not applicable for infants or children.
Source: WHO (39).

mercury. A population with an above normal seafood intake for the United Kingdom was chosen. Their fish consumption averaged 0.36 kg/person/week. During the study, the blood levels of mercury in this population rarely exceeded the 20 mg/ℓ of mercury, one tenth of the level that the an World Health Organization has reported is the minimum necessary to cause an adverse effect.

Although toxic metals other than mercury in seafood are believed to be a lesser hazard to human health, another recent study indicated that safe intake levels of cadmium may occasionally be exceeded. In Hong Kong, where both concentrations of metals in seafood and seafood consumption are high, a sampling of retail fish indicated that cadmium levels were approaching safety limit levels (31).

A number of naturally occurring interactions between heavy metals and other metals or compounds take place in the water, sediments, or marine organisms. Some of these interactions significantly increase the toxicity of the metals; others reduce it. For example, selenium and mercury, both of which are highly toxic, exist as trace elements in the sea and in fish, function as antagonists; their interaction decreases the toxicity of either one or both (8). Vitamin E, a nutrient found in fish, is effective in decreasing the toxicity of both mercury and silver. In contrast, toxicity is enhanced in combinations of zinc and cadmium, zinc and cadmium and calcium, lead and barium, etc. (see Section 2.3.2.)

Operationally, both subclinical and acute health effects of heavy metals or synthetic organics on marine organisms and their predators (including man) are unlikely if outfalls are located away from shellfish beds and contaminants are controlled at their points of manufacture and utilization. Where contaminants have entered the ecosystem, shellfish depuration or relaying in clean water is often an effective tool for removing pathogens and some mineral toxicants.

3.2 Design for Marine Ecosystems

It is widely recognized that ecological systems must be protected if we are to ensure the health and economic well-being of future generations. Evidence of the need for remedial and preventive environmental programs comes from natural and man-made ecological disasters such as that at Bhopal, India (isocyanates), Seveso,

Italy (dioxin), Minamata, Japan (methyl-mercury), Kesterson, California (selenium), and from longer-term conditions arising from acid rain, overfishing (Paragraph 2.3.7), ozone layer depletion, and global warming. Although generic command and control measures tend to be scientifically opaque and technologically excessive, they are legally attractive with their appearance of near-term simplicity and, once in place, their resistance to political change. AGENDA 21 and the Convention on Biodiversity adopted by the 1992 United Nations Conference on Environment and Development represent current efforts to set long-term goals.

Meanwhile, uncertainties abound in ecological protection measures. There are no accepted numerical standards or criteria for marine ecological protection from waste discharges as there are in public health, nor are sufficient integrated research and development programs under way to derive them (49, 47, 59). Ecological assessment procedures are based on nominal (equal or unequal) or ordinal (adding more than or less than) data that require analysis by non-parametric statistics. These do not suffice for engineering designs in marine ecosystem protection that require constant interval values which can be added, subtracted and by including a zero, multiplied or divided to form ratios. The latter evolve from a mix of engineering judgment, long-term ecological concerns, and increasingly from ethical and religious values so that near term probabilities can be assigned to the outcomes of physical works. Controversy and competition are assured.

3.2.1. Eutrophication

Eutrophication has been defined by Likens (21) as "the nutrient enrichment of waters which results in stimulation of an array of symptomatic changes, among which, increased production of algae and macrophytes, deterioration of water quality and other symptomatic changes are found to be undesirable and interfere with water uses". This widely accepted definition focuses on the problems of eutrophication.

Some examples of acute eutrophication and anoxia are found in marine waters where restricted water circulation or strong stratification limits the dilution of nutrients and oxygen-demanding material and the resupply of dissolved oxygen. These areas include estuaries and semi-enclosed seas (e.g., the Baltic Sea), fjords, archipelagos, bays, lagoons, and some continental shelf regions. They are not known to ever extend to the open ocean (32).

When the enhanced plant production causes excessive levels of organic matter to concentrate in the water column and sea bottom, bacterial decomposition will locally deplete the oxygen levels. Mobile organisms may leave the low-oxygen water, and sedentary organisms usually die. Under conditions of more moderate nutrient enrichment, however, overall system productivity of some planktivorous, commercial fishes (e.g., menhaden) may benefit from the abundance of phytoplankton, while others may suffer declines.

River basin development with seasonal impoundments and releases of water for downstream irrigation and agricultural development cause eutrophication and anoxia in coastal areas. Tolmazin (56) has described the characteristics and effects of increased summer outflows from the Danube, Dniester, and Dnieper Rivers into the Black Sea. Excess runoff is stored during the rainy season, diverted later on for irrigation, enriched with agricultural chemicals, and returned to the river with increased dissolved solids concentrations due to evapotranspiration   Enriched fresh water outflows during the irrigation season float over the heavier salt waters of the Black Sea.. Here they are warmed and further stratified by solar radiation. They become natural oxidation ponds where algae bloom and die, sink, decompose, and remove dissolved oxygen from the lower layer. Although mobile organisms may leave, sedentary ones usually succumb and die. Under conditions of more moderate eutrophication, however, overall system productivity of some planktivorous, commercially important fish (e.g., menhaden) may benefit from the abundance of phytoplankton, while others decline.

Algae utilize carbon, nitrogen, and phosphorus in specific ratios that are related to their metabolic requirements. The molar (Redfield) ratio for C:N:P in mixed marine plankton is 100:15:1 , the routinely determined ionic ratio for N:P is 7:1. Algae production is regulated by the nutrient in least abundance, which is called the limiting nutrient. Carbon can be limiting in enriched estuaries. In open waters, nitrogen is not the limiting nutrient since some blue-green algae common to marine and fresh waters are able to fix (i.e., convert into a usable form) nitrogen from the atmosphere. Phosphorus may be a limiting nutrient since, in well-aerated waters, it is often bound to oxygen in the sediments in the form of phosphates and is not immediately available for use by the algae. As organic matter in the sediments is decomposed by bacteria, the phosphates may be released to combine with nitrogen and stimulate plankton growth. Other factors that can limit plankton growth are temperature and light. Light availability may be limiting when turbidity is high. The effect of nutrient enrichment on algal production depends on three factors:

1. Whether the added nutrient is limiting. If so, its addition will increase algal production.
2. Whether luxury uptake occurs. This occurs when a nutrient is taken up by an organism but is not used immediately. In this case, the addition of even a non limiting nutrient might increase production, if the luxury nutrient then becomes limiting.
3. Whether the nutrient is in a preferred form. For example, an organism expends more energy in the taking up nitrogen from nitrates or nitrites than from ammonium.

Blue-green algae dominate when phosphate enrichment occurs, particularly in the fresher regions of an estuary, but blue-green dominance is less common in areas of higher salinity. When blue-green algae are dominant, the diversity and stability of the normal phytoplankton community decline. Blue-green algae are inedible to many plankton-feeders.

Attempts to control eutrophication should begin with determining whether eutrophication is natural or artificial. This is a necessary first step, since little can be done to prevent natural eutrophication.   Anthropogenic eutrophication generally is due to runoff from fertilized agricultural and urban areas, and / or nutrient-rich wastes discharged to sensitive waters.

The first pathway may be controlled by improved land use, vegetation buffer zones or engineered land-treatment systems. The second is likely to require higher levels of treatment before discharge or moving the discharge location. Both oxidation pond systems and tertiary sewage treatment can remove large amounts of nutrients from waste discharges.

The natural concentration of dissolved oxygen in coastal waters is generally adequate to support marine life. Hypoxia occurs when the dissolved oxygen diminishes to a point at which the organisms of interest can no longer survive or reproduce. In marine waters, this is at about 2 mg/ℓ. Hypoxic conditions often degenerate to anoxic conditions. Anoxia occurs below the level at which oxygen can be measured, As conditions become anoxic, sulfate-reducing bacteria digest organic material and release toxic hydrogen sulfide into the water. A number of basins in the world such as the Black Sea are characterized by naturally occurring anoxic conditions.

Some fjords have natural anoxic basins owing to reduced circulation and extended residence times below the depth of the sill at the mouth of the fjord. The anoxic bottom waters occasionally affect the oxygenated surface waters when coastal storm surges cause large intrusions of oxygenated seawater to flow over the sill. This forces toxic, deoxygenated bottom waters rich in hydrogen sulfide toward the surface, which may kill organisms living there. In certain temperate locations, anoxia occurs seasonally. Coastal waters in temperate climates become stratified only in summer. Phytoplankton productivity increases in the spring as a result of increased temperatures and solar radiation, and an abundance of nutrients. Subsequently, organic matter accumulates on the bottom, and as it decomposes oxygen concentrations in the bottom waters decrease. In severe cases, this gives rise to restricted circulation, and to anoxic conditions. Nutrient and organic inputs from waste discharges and land runoff can contribute to the formation of anoxic conditions. As in the case of eutrophication, it is difficult to distinguish between anoxic conditions caused by waste discharges and runoff from those that occur naturally since waste discharges and runoff may in fact contribute monotonically to naturally existing hypoxic or anoxic conditions or tendencies (see Section 15.2).

Whereas eutrophication and anoxia may cause permanent damage to the ecosystems of small, land-locked freshwater bodies, these conditions are more likely to be transient in coastal waters. In many areas where eutrophic and hypoxic conditions have led to deterioration in water quality, attempts have been made to correct these problems by controlling nutrient inputs to each of the various segments of the water body. Segmentation can be used as an analytical tool since it treats the water body as an interrelated ecosystem, (45, 50). This concept of segmentation is being applied to water-quality management in, among other areas, (1) the Great Lakes, which have been divided into zones having similar nutrient and chlorophyll a levels for the purpose of monitoring eutrophication, (2) San Francisco Bay, which was separated into six major areas according to flushing characteristics, to determine acceptable sites for discharging treated sewage effluents, and (3) Chesapeake Bay, which has been divided into forty-five segments on the basis of circulation, salinity, and geomorphologic characteristics. The segmentation concept has also been applied to the tidal portion of the Thames River, and it can be used in bays, estuaries, and other coastal waters where circulation is restricted.

Natural systems, particularly tidal and estuarine marshes, are important in removing nutrients from receiving waters. Such areas are characterized by high plant productivity and incorporate the wastes into their yield of organic plant material that supports estuarine species and provides food and cover for waterfowl and mammals. Because of the special value of these natural systems in the role of pollution assimilation, it is important to avoid expedient filling of these systems.

3.2.2 Toxic Wastes

Waste discharges include inorganic and organic compounds that are either suspended in the water column or in the sediments. Some of these chemicals, particularly certain trace metals and organic compounds, are toxic to marine life. As yet, little is known about the physical, chemical, and biochemical processes controlling transfers of toxic compounds within marine ecosystems (28). It is known, however, that the concentrations of most toxic metals do not increase as they move upward through the food web (biomagnification), but that the concentrations of many organic chemicals do. In general, polythlorinated biphenys (PCBs), polyaromatic hydrocarbons (PAHs), dioxins, furans, and some pesticides are presently of greatest concern (2). The uptake of contaminants by organisms does not necessarily harm them nor does increased time in the discharge area always result in elevated contaminant levels. Many marine organisms have developed detoxification mechanisms for metals in their environment, since metals occur in the marine environment naturally, as well as being transported there by waste discharges. Few marine organisms have mechanisms to detoxify synthetic organic compounds, although some may excrete organics and others may detoxify them by various metabolic processes. However, some organics increase in toxicity as they are metabolized and converted to other compounds.

Benthic organisms immediately adjacent to an outfall are most likely to be affected. The size of the area affected depends on the degree of flushing and dispersion in the discharge area, the topography of the sea floor, the sedimentation characteristics of the waste, and the quantity of waste discharged. As the deposited organic matter degrades, dissolved oxygen decreases in the pore waters of the surface sediments. Species of benthos unable to adapt to the altered substrate are replaced by opportunistic species that prefer more organic-rich sediments. In situations where an existing low organic substrate is replaced by organic-rich sediments, the diversity of benthic species decreases, whereas the abundance and biomass of opportunistic species increase. The overall effect of such an alteration of the benthic community is generally unknown. Exclusion of some benthic species may adversely or beneficially affect food-specific carnivores at higher trophic levels, whereas other carnivores may not be affected at all. Researchers in Southern California have observed some changes in the species composition of predatory bottom fish in the vicinity of a treated sewage discharge; this may be a result of a change in benthic species (26). Whether this change was beneficial or detrimental could not be determined.

Finfish can swim away from discharge areas and avoid irritants. However, since waste discharges typically offer readily available food sources, many finfish are usually attracted to the discharge area. Productivity and growth of certain finfish species may actually increase as they expend less energy in obtaining food.

Measurable effects of waste discharges are limited to the immediate area around an outfall. Far-field impacts are rare and any such effects are likely to occur among organisms that have spent time in the discharge area. For example, migratory fish may accumulate toxic substances in a discharge area and transport trace amounts of them elsewhere. Any effect that this has on the fishery, and subsequently on ecosystems, is almost certainly negligible where only a single outfall is concerned. However, multiple outfalls in an enclosed region might give rise to a more significant broad-scale effect. Fin-rot, for example, occurs over large areas.

Despite the concern regarding the toxicity of heavy metals, there is little documented evidence from field observations that metals have a significant adverse impact on marine organisms or their predators (including man), except in the extreme hydrographically and geographically unique case of methyl-mercury poisoning at Minimata, Japan. The only evidence that correlates marine species mortality with metal concentrations in discharge plumes is laboratory data that have not been, and probably cannot be, normalized to actual marine environmental conditions. A heavy metal (indeed, any substance) is toxic to an aquatic organism when present in levels in excess of an organism's tolerance. These levels are ordinarily greater than those found even in waters adjacent to point sources of pollution (1). Tolerance levels of organisms are highly variable between and within species. In any event, neither synergistic nor antagonistic effects of multiple toxicants at low levels are understood.

3.2.3   Minamata – A Special Case

The potential for ecological problems due to industrial releases of toxic heavy metal compounds into the marine environment first gained worldwide public attention when 1n 1951 a number of deaths in Japan.. Human consumption of sea food contaminated with methyl mercury from industrial wastes discharged to Minamata Bay (51).The cause of symptoms that had earlier been observed in the marine ecology and among domestic animals was confirmed by 1968 A second outbreak occurred at Nigata, Japan, between 1965 and 1970. There followed outbreaks in Sweden of mercury poisoning of predatory birds during the early 1960s, discovery in the late 1960s of mercury in fish in Lake St. Claire, and in Lake Erie. fish 8n 1970. The U.S. Secretary of the Interior declared mercury pollution as "intolerable." In 1970 the U.S. Food and Drug Administration set a limit of 0.5 ppm for mercury and began surveillance of fish. One million cans of tuna were withdrawn from the market and the importing and interstate movement of frozen swordfish was banned. Then in 1972 it was reported that analyses of museum samples showed similar concentrations in both tuna and swordfish. from around the world (23). The ban on the 459 million pound per year tuna market was lifted and the one on the 25 million pound swordfish market was retained. These records reveal the difficulties both in setting standards and in changing them.

3.2.4   Synthetic Organics

There is increasing evidence that man-made synthetic organic chemicals pose greater health risks than metals. Many of these compounds persist in the marine environment and tend to accumulate in the fatty tissues of organisms. Although synthetic organics include an enormous variety of compounds, pesticides, polychlorinated biphenyls (PCBs) polyaromatic hydrocarbons (PAHs), dioxins, have received the most attention. Most of these chemicals enter the marine environment in agricultural runoffs of and industrial discharges. Only a small fraction derive from domestic discharges, although effluent chlorination for public health reasons can affect receiving water biota so that dechlorination may be required (50). (see Chapter 16).

Reports of mortalities of marine species caused by organic chemicals are rare, with the notable exception of the death of some captive seabirds, later ascribed to the consumption of fish with a high DDT content. These substances can produce sublethal effects, however, which may alter metabolic, reproductive, physiological, and behavioral responses in marine species. Fin rot disease, which occurs primarily in bottom fishes, has been observed in some discharge areas. The distribution and prevalence of this disease is highly variable and it has been found in fish far removed from discharge areas. There is some evidence that the increased prevalence may be correlated with the presence of chlorinated hydrocarbons and high levels of toxic metals in sediments (28).

Petroleum hydrocarbons, some of which are carcinogenic, continue to increase in the marine environment. Major sources of these compounds are shipping operations and land runoff, rather than municipal discharges.

3.3   Ocean Dumping

For over a century the coastal oceans have been used for planned disposal of dredged material, industrial wastes and sewage sludge. Although ocean waste disposal practices are largely up to the discretion of individual nations, the London Dumping Convention was adopted in 1972 to control ocean dumping of certain wastes from vessels. Regional agreements are found throughout the world (18). Signatories have agreed not to dump certain substances in the sea, and to dump others only with special care. Under the convention, permits for ocean dumping off their shores are issued by the participating states under the auspices of the International Maritime Organization (IMO). The ban was extended in 1985, by the London Dumping Convention Meeting of Parties, to include dumping of low-level radioactive wastes until it is "proven safe." International actions on related issues include the 1973 Convention on Prevention of Pollution from Ships (MARPOL), the 1989 Basel Convention on the Control of Transboundary movements of Hazardous Wastes and their Disposal, and the 1989 agreement between the European Community and ACP (Africa, Caribbean, Pacific) states banning exports of hazardous to countries that lack proper controls. The last two are listed because of the historical risks of accidental discharges of these materials at sea.

Meanwhile, costs of disposing of dredged spoil are escalating in industrial countries, particularly where litigation is attractive. This has led the Port

Authority of New York City to consider reducing the potential of suits over contaminated material by hauling it over 3,000 km to the State of Utah, whose citizens' views are not reported (New York Times 4/1/95).

When dumped at sea, sewage sludge and other wastes are typically released in the wake of a moving vessel, which promotes rapid initial dispersion and dilution. Industrial and acid wastes are normally towed in special rubber lined barges and, similarly, are pumped through discharge pipes at keel level, to provide rapid dilution in the turbulent wake. Dredge spoils, the largest source of solid wastes to the ocean on a mass basis, are usually released *en masse* through doors in the bottom of the dumping vessel. These dredged materials often contain large quantities of coarse sand that accumulate in a pile at the dumpsite, while fine-grained, contaminant-rich particles tend to be dispersed during disposal and through later resuspension.

In general, it is accepted that the effects of dumping onto the continental shelf of sewage sludge are similar to those observed in the vicinity of sewage effluent and sludge outfalls.. Wastes are swept away from a dumpsite or outfall location in directions and at speeds determined by the mean flow of the long-shore current. Dispersion of the sewage particles is also controlled by tidal currents and wave-induced turbulence. In areas where tidal velocities or wave action, or both, are strong, dispersion is more rapid, and any particles in the sewage reaching the sediments near the outfall or at the dumpsite are effectively resuspended and transported away from the discharge site. Sites with strong tidal and wave-induced mixing are often referred to as dispersive sites since little or no accumulation of organic particles occurs in the sediments. In contrast, sites with poor mixing such as those in bathymetric lows are often referred to as containment sites since particles of the sewage tend to accumulate in the sediments at the discharge site. Areas that receive the largest loadings of contaminants in the sediments near the discharge site retain only a very small fraction of the waste, and the sewage particles are widely dispersed and incorporated into the sediments. The current practice in selection of disposal sites is to provide the best dispersion characteristics and, therefore, to minimize accumulation of sewage particulates in the sediments at the discharge site.

Note added in proof. The high and increasing costs and uncertain long-term environmental effects of processing and/or disposing of sewage sludge onto land are strong incentives for research into new technologies that promise financially and environmentally sustainable means for ocean disposal. The spectrum of solid materials ranges from very hazardous radioactive wastes to sludges from domestic sewage treatment An early suggestion was for placing (permanent) containers of radioactive wastes in subduction zones where oceanic plates and their overlying sediments are slowly sequestered underneath continental plates; this idea was abandoned when estimated times for subduction were much greater than the half lives of the wastes.

More recently, concepts for abyssal isolation, that is distinguished from ocean dumping, of sewage sludges (20% solids), municipal incinerator fly ash (85% solids), and contaminated dredge spoils (32% solids) have been the subject of research. Conceptual designs have been made by the U.S. Nany Research Laboratory of systems using very large transport vessels of 25,000 dead weight tons discharging directly to the sea floor through tethered pipes, or released in geosynthetic fabric containers from the surface or from submarine gliders to bottom depths of at least 3,000 m. Contingent financial costs of the ship-borne portion of the costs for comparing the three concepts are from about $15 to $35/ton for large service areas and land transport systems These costs have been considered promising enough to continue the investigation.

The research is continuing, particularly into hydrodynamics of settling particles and containers, and into distributions of bottom currents and dispersion patterns. Questions relating to marine snow, the extent of abyssal storms seaward of the western boundary current regions, and other environmental factors are being considered. The permanence of the geosynthetic fabric containers is a matter for future research, since this predicates the evolution of steady-state conditions.

(Source: Valent, P.J., Young, D.K. and Palowitch, A.W. 1996. Technical and economic assessment of waste isolation on the abyssal seafloor. Oceanology International 96, Conference on the Global Ocean, Proceedings, Ocean House, 560 Kingston Road, New Malden, Surrey KT3 3LZ, UK.. Vol. 1, pp 305 - 315.)

3.4　　　References

1. Boesch, D. F. and M. H. Roberts. 1983. Biological effects. In Myers, ref. (27) 425-517.
2. Breteler, R. J. 1984. Chemical pollution of the Hudson Raritan Estuary. NOAA Tech. Memo.. National Oceanic and Atmospheric Administration, Washington.
3. Cabelli, V. J. 1983. Health Effects of Criteria for Marine Recreational Waters. Pub. no. EPA-600/1-80-031. U.S. Environmental Protection Agency, Washington.
4. Cabelli, V. S., M. A. Levin, and A. P. Dufour. 1983. Public health consequences of coastal and cstuarine pollution. In Myers, ref. (27), 519-5755. Duedall, I. W., B. H. Ketchum, P. K. Park, and D. R. Kester. 1983. Global inputs, characteristics, and fates of ocean-dumped industrial and sewage wastes. In I.W. Duedall, B. H. Ketchum, P. K. Park, and D. R. Kester, eds., Wastes in the Ocean: Industrial and Sewage Wastes in the Ocean, vol. 1. Wiley-Interscience, New York, 3-45.
6. Feachem, R. G., D. J. Bradley, H. Garelick, and D. D. Mara. 1983. Sanitation and Disease: Health Aspects of Excreta and Wastewater Management. John Wiley & Sons, Chichester, England.
7. Gameson, A. L. H., ed. 1975. Discharge of Sewage from Sea Outfalls. Pergamon, Oxford, U.K.
8. Ganther, H.E. 1980. Interactions of Vitamin E and selenium with mercury
9. Goldberg, E. D., ed. 1979. Assimilative Capacity of U.S. Coastal Waters for Pollutants. Proceedings of a Workshop, Crystal Mountain, Washington, July 29-August 4, 1979. National Oceanic and Atmospheric Administration, Boulder, Colorado.
10. Gunnerson, C. G. 1958. Sewage disposal in Snnta Monica Bay. Trans. Amer. Soc. Civil Engrs., v. 124, 823-850.

11. Gunnerson, C. G. 1961. Discussion on settling properties of suspension; Trans. Am. Soc. Civ. Engrs., v. 126, 1772-1775.
12. Gunnerson, C. G. 1978. Discharge of sewage from sea outfalls. In Gameson, ref. (7), pp. 415-525. In W. Bascom, ed. Annual Report, 1978. Southern California Coastal Water Research Project, Long Beach, California.
14. Houser, L. S., ed. 1968. National Shellfish Sanitation Program Manual of Operation, Part I. U.S. Public Health Service, Washington.
15. Huggett, R. J., and M. Bender. 1980. Kepone in the James River. Environmental Science and Technology, 14:919.
16. Hunt, J. R., and J. D. Pandya. 1984. Sewage sludge coagulation and settling in seawater, Environ. Sci. Technol., v. 18(, n. 2, 119-121.
17. Institute of Civil Engineers. 1981. Coastal Discharges. Thomas Telford, Ltd., London.
18. International Maritime Organization. 1980. Internal report, London.
19. Kneip, T. J. 1983. Public health risks of toxic substances. In Myers, ref. (27), 577-658.
20. Koh, R.C.Y. 1982. Initial sedimentation of waste particulates discharged from ocean outfalls. Environ. Sci. Technol., v. 16, 757-763.
21. Likens, G. E., ed. 1972. Panel: Nutrients and Eutrophication: Prospects and Options for the Future.   In Nutrients and Eutrophication, Special Symposium, Vol. 1. American Society of Limnology and Oceanography, Lawrence, Kansas 297-310.
22. McLaughlin, R. T. 1961. Settling properties of suspensions. Trans. Amer. Soc. Civ. Engrs. v.126, 1734-1766.
23. Miller, G. E., P. M. Grant, R. Kinshore, F. J. Steinkruger, F. S. Roland, and V. P. Guinn. 1972. Mercury concentrations in museum specimens of tuna and swordfish. Science, no. 175, 1121-1122.
24. Mitchell, R., and C. Chamberlain. 1975. Factors Influencing the Survival of Enteric Microorganisms   in the Sea: An Overview. In ref. (7), 237-247.
25. Moore, B. 1975. The case against microbial standards for bathing waters. In ref. (7), 110-114.
26. Moore, M. D., and A. J. Mearns. 1980. Changes in bottom fish population, 1975-1980. In W. F. Bascom, ed., Biennial Report 1979-1980. Southern California Coastal Water Research Project,   Long Beach, California.
27. Myers, E. P., ed.   1983.   Ocean Disposal of Municipal Wastewater, 2 vols. MIT Sea Grant   Program, Massachusetts Institute of Technology, Cambridge, Massachusetts.
28. National Advisory Committee on Oceans and Atmosphere. 1981. The Role of the Ocean in a Waste Management Strategy. Government Printing Office, Washington,.
29. Pearson, E. A., ed. Proceedings, First International Conference on Waste Disposal in the Marine Environment. Pergamon, London.
30. Pearson, E. A., and E. F. Frangipani, eds. 1975. Proceedings, Second International Conference on Waste Disposal in Marine Waters. Pergamon, London.
31. Phillips, D. H. 1982. Trace metals of toxicological significance to man in Hong Kong seafood. Environmental Pollution, Series B, 3:27-45.

32. Riley, J. P., K. Grasshoff, and A. Vipio. 1972. Nutrient chemicals. In E. D. Goldberg, ed., A Guide to Marine Pollution. Gordon and Beach, New York, 81-110.

33. Segan, D. A., and D. G. Davis. 1984. Contamination of populated estuaries. NOAA Tech. Memo.. NOS OMA 11. National Oceanic and Atmospheric Administration, Washington.

34. Sherlock, J. C. 1982. Duplication diet study on mercury intake by fish consumers in the U.S.. Arch. Env. Health, v.37, 5: 272.

35. Shuval, H. I., A. Adin, B. Fattal, E. Ravitz, and P. Yekutiel. 1986. Health Effects of Wastewater Irrigation. Technical Paper 51. World Bank, Washington;

36. Sternberg, R. W. 1972. Predicting initial motion and bedload transport of sediment particles in the shallow marine environment. In D. J. P. Swift, D. B. Duane and O. H. Pilkey. 1972. Shelf Sediment Transport. Dowden, Hutchinson & Ross, Inc. Stroudsburg, Pennsylvania.

37 U.S. Environmental Protection Agency. 1982. Chesapeake Bay Program Technical Studies: a Synthesis. Government Printing Office, Washington.

38. U.S. Environmental Protection Agency. 1983. Chesapeake Bay: a profile of environmental change. Government Printing Office, Washington.

39. World Health Organization. 1972. Foot Additive Series (no. 4). Technical Series no. 505, Geneva.

40. Wang, T. H., R. C.Y. Koh, and N. B. Brooks. 1985. Interpretation of Sludge Sedimentation Measurements. In D. J. Baumgartner and I. W. Duedall, eds., Oceanic Processes in Marine Pollution. Vol. 6, Physical and Chemical Processes: Transport and Transformation. Krieger, Malabar, Florida.

41. Daly, H.E. and Townsend, K.N. 1883. Valuing the Earth: Economics, Ecology, and Ethics. 1993, .MIT Press, Cambridge, MA.

42. Degrément 1991. Water Treatment Handbook, 6th edition. 2 vols. Lavoisier Publishing, Inc. ,with   Springer Verlag, Paris

43. Fresenius, W. and Deutche Gesellschaft for Zusammenarbeit 1989. Waste Water Technology ,   Springer-Verlag, Heidelberg.

44. French, J.A.1978. Alexandria Wastewater Master Plan Study, Vol. IV, Marine Studies, Camp, Dresser, & McKee, International Div., Cambridge, Massachusetts.

45. Jordan, S., Stenger, C., Olson, M., Batnik, R., and Mountford, M. 1992. Chesapeake Bay Dissolved Oxygen Goal for Restoration of Living Resource Habitats. Maryland Department of Natural Resources, Annapolis, MD.

46. Kalbermatten, J.M., Julius, D.S., and Gunnerson, C.G. 1982.. Appropriate Sanitation Alternatives: A Technical and Economic Appraisal.. World Bank Studies in Water Supply and Sanitation 1. Johns Hopkins, Baltimore

47. Malone, T.C. 1992. Effects of water column processes on dissolved oxygen, nutrients, phytoplankton and zooplankton. In . Smith, D.E., Leffler, M., and Mackiernan, G., Editors. 1992) Oxygen Dynamics in the Chesapeake Bay. Maryland Sea Grant Program, College Park, MD.

48. Mearns, A.J. 1994. Requirements for ecological models and post-audits of marine waste disposal sites, Personal communication.

49. National Research Council. 1993. Managing Wastewater in Coastal Urban Areas. National Academy Press, Washington.
50. Nemerow, N.L.1991. Stream, Lake, Estuary, and Ocean Pollution. Van Nostrand Reinhold, New York.
51. Officer, C.B., and Ryther, J.H. 1981. Swordfish and mercury: a case history. Oceanus, v.24, 34-43.
52. Pearson, E.A., Ed. 1960. Waste Disposal in the Marine Environment. Pergamon, New York; Pearson, E.A., and Frangipani,E.F., Eds. 1975.. Marine pollution and marine waste disposal. Suppl. Wastewater. Water Sci and Technol., v.18, n. 11, Pergamon; Özturk, I. Ed. 1994. International Specialized Conference on Marine Disposal Systems. Iller Bankasi, Istanbul.
53. Salas, H.J. 1986. History and application of microbiological quality standards in the marine environment. Water Sci. and Technol. v.18, 47-58. Ms. revision, Pan American Health Organization (1991).
54. Strauss, M., and Blumenthal, U.J. 1990. Use of Human Wastes in Agriculture and Aquaculture. IRCWD, Ueberlandstrasse 133, CH-8600, Duebendorf, Switzerland.
55. Tschobanoglous, G., and Schroeder, E.D. 1985. Water Quality: Characteristics, Modeling, Modification. Addison-Wesley, Reading, Mass
56. Tolmazin, D. 1985. The changing oceanography of the Black Sea. 1: northwestern shelf. Progress in Oceanography v. 15, n. 4, 217-276.
57. World Bank 1991. Environmental Assessment Sourcebook. Volume 1. Policies, Procedures, and Cross-Sectoral Issues. Tech. Pap 139. Washington.
58. World Health Organization, Regional Office for Europe. 1979. Principles and Guidelines for Discharges into the Marine Environment, Copenhagen.
59. Gunnerson, C.G., Shuval, H.I., and Arlosoroff, S. 1984. Health effects of wastewater irrigation and their control in developing countries. Proceedings, Water Reuse Symposium III, San Diego, California, American Water Works Association, Denver, CO.
60. Salas, H.J. 1985. A manural for planning and conceptual design of submarine outfalls in Latin America and the Caribbean. PAHO, Lima.
61. Wilcocks, C., and Manson-Bahr, P.E.C. 1972. Manson'a TRopical Diseases. 17th Edition. Williams and Wilkins, Baaltimore, MD

4 Hydraulic Design

This chapter explains the processes by which an appropriate discharge site is selected and then the hydraulic design of the outlet and other parts of the system.Sections 4.2 and 4.3 introduce the terms and conditions, while Section 4.4 describes analytical tools and approaches and presents some basic formulas.

The central questions of outfall siting are addressed in Section 4.5. The techniques used to answer these questions and to arrive at the engineering decisions as to where the outfall is to lie are explained in terms of the concepts and techniques introduced in the earlier sections.

Outlet design and initial dilution computations are presented in Section 4.6. Other important hydraulic considerations, including internal hydraulics upstream of the outlet section and external hydrodynamics forces due to waves and earthquakes, are presented in Section 4.7. This order of presentation follows the principal sequence of steps of proper ocean outfall design.

However, the design process is iterative, in that it generally involves compromise among design factors, and one may have to consider siting, outlet design, and other hydraulic and non-hydraulic factors (such as structural, financial, or public pressure) several times in turn before arriving at a final design.

4.1   Concepts and Definitions

Listed below are brief definitions of the most common terms and concepts encountered in outfall siting and hydraulic design. Further analytical discussion of some of the concepts is presented in later sections.

Dilution and relative concentration. The dilution, S, is the ratio of the total volume of seawater-effluent mixture to the volume of effluent that would be found in a sample of seawater-effluent mixture at a particular location. The reciprocal of the dilution ratio is the relative volumetric concentration, $p = 1/S$

Hydraulic performance of an outfall is usually quantified in terms of the dilution, S, with a large number such as $S = 100$ or $S = 500$ implying more complete mixing of effluent with ocean water than a low number such as $S = 10$. To compute the effects of successive dilutions and concurrent decay of an effluent constituent, it is usually more convenient to work with relative concentration, p.

Diffuser. This is the outlet structure of the outfall that is usually, but not always, employed to begin the process of dispersing the effluent into the marine environment. Most often it consists of the final several hundred meters of the outfall, in the side wall of which effluent ports are often placed, so that effluent is introduced into the sea by many smaller distributed plumes, instead of a big one.

Effluent plume. Wastewater discharged from an ocean outfall mixes with surrounding water and is swept away by ambient currents. When the effluent discharge is continuous, the mixing and drifting effluent appears as a continuous plume, much like a plume of smoke from a chimney. Effluent dilution increases and effluent concentration decreases with increasing distance from the discharge point. At any one instant, the effluent concentration in a plume cross section roughly follows a normal (or Gaussian) distribution about the centerline, modified by a patchiness due to turbulence. On a larger scale, the plume path also fluctuates because of turbulence in the same manner that an atmospheric smoke plume from an open fire surrounded by seated people blows in the face of first one person and then another, although not in the faces of all people at once.

Buoyant plume; initial dilution. In many cases, an essentially freshwater effluent discharged into seawater will form a plume of water less dense than the surrounding water and will rise toward the surface in a buoyant plume. After the effluent reaches the sea surface (or other terminal rise height if kept submerged by stratification)., the initial dilution phase ends. It is followed by a gravitational spreading phase with dispersion by the turbulence and the shear flow of the ambient current.

Disappearance/Decay/Dieaway; T_{90}. Coliform bacteria are the most common non conservative sewage effluent constituents considered in outfall design. Bacterial concentrations in an effluent plume decrease with increasing time and distance from the discharge point. Bacteria concentrations are reduced not only by physical dilution, but also by mortality and by sedimentation, as discussed in Section 3.1.3. The rate of reduction in numbers is usually described by a first-order decay function of the form:

$$C_t = C_o e^{-kt} = C_o\, 10^{-t/T_{90}} \tag{4.1}$$

where C_t = concentration at time, t, C_o = initial concentration, k = decay coefficient, and T_{90} = time for 90% disappearance of bacteria

It is customary to express the decay rate as the time required in hours for 90 percent of bacteria in a sample to disappear; where $k_e = \ln 10/T_{90}$. For a T_{90} of

0.5 hours, a bacteria concentration at time zero of 10,000,000 per 100 mℓ will be reduced to 1 per 100 mℓ in 3.5 hours.

Advection, dispersion, shear (stirring), and diffusion (mixing). These are described in Chapter 2 and defined analytically in subsequent sections of this chapter. Advection patterns for outfall effluents are shown in Figure 4.1.

Density stratification. Density stratification (see Chapter 2) often blocks the rise of a buoyant effluent plume, forcing it to spread laterally in a subsurface water layer (Figure 4.2). This will keep a wastewater plume away from recreational surface waters. On the other hand, it may be a condition to avoid if oxygen depletion of subsurface waters is an essential design concern.

Particle settling. Most wastewater effluents contain solid particles that flocculate in seawater and settle at rates controlled by density, currents, and turbulence. Some particles remain on the bottom while others are moved by saltation or resuspension. When accumulation is sufficiently rapid, sediments become anaerobic and support an abundance of specialized pollution-tolerant species that feed on the detritus. Effects of sediment oxygen demand are ordinarily negligible, except, for example, when an untreated paper mill waste with a suspended solids concentration of thousands of mg/ℓ discharges to a shallow, sheltered bay with little current or wave action.

Professional Issues. Individual civil engineers, scientists, and others engaged in an outfall study are seldom the first and never the last to study a given site. It is arguably a professional obligation to report study results so that they may be easily found and completely understood by those who follow. The results and conclusions from such studies--including details on construction methods and costs, particularly on operating experience and costs--should be made widely available. This can be done by publishing a short article in an abstracted journal, so that the article can be found in a subsequent database search.

4.2 Qualitative Descriptions of Receiving Waters

Receiving water characteristics described in Chapter 2 are summarized below with respect to outfall siting and design. Quantitative analytical and engineering procedures for outfall siting and hydraulic design are presented in the sections that follow.

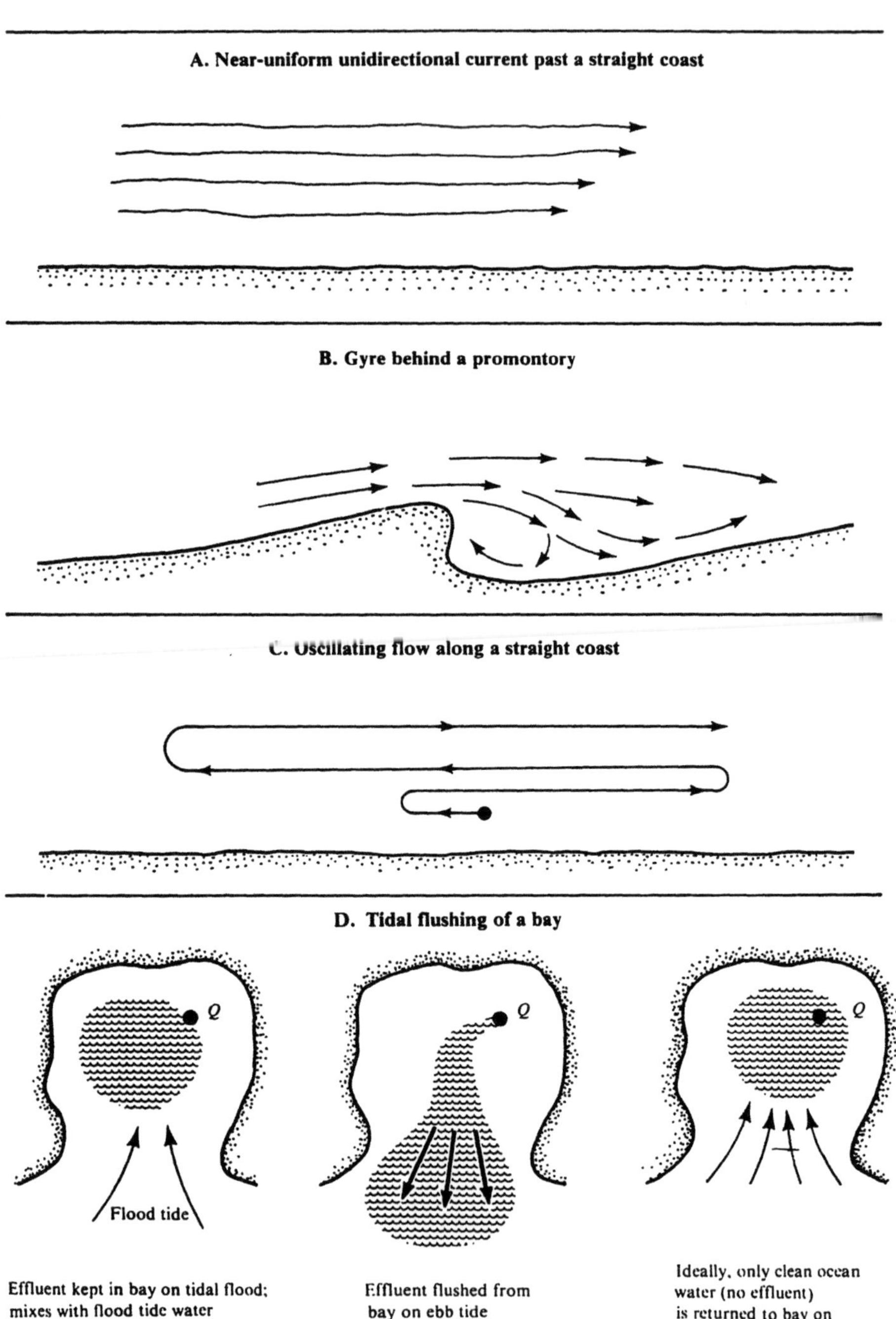

Figure 4.1   Advection and dispersion of effluent discharges.

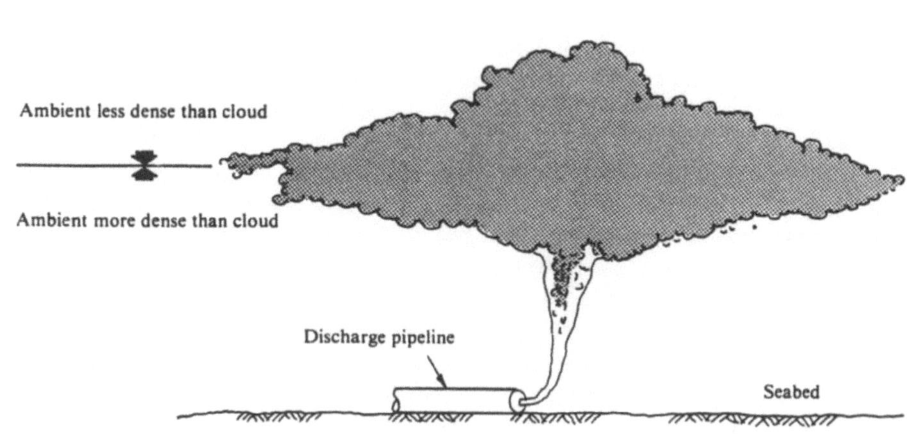

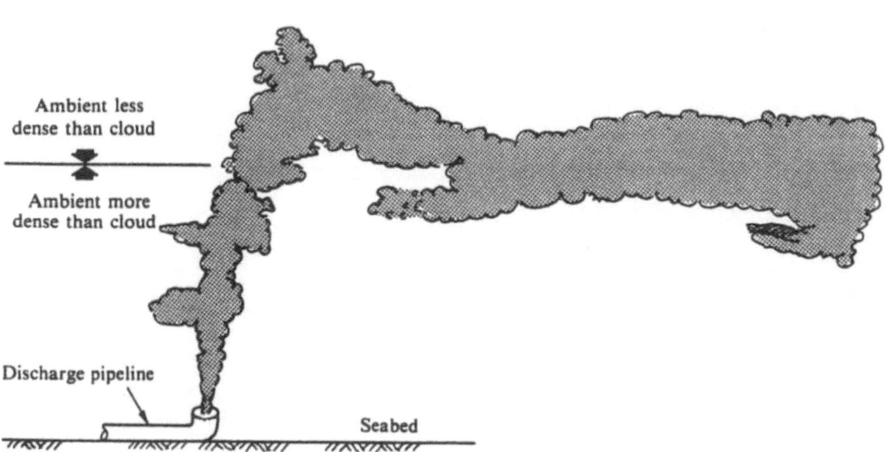

Figure 4.2. Buoyant plumes kept submerged by ambient density stratification. Note that submergence is independent of angle of jet. Sources: top - Fan (5), bottom - Wrignt (15).

4.2.1 Coastal Waters

Whatever the source of current motion, average currents flow parallel to the coast line and to depth contours, but onshore-offshore components arise periodically owing to waves, tides, and wind, including storms and diurnal land-sea breezes. This means that effluent plumes discharged offshore are seldom carried directly out to sea or directly onshore, but usually tend to drift parallel to the shore.

Shallow coastal embayments often attract development because they offer a protected anchorage and because the adjacent land is usually flat and is easy to build upon. Many such bays are now heavily polluted by industrial and domestic effluents. Unfortunately, the flushing capacity of embayments is less (sometimes by orders of magnitude) than that of the adjacent sea. New York Bight (United States), Izmir Bay (Turkey), Abu Kir Bay (Egypt), and Cockburn Sound (Australia) are examples of such bays. There are structural advantages in building outfalls discharging to such bays. The wastewater collection points are on the shore, bottom conditions are often structurally favorable, and wave action is minimal. However, their assimilative capacities are limited.

Coastal lagoons shoreward of barrier beaches occur along much of the world's coastlines. Here, brackish water bodies connect to the sea via occasional inlets through the barrier beach. Cities and towns are often built on the barrier beaches themselves (e.g. Venice, Lagos, Alexandria, and Miami Beach), or on the mainland shoreward of the lagoons (e.g. Titusville, Miami, Maestre). Sewage from such cities is often discharged to the lagoons through outfalls easily constructed in calm and shallow water.

However, the environmental balance of coastal lagoons is even more sensitive than that of the less constricted coastal embayment. Lagoons also tend to harbor intense ecological activity, and at the same time have much less dispersive capability than the open ocean.

4.2.2 Rivers

This book is concerned primarily with ocean disposal of wastes, and secondarily with estuarine disposal. River disposal is included for purposes of comparison. A river is a freshwater stream with unidirectional flow and with velocities that are affected by tides in the lower reaches. The plume from a wastewater outfall will be diluted and dispersed much as in the open ocean, except that (1) buoyancy in the plume is less because, unlike the ocean, the stream is not more saline and hence significantly heavier than the effluent, and (2) because dispersion due to turbulent diffusion may be enhanced by the shear flow because of the channel curvature.

Plume analysis in rivers attempts to determine how far downstream the plume will be advected before it is more or less completely mixed across the river cross-section. Complete mixing across the cross-section achieves the maximum dilution, and justifies one-dimensional analyses of chemistry or biochemistry.

Where there is a water supply intake on the opposite shore downstream of an outfall, however, a design criterion may be that the wastewater plume should not be fully mixed across the stream at the water intake. Some computational techniques for plume dispersion in rivers are presented in Section 4.4.5.

4.2.3 Estuaries

Most large coastal cities are located on estuaries that provide harbors for naval and commercial shipping, water transport routes to the interior, and, unless they are too badly polluted, valuable fisheries. Wastewaters are almost always discharged to these estuaries, unless major engineering measures have been undertaken to discharge the water elsewhere.

Islands in estuaries strongly affect tidal flows. Sewage discharged at a given point near an island may be pumped entirely around an island by tidal action rather than flushed directly out to sea. Similarly, circulatory pumping around a submerged shoal will cause a net upstream flow in one channel and a net downstream flow in the other.

4.3   Methods of Analysis for Outfall Siting

In outfall design, there is no decision of greater strategic importance than selecting site or sites where effluent is to be released to the marine environment. The well-chosen site puts effluent sufficiently far to sea that it can be dispersed without demonstrable harm to sensitive receptor areas such as beaches, shellfish beds and coral reefs. It limits discharge to bays, recirculation zones or other zones from which the effluent will not be sufficiently dispersed.

4.3.1 First Questions

Initial planning for ocean disposal systems requires answers to the following questions in order to scale the project and to guide field studies:

(1) How do outfall alternatives relate to the sewage collection and treatment options, present and future?
(2) How many outfalls should there be?
(3) Should the discharge be to the estuary or to the ocean?
(4) How far offshore and to what depth should the outfall extend?
(5) Should a diffuser be provided? If so, what should be its size and location?
(6) Is plume submergence likely? If so, is this desirable, or is it undesirable?

It usually takes months of study to arrive at the final answers. In the meantime, reasonable approximations may be provided after only a short period, perhaps a few weeks, of study to determine what will be needed for scoping, scheduling, budgeting, and conducting the full study program.

Information at hand. The first step is to collect all the readily available reports, tables, charts, and atlases pertaining to the area's hydrography, topography, geology, meteorology, environmental standards for land use and their background, and fisheries (see Chapter 2). Previous engineering feasibility studies, of which there are often more than one, are particularly useful. Other sources include engineering, earth sciences, and geography departments of local universities; ship supply stores, nearby airports, harbormasters, and government agencies of maritime nations. Many of the latter publish their own versions of nautical charts, borrowing from one another for correcting and updating.

Previous engineering feasibility studies, plans, and reports for water supply, wastewater disposal, cooling water discharges, harbor or shoreline development, or oil pipelines are particularly valuable.

Size and strength of discharge. The next step is to determine the design flow rate, by determining the tributary areas, populations and design flows for the outfall system. Municipal sewage flows may range from 100 to 600 liters per capita per day, depending upon economic development, water service levels, land use, industrialization, water reclamation, and waste. Additional contributions from groundwater infiltration and storm water should be accounted for.

Advection. To determine where the effluent from an outfall will go, it is necessary to obtain all previous current data. If conditions permit, estimate current speeds from tidal prism considerations. Apply information identified in Chapter 2 and Section 4.3. Query local boatmen.

Approximate initial dilution. For a simple open-ended pipe of diameter d, discharging into water of depth D, the initial dilution of the resultant plume is of the order of D/d, assuming a discharge velocity of the order of 1 m/sec (2).

For a multiport diffuser of length b aligned perpendicular to an ambient current of velocity U, in water of depth D, the ambient water flow available for diluting the effluent is of the order of UbD, i.e. a flow velocity multiplied by an approximate flow depth multiplied by an approximate flow width. The initial dilution is therefore of the order of UbD/Q_e, in which Q_e is the effluent flow rate.

These rough approximations may be compared with the more precise relationships given in Section 4.6.

Approximate ultimate dilution. In a river, the ultimate physical dilution of the discharge cannot exceed Q_r/Q_e, the ratio of the river's flow rate to the effluent flow rate. In an estuary, the ultimate dilution cannot exceed $(Q_r + Q_t)/Q_e$, where Q_t equals the volume of the tidal prism divided by the ebb flow period.

In unidirectional flow along an open coast, the estimate of ultimate dilution is similar to that of a river, but is somewhat more complicated. An effluent plume flowing parallel to a coast will be dispersed across a cross-sectional area that gradually increases with distance and time. The diffuser is located x meters offshore, in y meters depth of water (Figure 4.3). The plume is advected parallel to the shoreline, and spreads with respect to time and distance from the discharge point. At some point, the plume edge will reach the shore, and the shore will be lapped by waters containing a seawater/effluent ratio whose value one wishes to know. That ratio is of the order of UA/Q_e, in which u is the ambient current speed, A is the plume cross-sectional area at the point of interest, and Qe is the discharge flow rate. The approximate value of A is likely to be of the same magnitude as the cross-sectional area of flow between the shore and the discharge point, thus: $A = \int y(x)\, dx$, where y is the water depth at a distance x from shore.

4.3.2   Worked example of Near-Field Dilution

If the effluent discharge, Q_e, equals 10 m^3/sec and the current speed u is 0.1 m/sec, roughly how far offshore must be the discharge site be to ensure dilution of the order of 1,000 by the time the effluent plume reaches the beach? The shore and bathymetric contours are straight and parallel, and the bottom slope is constant at 1:50. (There is nothing magic about the ratio, 1,000 : 1. It corresponds to both a 99.9 percent removal, and one cupful in a bathtub.)

Try an offshore distance of 1 km. For this trial, A is on the order of 1,000 m x 10 m average depth, or 10,000 m^2. Thus UA/Qe is 0.1 m/sec x 10,000 m^2/(10 m3/sec), or only 100. This will not provide a dilution of 1,000. This result has nothing to do with diffuser design; changing the diffuser design will do nothing to alter the fact that there is simply inadequate dilution water available at that site.

Try an offshore distance of 4 km. For this trial, A is of the order of 4,000 m x 40 m average depth, or 160,000 m^2. Thus uA/Qe is approximately 1,600, which is of the order of magnitude desired.

4.3.3   Field Surveys

For outfall siting and design, field studies most commonly include current measurements, density profiles, bathymetry and other seabed surveys, water quality sampling, sediment sampling, wave climate measurements, and diver surveys.

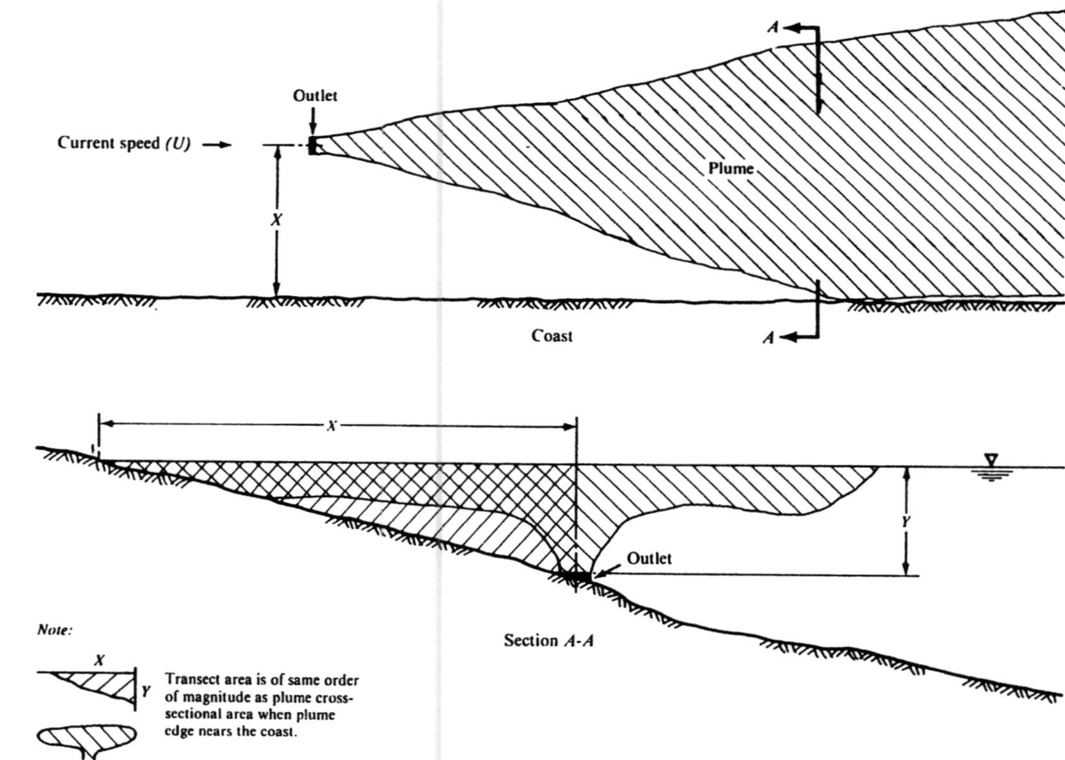

Figure 4.3. Estimate of significant far-field dilution along an open, straight coastline.

The last category routinely includes photographs and narrated video. Any underwater camera or video camera is useful. Vocal narration through a SCUBA regulator is difficult to understand, yet feasible to use.

For many coastal cities, wind data routinely collected at their airports are useful in evaluating wave climates and land-sea breeze regimes. Wind, water level and wave data are often measured by port or harbor authorities. These may be validated or extended by short-term measurements in potential outfall areas.

Current Measurements. To site an outfall properly, one needs to understand the typical water movement throughout the region between the point of discharge and sensitive sites such as beaches and shellfish beds. This requires measurement of current vectors throughout the length, breadth, and depth of the study area sea, for 13 or more months to include all seasons, concurrent and subsequent data analysis, recognition of the dominant patterns; and application of this understanding to design of an outfall. Since such a program will usually be impractical, one instead hypothesizes what the dominant current behavior will be, then samples the current to test the hypothesis. Interpretation of the current data is often extended by the use of circulation models.

Models and data are interactive. The models greatly extend the understanding and interpretation of some of the data, while other data serve to validate the model. For example, a model may take the record of water levels at the mouth of a semi-enclosed bay, together with the bathymetry of the bay, to predict the time-varying current pattern throughout the bay. How valid is this prediction? If there is also a measurement of current for the same time period or conditions as for the water level record, and the model can predict the same current pattern as is measured, the model is at least partially validated. Current measurements are discussed in this section. Models are discussed below..

Current measurements can be classified as either "Eulerian" or "Lagrangian". Each type of measurement has its place in a complete set of field studies. Lagrangian measurements began with observations of ship and small craft movements by ancient Greek and Polynesian sailors They were first charted in relation to water masses in the Gulf Stream by Benjamin Franklin in the 1770s because it took mail -packets from England twice as long to go from east to west as it did from west to east. (Note that mariner's conventions still define a westerly current as flowing to the west, while a west wind comes from the west) Small-craft movements are still useful in revealing unexpected transients particularly in stratified waters where they are marine equivalents of atmospheric jet streams.

Lagrangian studies are made with drogues (underwater weighted sails, crosses, or sea anchors tethered to floating markers or radar targets), surface or weighted drifters with neutral buoyancy, or dye in which the path and velocity of a particle are traced in the flow field.

Eulerian measurements provide large quantities of data from current meters at fixed points in space which are then translated into Lagrangian circulation patterns. Since the 1940's, current meters have evolved from the classical Ekman mechanical meter that measured average velocities in different directions at selected depths over periods of minutes or hours. It worked by releasing lead shot through a magnetized trough into one of 36 chambers after a fixed number of revolutions of an impeller lined up with the current by a large vane   A series of improvements depend on electrical or electronic substitutes for mechanical works.

Measurements are made in one place for a at least a fortnight or moved from place to place in the velocity field. They may be augmented sequential aerial surveys for qualitative insight into flow patterns of converging water bodies with differing densities or turbidities. Tidal outflows, inlet jets, and promontory gyres may stand out sharply.   Photos from a series of flights at two-hour intervals over a tidal cycle provide a useful qualitative understanding of surface flows. However, limited funds will usually be better spent on quantitative measurements *in situ.*.

Acoustic Doppler Current Profilers (ADCP). These state-of-the-art systems are appropriate for large-scale investigations. To measure velocity at, say, mid-depth, any other current meter must itself be moored at mid-depth; and there it can measure only the velocity at mid-depth. An acoustic Doppler meter, on the other hand, can be placed on the sea bed and sample the current velocity at each of several layers above it. Typically, the layers, or "bins", are each 1 m thick. The top two m and the bottom two m of the water column above the meter cannot be sampled, but all intermediate depths can. The advantages of this instrument are its multi-depth sampling ability, and its seabed location, protecting it relatively well from being snagged and dragged by passing vessels or fish nets.

Vessel-mounted ADCP. Here the ADCP can be mounted on a survey vessel and point down from the surface as the vessel proceeds. It measures both the velocity of the water in each "bin" below the vessel , and of the velocity of the seabed with respect to the vessel from which the seawater velocity is calculated. Together with a time series record of vessel location presumably a close approximation to the entire velocity field is obtained.

This technique is useful for detailed understanding of complex flow fields, such as through groups of islands. When the survey vessel is fast enough (or the survey area small enough) the vessel can complete the survey course several times within a tidal cycle, one can learn better than ever before the details of how flow passes among the islands and whether there is a net "pumping" circulation superimposed on the tidal oscillatory flow. Since the data collection capabilities are substantial and corresponding data reduction requirements are exhaustive, optimum resolution of sampling and digitizing intervals in time and space is an essential component in project planning. (See Section 10.4).

Wave Climate Measurements.   It is often critical to make a good estimate of the greatest storm waves the outfall must withstand during its design life.

Direct wave and sea level measurements can of course be made, but the span of a typical planning and design project is far too short to estimate the "100-year storm wave" with confidence. One should instead obtain the longest local record of wind speed and direction up to, say, 20 or 30 years and almost certainly longer than one or two years.during an outfall design period.

Local long-record wave measurements obtained by other projects are welcome, of course. From these, estimates of extreme-event wave height at the outfall site can be made, and usually at far less cost than deploying a wave gauge. Wave estimation techniques are explained in detail in the Shore Protection Manual (18). General principles of wave and sea level mechanics and observations are presented in Sections 2.2.3 and 2.4.3

Temperature, salinity, and density profiles. Seawater density is determined by temperature and salinity (T and S) from probes lowered from a survey boat. Continuous-profile probes record T, S. and depth from the surface to the sea bed. However, for outfall planning and design, an instrument that merely indicates T and S is adequate. Readings are made as the probe is lowered manually from the surface to intermediate levels with T or S gradients, and to the bottom. Density is conventionally indicated as σ_t, where $\sigma_t = 1,000 (\rho - 1)$ and may be determined from nomograph or, if greater accuracy is required, published tables.

Section 4.6.4 describes how density profiles are related to effluent plume submergence. The precision of density measurements should be within one-tenth of the surface-to-bottom density difference that is likely to keep a proposed effluent plume submerged. The salinity probe should periodically (if possible, before and after every voyage) be standardized against a laboratory-grade salinometer which is in turn calibrated with standard Copenhagen seawater.

The probe cable should be weighted sufficiently to hang essentially vertically from the survey vessel, whether drifting or moored. The weights should be ≥ 1 m from the probes to avoid electrical interference with the salinity probe.

Seabed Surveys. Seabed surveys are made to determine bathymetry (topography), texture, and soil characteristics. Four classes of survey are: (i) preliminary reconnaissance, (ii) systematic broad surveys to select alignment corridors, (iii) detailed studies for the final alignment, pipeline design, and construction estimates, and (iv) construction control.

A preliminary reconnaissance is based on published nautical charts showing depths with some indications of bottom conditions (rocky, sandy, soft, shells, hard, sticky). A recording fathometer is used to locate sampling or current meter stations and to reveal unsuspected bathymetric or soil conditions.

Systematic surveys along alternate alignments are used to confirm bathymetry and sub-bottom soil conditions and probable construction requirements. A "corridor" is 100 to 200 m wide. An ideal survey mode is to proceed along the alignment at a slow, steady speed while recording bathymetry, seismic sub-bottom profiles, side-scan sonar, and vessel position. Tide elevation must also be noted. Bathymetric surveys should be made shortly before the winter storm season and immediately after a major storm period to determine the range of scour and fill.

Seismic sub-bottom profiling should give good resolution at sub-bottom depths up to 10 m. Deeper penetration is not necessary and loses resolution at shallower depths. Side-scan sonar provides a continuous acoustical image on a wide strip chart of seabed surface features such as rocks, cables, sand ripples, pipelines, or shipwrecks within several tens of meters to either side of the track.

The systematic surveys should include bottom sediment sampling with a gravity corer for silts, clays, or soft rock. Sand and rock samples can be taken with a bottom grab sampler, or they can be collected by hand by a diver.

With sufficient light and water clarity, a diver with an underwater camera and color film can record typical rock outcrops, sand ripples and waves, or reef structure. A series of photographs taken in a slow panoramic sweep are often more useful that an equal number of stand-alone photos.

Detailed studies along two or more corridors can indicate the best choice of alignment to within 2 m. Uniform seabed conditions are preferred. Rock bottoms, cliffs, and ravines present difficult construction problems. Cables, other pipelines, and shipwrecks should be avoided entirely.

Detailed studies, using fathometer, sub-bottom profiler, and side-scan sonar are made on a grid spacing of 25 to 50 m and cross-tie runs perpendicular to the corridor at intervals of 50 to 100 m. Gross inconsistencies between corridor and cross-tie data indicate positioning error.

The detailed studies should also including borings or corings at frequent intervals to provide ground truth for interpretation of stratigraphic profiles from the sub-bottom profiler, as well as samples for grain-size analysis and dynamic analysis for structural design. Intervals of borings vary from 10 to 100 m, depending upon bottom variability. Construction control consists of high-resolution bathymetric mapping capable of monitoring, say, the completeness of excavation and backfill. Soils data may seem expensive but are worth many wimes their costs when they correctly foretell difficult soil conditions. Many, many contractors have come to regret inadequate soils data.

4.3.4   Numerical Modeling

High-speed computers have made practical the numerical simulation of current patterns and other physical processes for outfall studies. Programs have been developed for thermal discharges, sewage discharges, and currents.

At the core of any mathematical analysis of ocean outfall discharges is a series of equations that describe the pertinent physical, chemical, and biological processes of a study area. For example, physical processes that define circulation or current patterns are described by differential equatgions for conservation of water mass and momentum in three spatial dimensions:

$$\nabla \cdot \bar{u} = 0; \text{ and } \frac{\partial \bar{u}}{\partial t} + \bar{u} \cdot \nabla \bar{u} + \frac{1}{\rho} \nabla \cdot \bar{P} + f \cdot \bar{u} = \frac{1}{\rho} \nabla \cdot \bar{\tau} \tag{4.2}$$

Where u = mean velocity, P = mean pressure, f = Coriolis acceleration, τ = stress tensor, t = time, and ρ = density,

`Advective and dispersive processes are described by second-order partial differential equations that conserve constituent mass, called mass transport equations, of the form:

$$\frac{\partial C}{\partial t} + \bar{u} \bullet \nabla C = \nabla \bullet \left(\overline{E} \nabla C \right) \tau \tag{4.3}$$

where C = constituent concentration, E = diffusion tensor, and r = source/sink terms which describe constituent addition, loss, and reaction kinetics.

The techniques needed to use such systems of equations may deal with the three spatial dimensions explicitly, but very often the systems are simplified by integrating the equations over one or two spatial dimensions. For example, transverse and vertical movement and diffusion may be considered negligible compared to longitudinal movements, and thus may be said to integrate the governing equations over the transverse and vertical dimensions to obtain a one-dimensional mathematical description. In a shallow estuary, one might choose to neglect vertical variations and develop a vertically integrated two-dimensional model that describes movements in the two horizontal dimensions.

Solution techniques are analytical, empirical, or numerical. The first use closed-form analytical solutions to the equations, such as tidal prism analysis and steady-state mass transport in a uniform flow field. Empirical techniques provide regression analyses of field or laboratory data which reveal functional relationships between a dependent variable, such as initial dilution, and one or more independent variables such as outfall discharge or local depth. The equations of Section 4.4.5, 4.6.1, and 4.6.3 represent analytical techniques using empirical coefficients.

Numerical approximation techniques render the governing differential equations in an approximate finite-difference form, and solve repeatedly through time and space taking into account all the variability of, say, depth or shoreline geometry. These types of numerical approximation often used in circulation studies are called link-node, finite-difference, and finite-element techniques. In a link-node model, the study area is divided into a series of storage areas called nodes, at which water volumes and constituent concentrations are determined. These storage areas are interconnected by flow channels called links, along which constituent mass and ambient flow are transported. In Figure 4.4a, a one-dimensional analysis of a narrow channel is represented by a single strand of links and nodes, while a wide bay is represented by a two-dimensional link-node matrix.

The finite-difference technique is the oldest and most widely-used scheme. The water body under study is divided into a matrix of quadrilateral cells, usually square or rectangular as shown on Figure 4.4b. Water surface elevations commonly

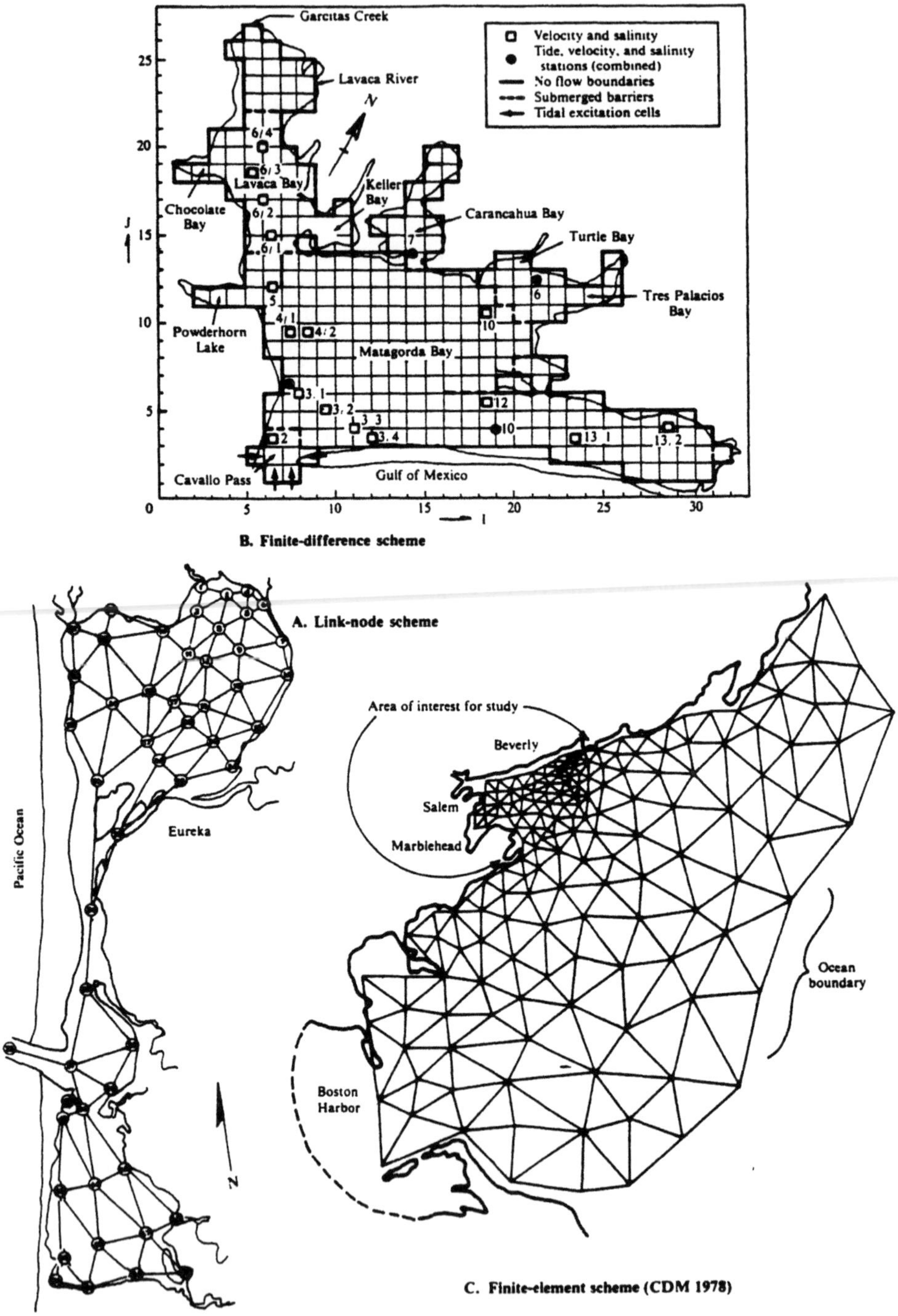

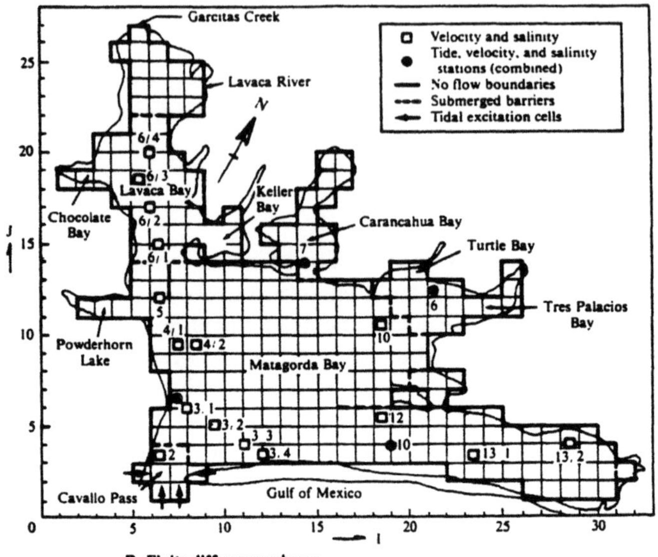

Figure 4.4 Numerical models for estuaries.

defined at the center of each cell, and normal velocity or inflow per unit width is defined for the side wall of each cell. An advantage of this technique is that it is conceptually straightforward and easy to program. A disadvantage is that a grid of rectangles my permit only a crude representation of a complex shoreline with many promontories or dendritic areas; this disadvantage can be minimized by a nested-grid model.

The finite-element technique common to structural analysis is also used in circulation analysis. The water body under study is divided into a series of elements which may be triangles, rectangles, or even higher-order polygons (Figure 4.4c). Within each element, solution variables are defined as nodes, usually at element vertices or midsides. The finite element technique employs a method of weighted residuals defining the distributions of the independent variable within each element as a particularly efficient computational procedure for solving the governing equations. The triangles can be designed to simulate a tortuous coastline much more gracefully and faithfully than can the normal rectangles of the finite-difference technique.

Many numerical models are available. To select the most appropriate numerical approach for a particular study, one must consider water quality standards, scales of the analyses, characterization of the physical system, data availability, dominant physical processes, and model availability. Where there are regulatory standards for water quality, the degree of required compliance within narrowly defined geographical areas determines the level of analytical effort and the choice of model. Most standards are written in terms of near-field and far-field regulations, and these define the scales of analyses to be performed and physical processes considered. Another scale is intermediate. These scales are defined below.

The near-field region is immediately adjacent in which the hydraulics of the outfall jet dominate ambient circulation. It is sometimes referred as the zone of initial dilution or the region of advective stirring (Section 2.2.6). The region is characterized by jet momentum effects, buoyancy due to thermal or salinity differences, and entrainment

The intermediate-field region lies immediately beyond the near-field region. Ambient processes control advection and spatial mixing over one or more directions, vertically in shallow estuaries and longitudinally in rivers.

Model characteristics often indicate type of model should be used, even before other aspects are considered. These include geometric length scales, complexity of the land boundaries, and extent of the open water boundary. The ratio of length scales in the physical system indicates the dimensions of model required. In a river, for example, after initial dilution, the plume will generally become well-mixed booth laterally and vertically within a short distance downstream, a factor of particular interest to riparian users drawing water from either bank. Dominant variation occurs only in the longitudinal direction and a one-dimensional model best simulates far-field concentrations. In a shallow, wide estuary such as Tampa Bay, Florida, a plume is quickly vertically well-mixed, and in the far-field, concentrations are determined by horizontal processes. Here, a two-dimensional (longitudinal and transverse) model should be used. A fjord, by contrast, may first become laterally well mixed, and be best modeled using a longitudinal and vertical model. Finally, in a deep, open water body such as Monterey Bay, California, processes operate in all spatial dimensions, and thus requires a three-dimensional model.

For very complex land boundaries such as Chesapeake Bay, it is often desirable to use a finite-element representation or a nested-grid model with finite-difference rather than the conventional finite-difference Model (Figure 4.4c). If the shoreline is more regular, the normal finite-difference grid may be adequate with the added advantage that this model is simpler to develop and use.

In modeling hydrodynamic and transport processes, special care must be taken in considering boundary conditions in open-water regions. It is relatively straight forward to model these processes in a water that is almost entirely enclosed (Figure 4.5a) such as Chesapeake Bay where boundary conditions are established at the entrance. Usually a stage (water-surface elevation) boundary situation is imposed with the implicit assumption that momentum fluxes are reflected at the entrance. Internal tidal hydrodynamics are driven by gravitational forcing.

When portions of a coastal region (Figure 4.5b) such as the southern California Bight or Marmala Bay in Hawaii are being modeled, specifying open ocean conditions is much more complicated; and the resulting models and their validation are much more costly. Aside from gravitational forcing, momentum effects from eddies and density-driven currents can enter the region and dominate the local circulation patterns calculated from long-term current meter records. For such areas, programs can be used that interpolate current meter data into a numerical grid while maintaining mass continuity. For semi-enclosed study areas (Figure 4.5c), appropriate model selection and verification is also difficult. In both cases, a thorough field investigation and understanding are required before choosing and applying a model whose value for design and operation may be then have been pre-empted by the empirical observations and their statistics.

In all cases, prototype data are essential. For most sites, meteorological, stream flow, tide, and outfall discharge data are common while receiving water current and water quality data are sparse, although they may be sufficient to determine the dominant physical processes, and in some cases may even be adequate for model calibration. Usually, more synoptic data are needed for model validation. As noted above, there are also cases where the extensive empirical and statistical data needed for verification are in themselves sufficient for outfall design without further recourse to a numerical model.

In any investigation of initial and ultimate constituent data, it is necessary to understand the physical processes. This is done by reviewing available data, from experience with similar systems, or occasionally from a sensitivity analysis of a numerical model using some data to establish ranges of conditions. In the near-coastal ocean, circulation is due to a combination of waves, tides, winds, atmospheric pressure gradients, fresh water inflows, density induced currents, and Coriolis acceleration. In in-shore areas, certain process dominate so that the others

A. Mostly enclosed study area

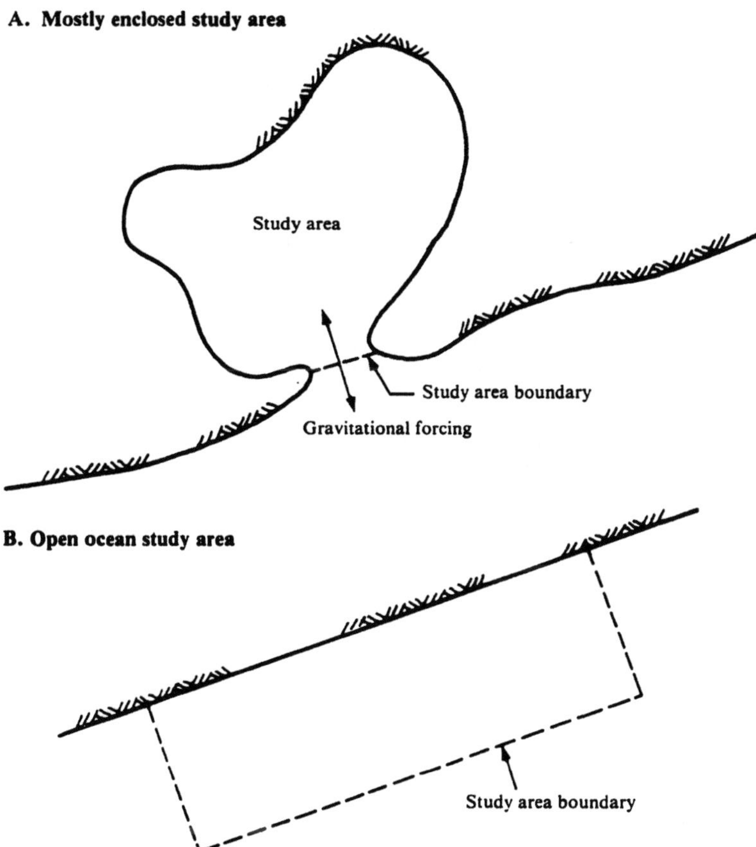

B. Open ocean study area

C. Semi-enclosed study area

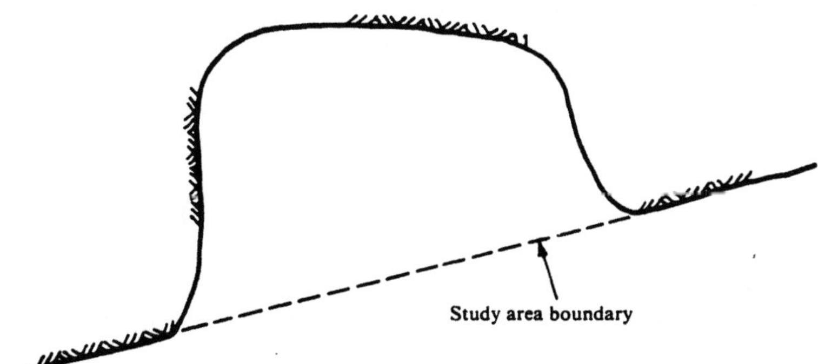

Figure 4.5. Types of ocean boundaries for numerical models.

may be neglected. In a fjord, tide and density forces dominate while in a wide, shallow estuary, Coriolis and wind effects control mixing and circulation.

Once the means to simulate and evaluate the hydrodynamic processes has been established, it must be determined what constituents and reaction kinetics should be included in the transport or water models. Most outfalls discharge heated water or freshwater sewage, so that initial momentum and buoyancy effects dominate, and constituents concentrations are large. First-order reaction kinetics that apply to chemical changes and to BOD, pathogen persistence, and other microbiological processes begin here. Their coefficients can be estimated from the literature and confirmed by limited field studies.

Having accomplished a preliminary investigation of the study area, the design engineer can select a numerical model to address both the geometric features and physical processes of the area.

For larger projects, highly technical numerical model approaches are usually justified on technological and institutional grounds. This involves several steps, some of which are optional, depending on data availability. These include model mobilization on the computer system, initial calibration of friction factors, dispersion coefficients, and first-order reaction rates, followed by sensitivity analyses, design of additional field programs to verify advection due to tidal pumping or gyres, final calibration, validation, and production runs.

Production runs indicate outfall locations that will minimize discharge impacts, predict constituent concentrations in critical areas, and their frequency distributions for comparison with environmental criteria and guidelines. Graphic summaries such as those shown on Figures 4.6 and 11.11 - 11.15 of current and water quality distributions are used by officials in authorizing and financing ocean outfalls.

4.3.5   Physical Hydraulic Models

Physical hydraulic model typically cost from US $30,000 to $100,000 for a pump intake or drop chamber study, to millions of dollars for a model of a major estuary. For major urban outfall systems, physical models have since the 1950s (11) sometimes been justified either as a guide in unusual small-scale internal hydraulic design problems such as unique outlet design, or as an aid in predicting dispersion of effluent through a complex system such as Puget Sound, San Francisco Bay, the Scheldt, or the Singapore Straits.

For smaller outfalls, the design and construction of a new physical model will seldom be justified. However, if the proposed outfall site is within the study area of an operating model, a small program of studies with that model can probably be justified. If the model has been decommissioned and the runs can not be easily made, the results from earlier tests may be useful.

Geographical physical models are most useful as indicators of mixing from point sources, of flow phenomena such as eddies shed by a promontory at a particular tide stage, or for the pattern of net tidal circulation around a large island

A. Velocity field plot (CDM 1978)

B. Concentration contours

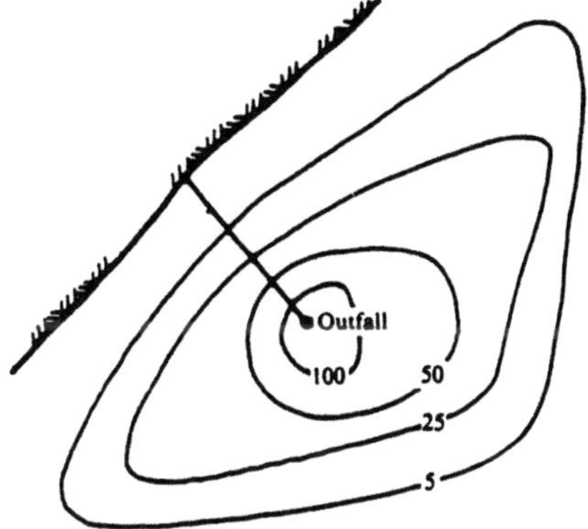

C. Progressive vector diagram (CDM 1978)

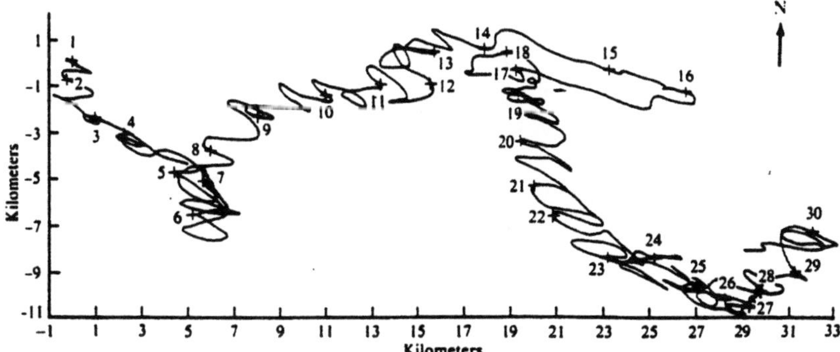

Figure 4.6. Representative types of computer output.

in a braided estuary. As with numerical models, phenomena observed in the hydraulic model must be validated by field tests.

The advent of fully three-dimensional numerical modeling, with user-friendly input, rapid computation, and automated graphical display, has in most cases replaced the labor-intensive, space-extensive geographical physical model with its complications of Froude/Reynolds incompatibility, and vertical/horizontal scale distortion. Recent physical modeling for outfall internal hydraulics includes purging studies for the Sydney outfalls and for Boston's Deer Island Outfall. Other model studies for the Deer Island Outfall include a study of near-field dilution, to determine an appropriate spacing for the rose-capped diffuser risers

4.4   Equations for Estimating Turbulent Diffusion

Several practical equations for calculating plume widths and centerline dilution of sewage from an ocean outfall with increasing time from the point of discharge, together with their appropriate application and shortcomings are presented in the following paragraphs. For a definitive discussion, see Fischer, et al (6)..

Lateral dispersion is the principal factor in turbulent diffusion. Vertical diffusion is of limited interest as it can only increase dilution by no more than a factor of three because the initial plume occupies one-third or more of the water column. Similarly, longitudinal dispersion in a continuous plume succeeds only in mixing the plume with itself. The theoretical analysis is based on the assumption that particles in a turbulent flow will tend to have a transverse (y-direction, in Figure 4.7a) distribution that is Gauassian:

$$p(x,y) = \frac{c}{c_0} = \frac{Q}{dus\sqrt{2\pi}} \exp\left(\frac{-y^2}{2s}\right) \tag{4.4}$$

in which p = volumetric effluent concentration at distance, x; Q_e = effluent flow rate; d = water depth; s = standard deviation of plume cross-section distribution, and Qe/u volume of effluent per unit length along the plume x-axis.

If c_0 mg/ℓ is the mass concentration of a contaminant in the undiluted effluent, its concentration at point (x,y) is pc_0 mg/ℓ if the constituent is conservative, or pC_0 e $^{-kt}$ for a decay rate, k, with respect to time, t.   The remainder of this section will be in terms of the volumetric concentration, p, which will simply be called the concentration.

The centerline concentration is

$$p(x.0) = \frac{Q_e}{dus\sqrt{2\pi}} \tag{4.5}$$

A. Dispersion in a wide stream

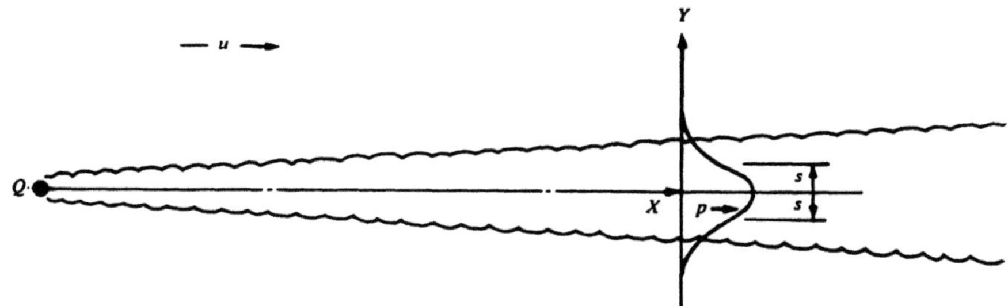

B. Wall effects in a narrow stream are modeled by image sources and plumes

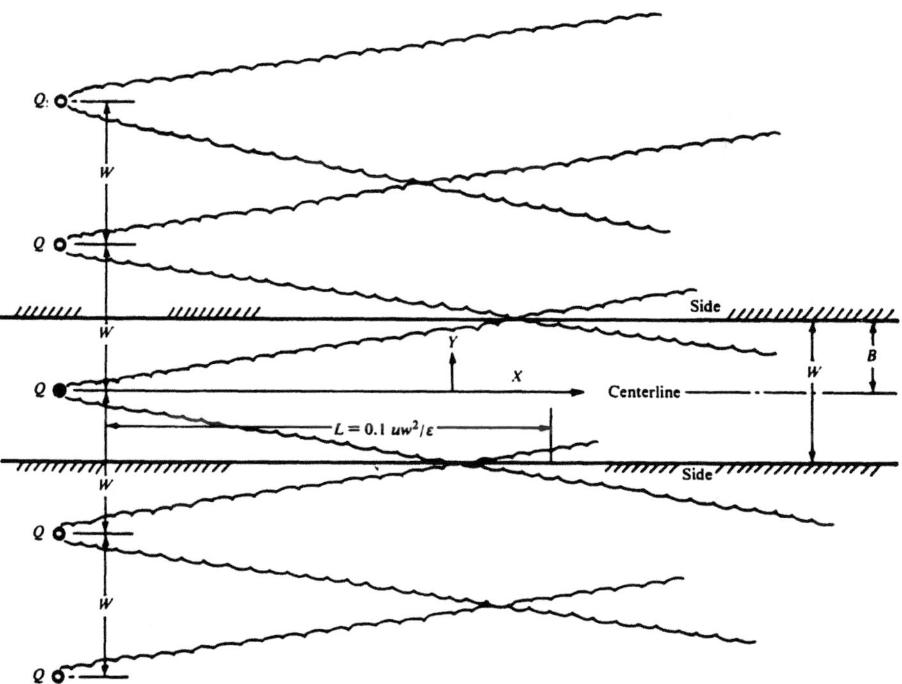

Figure 4.7. Plan views of point source plumes in streams.

The centerline concentration decreases with increasing "s". The rate at which s increases is of major interest in plume studies. It is customary to work with the variance, s^2. Its time rate of increase is defined as

$$\frac{ds^2}{dt} = 2\varepsilon \tag{4.6}$$

in which ϵ is the dispersion coefficient.

   Where the plume size, as characterized by s, is greater than the largest turbulent eddies promoting the dispersion, ϵ is a function of water depth, d, and shear velocity u*

$$\varepsilon = Kdu^* \tag{4.7}$$

in which u* is related to the bottom shear stress. This may be calculated in terms of the friction slope of the flow speed, u, and the Manning friction factor, n:

$$u^* = \sqrt{\tau Q\rho} \tag{4.8}$$

$$u^* = \sqrt{gd*slope} \tag{4.9}$$

$$u^* = ung^{1/2}d^{-1/6} \tag{4.10}$$

   Suggested values for K are listed in Table 4.1.   In straight, unstratified reaches of tidal estuaries, values of about 1.2 are appropriate (6).

Table 4.1. Coefficients for transverse mixing in open channels (6)

Channel	$K = \tau/(du^*)$
Straight uniform laboratory channels	0.1 to 0.2
Straight or meandering natural streams	0.4 to 0.8
Straight, unstratified reaches of tidal estuaries	1.0 TO 1.4

   Source: Fischer, et al. (6)

   Where the size a parcel of effluent, L, is smaller than largest turbulent eddy, as is the case for, ϵ increases with L according to the four thirds law:

$$\varepsilon = L^{4/3} \tag{4.11}$$

Applications. (1) The simplest application is for a point discharge to a wide river before the plume reaches the opposite bank:

$$p(x,y,t) = \frac{c}{c_0}(x,y,t) = \frac{Q}{d\sqrt{4\pi u \varepsilon x}}\exp\left(\frac{-uy^2}{4\varepsilon x}\right) \tag{4.12}$$

in which C/C_0 is the is the concentration of the effluent constituent, Q is the mass discharge rate of the constituent, d is water depth, and u is the mean advective current speed as illustrated in Figure 4.7. This simplest case is used for analyzing point discharges, and is a building block for more complex cases, using the principle of superposition and the method of images.

(2) For stream width, w, Equation 4.12 applies until the plume touches one of the banks. Downstream of that point, the density distribution within the stream behaves as if in an infinitely wide stream with an "image" source abreast of the true source. If the plume is at midstream, the method of superposition gives the downstream concentration distribution as:

$$p = \frac{c}{c_0} = \frac{Q}{d\sqrt{4\pi u \varepsilon x}}\sum_{n=-\infty}^{\infty}\exp\left(\frac{-u(y-nw)^2}{4\varepsilon x}\right) \tag{4.13}$$

This situation is illustrated in Figure 4.7. The plume trailing from the coordinate origin, confined between the sides w apart, is represented as a composition of plumes at $y = 0$, $y = +w$, $y = +2w$, ..., $y = +nw$ for n up to ∞.. Two results of this summation are shown on Figure 4.7a. The centerline concentrations at $y = 0$, initially very high near the source (small x) decreases gradually to $C = Q/(udw)$, again the value for complete mixing. The plot shows that the centerline and side concentrations differ by less than 10% when

$$x = \frac{uw^2}{\varepsilon} = L \tag{4.14}$$

Equation 4,14 may be used to calculate the downstream distance, $x = L$, which is required for complete mixture of centerline discharge across a stream.

(3) Essentially the same mathematics may be used to analyze the spread of a plume from a shoreline discharge across a stream. In Figure 4.7b imagine the "centerline" to be a channel side, the other channel side being the side shown at $y = w/2$. We can thus study the spread of a bankside discharge across a stream of breadth, $B = w/2$. the plot of Figure 4.8a is repeated in 4.8b, but it relabeled for the case of bankside rather than centerline discharge. The downstream distance required for complete across-stream mixing is:

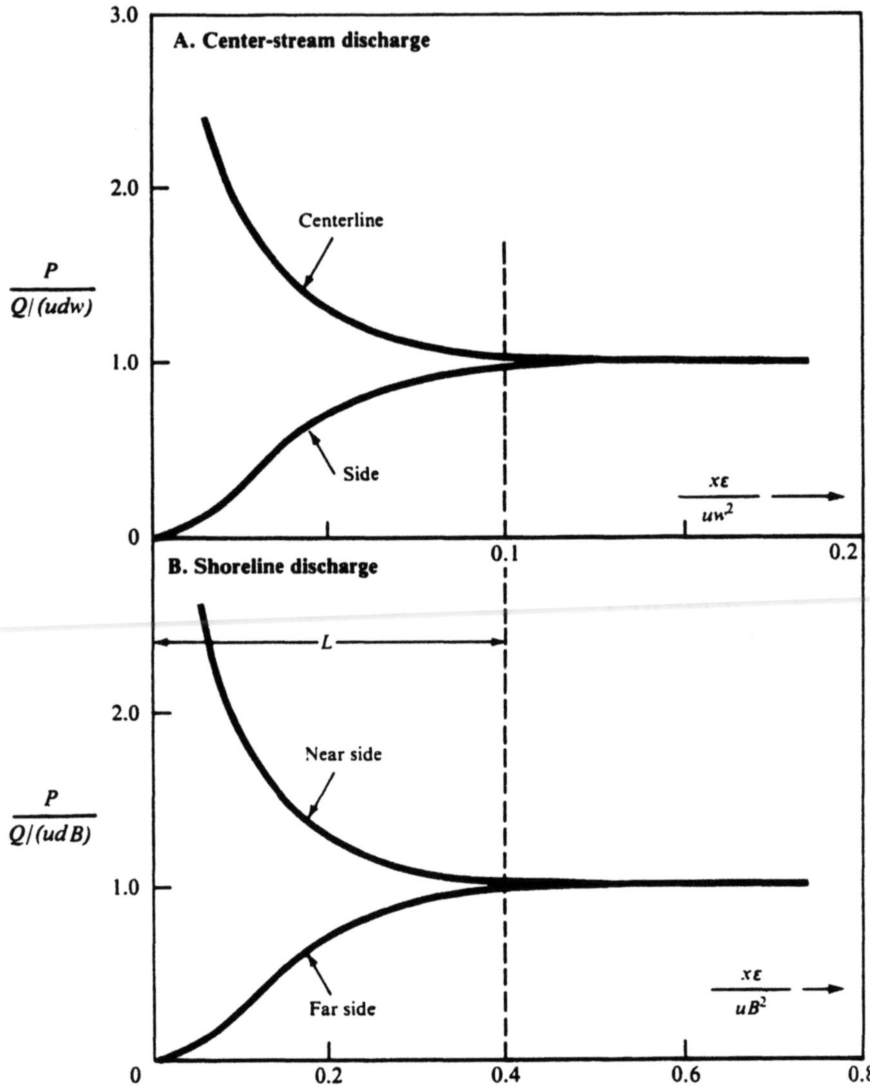

Figure 4.8   Effluent concentration as a function of downstream distance for point sources at midstream (A) and at one shore . Source: Fischer, et al (6).

$$x = \frac{0.4B^2}{\varepsilon} = L \tag{4.15}$$

(4) Discharge from a diffuser of finite length, b, in a river is computed in theory by superimposing the plumes, each described by Equation 4.12, from many small sources distributed along the diffuser. The mathematical expression for the plume centerline concentration for such a superposition is:

$$p = \frac{c}{c_0} = \frac{Q}{ubd} erf \sqrt{\frac{ub^2}{16\varepsilon x}} \tag{4.16}$$

for which erf is the standard error function for a Gaussian distribution.

$$erf(M) = \frac{2}{\sqrt{\pi}} \int_0^M \exp(-V^2 dV \tag{4.17}$$

Equation 4,16 is plotted in Figure 4.8 in normalized form.

Note that for large x (hence small M) far downstream from the diffuser, erf(M) approaches $2M / \sqrt{x}$, so the centerline concentration approaches:

$$\frac{c}{c_0} \rightarrow \frac{2Q}{ubd} \sqrt{\frac{ub^2}{16\pi\varepsilon x}} \tag{4.18}$$

$$\text{Or,} \quad \frac{c}{c_0} \rightarrow \frac{Q}{d\sqrt{4\pi\varepsilon ux}} \tag{4.19}$$

Note that this is identical to Equation 4.12 for plume centerline concentration (y=0). Figure 4,9 shows that Equation 4.16 can be approximated by Equation 4.19 for y = 0 whenever x exceeds about 0.5 ub^2/ϵ.

(3) Turbulent dispersion from a diffuser of length b in the open ocean where the "4/3rd law" applies, can be analyzed using the Brooks' (1):classic formula for centerline concentrations:

$$p = \frac{c}{c_0} = \frac{Q}{ubd} erf \sqrt{\frac{1.5}{(1+\frac{8Ex}{2})^3 - 1}} \tag{4.20}$$

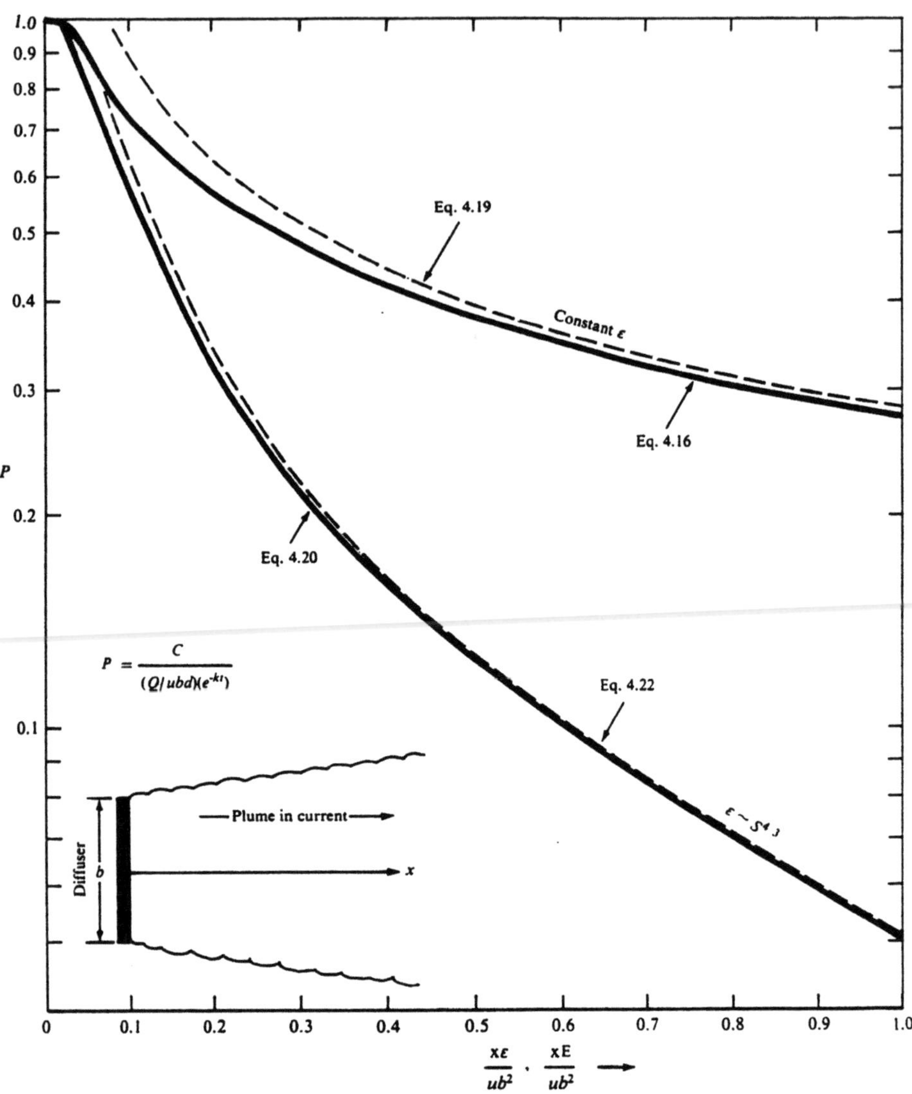

Figure 4.9. Relative concentration at the centerline downstream from a line source of length, b

This expression is plotted after French (7) in normalized form in Figure 4.9. The diffusivity, ϵ, increases with the width of the plume according to Equation 4.11. The initial value, E, is related to diffuser length:

$$E = 5x10^{-4}b^{4/3} \tag{4.21}$$

As with dispersion from a diffuser, the length of the diffuser becomes less and less important to the dilution the farther downstream one proceeds. For large x (small M), Equation 4.20 can be approximated by

$$p = \frac{c}{c_0} = \frac{Q}{ubd}\frac{2}{\sqrt{\pi}}\sqrt{\frac{1.5}{(1+\dfrac{8Ex}{2})^3 - 1}} \tag{4.22}$$

Equation 4.22 is plotted with dashed lines in Figure 4.9; it is a good approximation to Equation 4.20 for x larger than about 0,025 ub^2/g. Substituting Equation 4.21 into Equation 4.22 (using meter-kilogram-second units), the limit for large x is:

$$p = 5463\frac{Q}{d}u^{1/2}x^{-3/2} \tag{4.23}$$

in which the diffuser length does not appear at all. Thus diffuser length, which is a very important factor in initial dilution computations, is essentially unimportant in estimating dispersion further down-plume from the diffuser. An exception is the case in which there is potential for plume submergence (see Section 4.6.4)) with different advection-dispersion patterns beneath the surface and at the surface.

4.5   Comparisons of Results

Current patterns may be predicted by field studies using drogues or moored meters, by numerical models, or by physical models. Generally, there is little agreement between, say, drogue and meter values for instantaneous current velocities in one locality, but considerable agreement between velocity integrals, such as flux over a tidal cycle, or trends, such as seasonal speed direction histograms.

Example 1. A drogue deployed near and at the depth of a recording meter may show little agreement with the velocity measured simultaneously by the meter. However, after a year of nearly continuous current meter recordings and dozens of drogue deployments, the sum of the drogue measurements will show a range of current speeds and directions that is in close agreement with a similar summary of meter data.

Example 2. Tidal current charts for a major seaport (Figure 4.10) suggest that for three hours of a flood tide, effluent from a particular sewage outfall is entrained into a branch estuary (A); for the next three hours the effluent is carried up the stem of the main estuary (B);and during the six hours of ebb, the effluent is carried out of the harbor (C). A link-node numerical model of the same estuary system predicts that for each of the six hours of flood, roughly half the effluent flows to A, and half to B: and that during the ebb, all effluent goes toward C. Again, the charts and the model disagree on flood current patterns on an hour-by-hour basis, but both agree that over one tidal cycle, 35 percent of the effluent is drawn into branch estuary A.

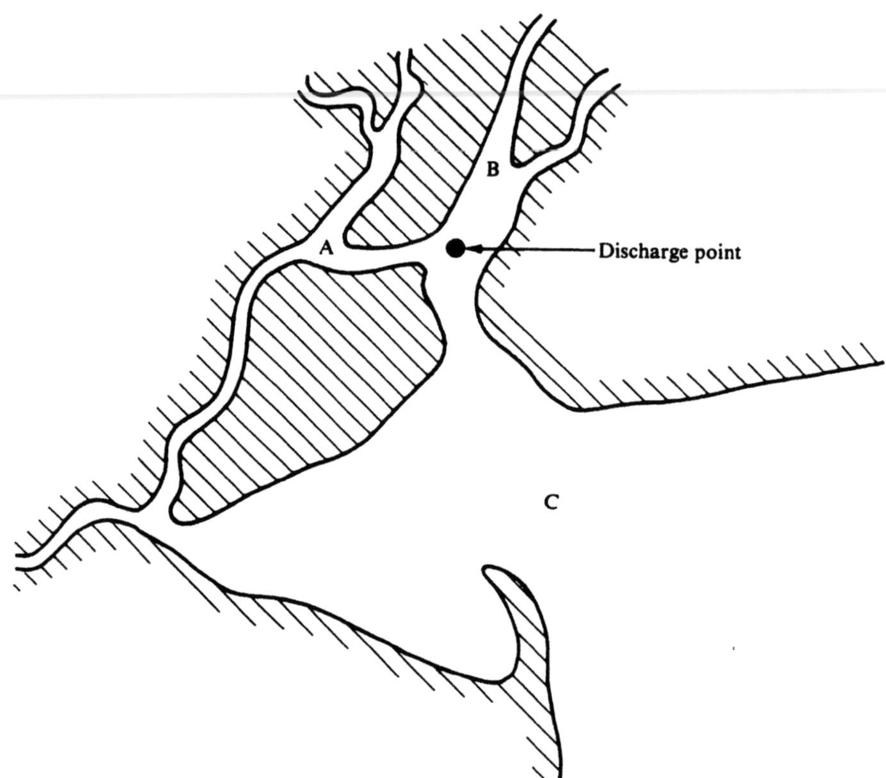

Figure 4.10   Hypothetical estuary for example 2, Section 4.4.6

4.6   Outfall and Outlet Location

Ocean outfalls complete coastal cities wastewater management system goals to remove people's wastes from their presence and their consciousness. This begins with buried collection systems on land and ends with submerged waste fields in the ocean. The geographic locations of outfalls are determined by the interactions of nearshore land use and topography, jurisdictional boundaries, offshore sediment type and bathymetry, waves, and water quality criteria.

4.6.1   Information Needs

The central questions of siting are (i) How can water quality criteria be met? (ii) Which alternative sites are most cost-effective? (iii) What level of wastewater treatment is required? (iv) How can future wastewater management system expansion flexibility be maintained? By staged construction?

The information needed to answer these questions includes: (i) present and probable future effluent quantity and characteristics, (ii) topography and land use in service area and bathymetry and sediments of the discharge area, (iii) locations of recreational, touristic, marine sanctuary, and commercial fishery and shellfish areas, and in estuaries, fish migration, (iv) local shipping, anchorage, and shore protection requirements, (v) local water quality criteria, (vi) local current and stratification patterns and variance, (vii) detailed seabed conditions along alternative outfall alignments, and (viii) wind and wave climate and variability with particular reference to construction (4,8).

A smooth, sandy bottom with good bearing capacity along a straight shoreline is preferred. Rock bottoms make construction more difficult and expensive. Very soft marine clays, particularly those with high organic and methane content are also difficult. Deeper discharges promote greater initial dilution so that if two sites provide necessary travel time to shore for advection, dispersion, and first-order decay of non-conservative constituents, choose the deeper . On the other hand, deeper construction is more expensive and plume submergence may be occasionally compromised by upwelling along the shoreline. In any event, it is necessary to compare in appropriate detail the marginal costs and benefits of all feasible locations.

4.6.2   Water Quality

Water quality criteria and standards recognize a wide distribution of values and are usually stated as a limiting value for some fraction of the samples. This is important, since there is no plausible outfall length within which discharge standards will either always or never be met. Rather, the frequency with which a limiting nearshore value is exceeded will decrease monotonically with increasing outfall length. For design purposes, the outfall length/violation frequency function, and the acceptable frequency of violation, each of which may vary seasonally, are determined.

Limiting concentrations for microbial constituents, identified in Chapter 3, are the most common criterion for determining the design travel time from the discharge area after initial dilution. Site-specific first-order decay factors for these and other non-conservative constituents require field studies since there are at present no general models for aggregating laboratory data on indicator or pathogen survival, sedimentation factors, and receiving water chemistry and biology with field data on transport, dispersion, and other physical factors. Relatively conservative constituents, some of them toxic, that can accumulate in marine life are then best controlled at the source or, at greater cost and with less success, at the municipal treatment plant. Surface sleeks of oil and grease that can blow ashore during an afternoon sea breeze can generally be removed at the plant.. Although empirical supporting data have not been found to date, some managers and designers have focused on nutrient removal from effluents before discharging them into areas where upwelling nutrient-rich waters are already supporting possibly limiting levels of desirable marine life. The design problems here lie in matching differences in scales of end-of-pipe, near- and far-field effects, and marine ecosystems monitoring (see Chapter 10).to the scales of the service area, contaminant source control, water conservation, and reclamation, and central wastewater treatment and disposal works. Meanwhile, the acceptable violation frequency establishes the required discharge distance from shore.

4.6.3 Worked Examples of Outfall Siting

Example 1. Consider a primary effluent discharge of 100 ℓ/s. with 100 mg/ℓ SS, 100 mg/ℓ BOD, and 1 million fecal coliforms /100 mℓ. Chlorination would reduce the fecal coliform count to 1,000/100 mℓ. The coastline is essentially straight, with a parallel coral reef 800 to 1,200 m offshore. The average current is 0.1 m/s out to the reef, and 0.5 m/s beyond the reef. At another site 100 km away with similar effluent and receiving water characteristics, *in-situ* bacterial disappearance tests indicated an overall T_{90} (see Section 3.2.1) of 1 hour.

Water quality criteria to be met at least 95 percent of the time are (i) settleable solids to accumulate not more than 0.5 kg/m^2/year, (ii) dissolved oxygen to be depressed no more than 1 mg/ℓ below ambient concentration, (iii) fecal coliform to be not more than 200/100 m/100 mℓ at shore or over the reef , and (iv)chlorine residual to be not more than 0.002 mg/ℓ over the reef.

With a straight coastline and reef, the position of the outfall on the shore will probably be near the treatment plant following a favorable routing over the seabed. The siting question is thereby reduced to, "How far offshore?" First, check the dilution flux inside the reef. The average current speed is 0.1 m/s; the cross-sectional area is about two-thirds of the width (1,000m) times the maximum depth, 8m, or 5,000m^2. The dilution flux is thus 500 m^3/s which, divided by the discharge rate of 100 ℓ/s, assuming complete mixing, gives a dilution ratio of 5,000 to 1. How far will dilution go towards meeting the criteria?

Constituent	Effluent, C_o	Criterion	Required dilution , C_0/C
SS	120 mg/ℓ	Study separately	
BOD	140 mg/ℓ	"1 mg/ℓ "	140
Cl	2 mg/ℓ	0.002 mg/ℓ	1,000
Fecal coli-	1,000,000	200	5,000
form/100 mℓ	1,000 if chlorinated	200	5

The required dilution ratios, except for unchlorinated effluent, are much less than the maximum attainable 5,000 to 1. As for fecal coliforms, they are assumed to be reduced by a factor of 10 during each hour of travel (this would be confirmed by parallel laboratory or field comparisons of the two effluents) . During this time, the plume will have traveled an average of 360 m.

The suspended solids mass discharge rate is 120 mg/ℓ x 100 ℓ/s = 378,000 kg/year.. For a first approximation, assume that 5 percent or 18,900 kg/year will have particle setting velocities $\geq$1 cm/s. They will reach the bottom within 8,000 seconds while the 0.1 m/s current will have carried them up to 800 m.

The particles will settle in a swath, the width of which depends initially on diffuser length but after 800 m of drift will be several times greater than the water depth and, in this case of unidirectional flow along a straight shoreline, will be an order of magnitude less than the travel distance. Thus it is assured that the plume path is much greater than 8 m and much less than 800 m. Use 80 m for the width, for order-of-magnitude computations.

For 18,900 kg/y settling in a swath 800 m long by 80 m wide, the average accumulation rate is 0.3 kg/m^2/year, well within a criterion of 0.5 kg/m^2/y. With current speeds averaging 0.1 m/s and with such wave action as will pass over the reef, frequent resuspension, advection, and resettling of these particles should be expected. For this small outfall, discharge shoreward of the reef is feasible. With an outlet at the deepest point between the reef and the shore, the above ratios are recalculated for slow (say, 10-percentile) current speeds and for the peak rate of discharge, to determine the limiting conditions under which the standards will still be met. With or without chlorination, coliform and chlorine standards will be met on the beach and over the reef.

Example 2. Assume the same conditions and contaminant concentrations as in Example 1, but with a tourist development that increases the discharge to 1000 ℓ/s

The dilution flux of 500 m^3/s, divided by the discharge gives a dilution ratio of 500. This is adequate for BOD dilution, but insufficient for chlorine dilution. Fecal coliform standards will be met because of bacterial disappearance. The suspended solids accumulation will now be 3 kg/m^2/year. Accordingly, it will be necessary to extend the outfall beyond the reef so the transect cross-section area of ocean between the reef and the chosen discharge point is greater than A, say 2A, in which A is the area required to obtain ocean water flux sufficient to provide adequate effluent dilution at the reef. If A = 1,250 m^2 with current speed 0.8 m/s, the ocean flux of 1,000 m^3/s will provide a dilution capacity of the order of 1,000. Doubling the area will double the dilution capacity.

Example 3. A 5m^3/s paper mill effluent has 3,000 mg/ℓ suspended solids. Sedimentation tests (14) have shown that 50 percent of the particles have settling rates in seawater in excess of 0.1 cm/s. Consider discharging this effluent to the sheltered bay shown in figure 4.5a, the average depth of which is 10 m and in which the 80-pctl current speed is 0.08 m/s. The bay area is about 100 km^2. Here, the question is whether there is a site such that the rate of accumulation of solids on the seabed does not average more than 1 kg/m^2/year. The bay will have too little wave action or currents exceeding an estimated 9 cm/s to resuspend sediments, What settles on the seabed will remain there.

What area is required to accommodate this discharge, at 1 m^3/year,?   5 m^3/s x 3,000 mg/ℓ x 86,400 s/day x 365 d/year = 4.73 x 10^8, requiring 473 km^2, far more area than is contained in the bay. Therefore discharge to the bay should not be permitted without treatment to reduce the high SS load. This conclusion is reached even without considering current speeds or particle settling velocity..

Example 4. An outfall is to be sited in a deep, seasonally stratified tidal estuary, near the treatment plant sited at T on Figure 4.11. Drogues deployed at point X tend to be entrained in a gyre induced during the seasonal flood flow. Those deployed at Y generally drift parallel to the coastline.

Transect A is typical of transacts near the point of land on which the treatment plant (T) is located. An effluent from a diffuser at X would float. A plume from Y would be below the pycnocline, or depth of strongest density gradient, and so would remain submerged. In any event, X is considered too close to shore because the effluent plume will be entrained by the gyre. If we discharge at point Y on Transect A, any effluent that reaches the surface will be efficiently advected away. If the plume is trapped below the pycnocline, it will be kept away from the shoreline while reducing oxygen levels at or below the pycnocline.

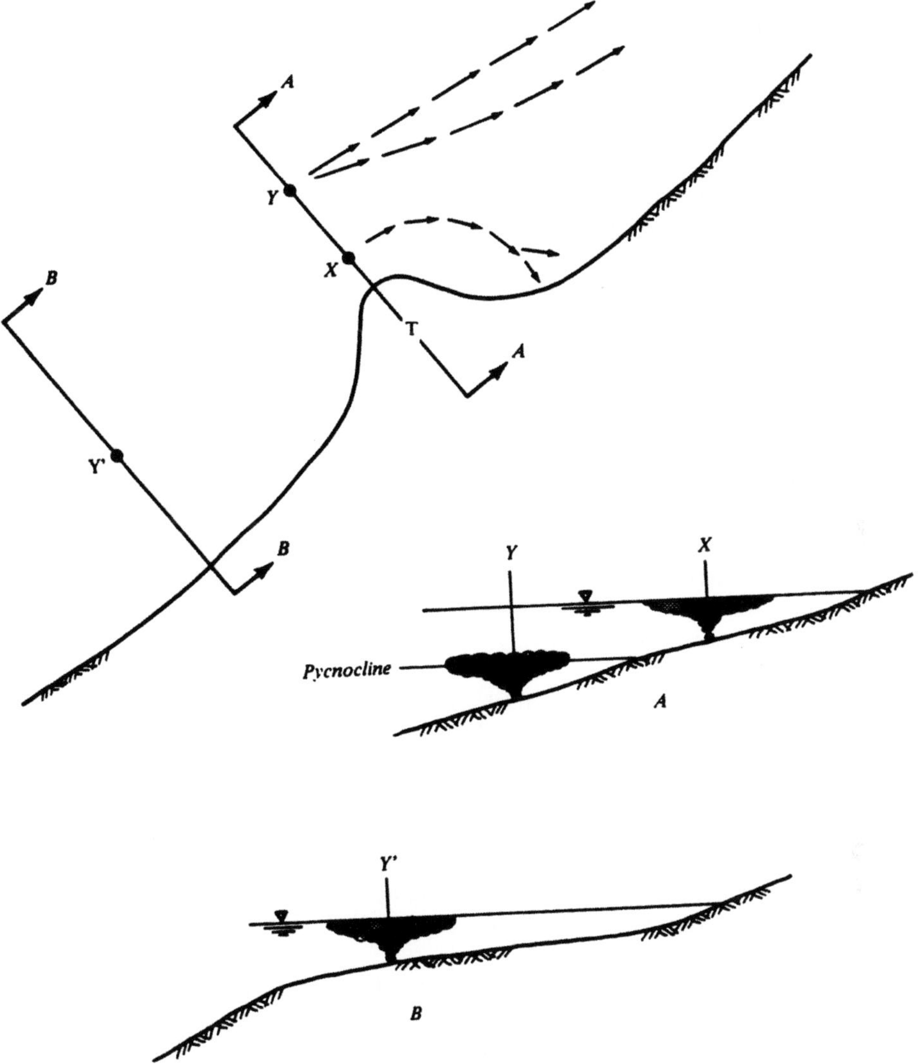

Note: Discharege at X: effluent plume rises to the surface and is entrained in the gyre downstream from the point. Discharge at Y: effluent remains submerged and is carried past the point. Discharge at Y': effluent plume rises to the surface and will be carried past the point.

Figure 4.11. Optimal discharge locations for a hypothetical outfall.

Check the likelihood of plume submergence by using the methods of Section 4.6.5: (i) obtain density profiles at all seasons for the area. (2) Measure the density difference between discharge depth and the water surface. (iii) For projected discharge rates, use equations 4.25 and 4.30 to determine the density difference required for plume submergence. (iv) If this density difference exists or is exceeded between discharge depth and the surface during any portion of the year, the plume will remain submerged for that time.

Determine the portion of the year for which such a density difference exists. (v).Repeat the last three steps but for the pycnocline rather than the sea surface to determine the seasons, if any, that the plume will be submerged.

If the plume will indeed be submerged a significant part of the year, determine the extent to which dilution alone will be sufficient to maintain acceptable dissolved oxygen levels. This may be accomplished with a sufficiently long diffuser (but check to see that any mandated initial dilution is obtained) or by going further offshore to deeper waters for greater initial dilution. Note increased construction costs for depths in excess of about 60 m.

An alternative is to seek another alignment. In this example, Transect B has a much flatter profile, so that Y' is as far offshore as point Y, but is in water shallow enough to be above the pycnocline. The extra costs of an outfall to point Y' must be justified by the extra benefits accruing when entrainment is avoided and the plume is submerged.

Example 5. Is it feasible to discharge 1 m^3/s of secondary effluent to the coastal lagoon shown in Figure 4.12 ? The mean tide range is 1 m. The area of the lagoon is 100 hectares. Tides are mixed semidiurnal and semidiurnal.

The tidal prism is 100 ha x 1 m = 1,000,000 m^3. Diurnal tides have a period of about 24.8 hours. At a very unlikely best, if the effluent is mixed uniformly throughout the lagoon, all of it will be exported on an ebb tide. With a sufficiently strong permanent longshore current, seawater entering the lagoon on the flood tide (not very likely) the water entering on the flood will have no sewage content.

The mass of sewage entering per tide cycle = Q x t = 1 m^3/s x 24.8 h x 3600 s/h = 89,300 m^3 The volume of seawater that leaves the inlet is the tidal prism, 1 million m^3· The ratio of these two numbers is 11.2. This is the theoretical maximum far-field tidal prism dilution, enough to affect the salinity and nutrient levelsand probably the biota of the lagoon, but of minor significance in public health.

4.7   Outlet Design and Initial Dilution

As of 1995, easy-to-use public-domain software is available for initial dilution computations for most diffuser configurations. Basic considerations are discussed below. Use of the models is discussed in Section 4.6.6.

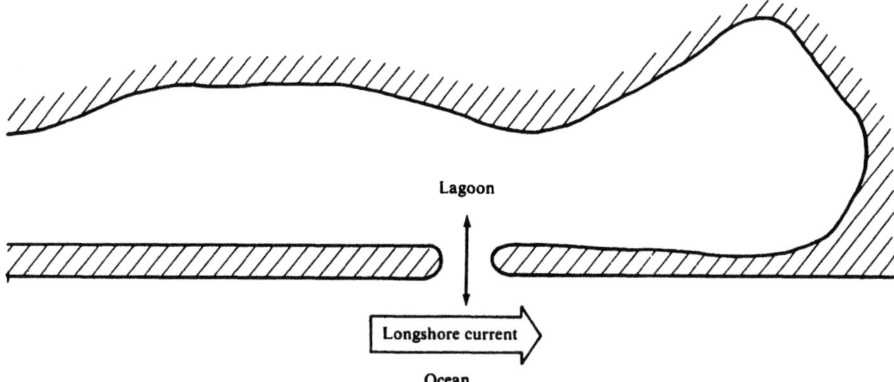

Figure 4.12   Hypothetical lagoon for sewage disposal (Example 5).

4.7.1   Single Open Ends, Rose Diffuser Caps, Multiport Diffusers

The outlet of an ocean outfall may consist of a single open end, perhaps with an elbow to conduct the flow to a point of release a meter or so above the seabed. The single end of the pipe may be fitted with a rose diffuser capital that directs the flow to discharge horizontally in many directions, much like a gas ring on a cooking stove. Or, the outlet may consist of a multiport diffuser, in which the last sections of pipe contain a regularly spaced line of relatively small ports, and the end of the outfall pipe is capped. Finally, the individual ports of some multiport diffusers may themselves be fitted with rose diffuser caps.

The purpose of a diffuser is to ensure a much greater initial interception of ambient dilution water by the effluent stream for greater initial dilution. While most small outfalls, as well as early larger ones, have single open ends, the multiport diffuser has become a conventional design feature for large-diameter outfalls for sewage and cooling water discharges.

Some multiport diffusers have been constructed with two or more branching arms. These are unnecessary; by the time the sewage field has moved one or two diffuser lengths downstream, average dilution is the same whether the current is normal or parallel to the diffuser section (6). Thus, most diffusers are simple extensions of the outfall pipe itself, sometimes with reductions in diameter progressing towards the end.

A multiport diffuser provides increased initial dilution within a small mixing zone near the diffuser, increasing the possibilities of plume submergence due to ambient stratification, whereas a plume from a single open end would be more likely to rise to the sea surface. By the time the diluted effluent has traveled a few diffuser lengths downstream, particularly when there is little stratification and no plume submergence, the plume dilution distribution becomes independent of the diffuser length (see Section 4.4.5).

Use of the single open end is recommended in cases where it will provide adequate initial dilution to meet water quality standards. It is also recommended in cases where plume submergence due to a diffuser is unattainable or undesirable. It is obviously the simplest terminus to build and maintain. For a single open end, the design considerations are:

- If an elbow is used to deflect the effluent upward, reaction to flow momentum within the bend will tend to drive the elbow into the seabed and pull it away from the next section of pipe. Provide thrust resistance as necessary.

- Provide a sufficiently constrictive bezel or nozzle (much like those for major smoke stacks) to increase the flow velocity, u, and decrease the port diameter, D_p, to ensure that the port emission Froude number, **F**, is always greater than 1, even at low flows, to prevent seawater intrusion into the pipeline. For u = exit velocity, ρ = ambient field density, $\Delta \rho$ = difference between effluent and ambient density, and g = gravitational acceleration.

$$\mathbf{F} = u/(g\, \Delta\rho/\rho\, D_p)^{1/2} \tag{4.24}$$

- If the natural seabed materials are uncohesive sands or silts, consider placing crushed rock around the outlet, to a depth of 10 to 15 cm, and within a distance of 2 to 3 m from the outlet pipe, to minimize erosion of the seabed near the pipe, due to the currents induced by the discharge itself. Erosion due to wave action may also be a concern (see Chapters 2 and 6).

Rose diffuser caps increase the initial dilution from a single port, or from each of the ports of a multiport diffuser, by a small amount. Where receiving waters are very shallow, the very evident surface boil from a single open ended discharge can be diminished--but not eliminated--by adding a rose diffuser cap.

The ports from buried multiport diffusers must consist of riser pipes extending from the buried diffuser header up through the sea bottom. To provide such a riser for each port can be very expensive. Fortunately, it has been found that the initial dilution from a diffuser of given overall length, flow rate, and port number is achieved nearly as well if the ports are not uniformly spaced in a line, but clustered, six or eight together, in rose diffusers caps on riser pipes totaling 1/6 to 1/8 the total number of ports. For the Deer Island tunneled outfall at Boston (see Section 11.6) where each riser would cost several million dollars, physical model tests were conducted to determine the minimum number of risers necessary for a diffuser section 2,000 m long, where the ocean was 31 m deep (P.J.W. Roberts, personal communication, 1993). Based on the tests, the number of risers was set at 55, implying a spacing of 37 m, slightly greater than the water depth.

4.7.2   Initial Dilution for Plumes from Single Round Ports

Over the past quarter century there has been extensive theoretical and experimental research on the movement and dilution of buoyant plumes from round discharge ports, such as an open pipe end. The trajectory and dilution pattern have been related to orientation of the port axis (vertical angle above horizontal plane; horizontal angle from direction of ambient current flow); to the density difference between the effluent and the ambient fluid; to the mass flux and momentum flux; and to the density stratification. The accuracy of dilution predictions by these analytical procedures is considered to be within 15 to 20 percent. Orientation (aim) of the discharge is an important factor for dilution when the discharge momentum is large and the discharge is jet-like; but orientation is of negligible importance when momentum is small and the plume motion is driven by its buoyancy. For a more complete discussion, see Fischer et al. (6).

In most practical cases encountered in ocean outfall diffuser design, the port emission Froude numbers are sufficiently small (4 or less) that the momentum can be neglected in the analysis. Here, u is the velocity of flow leaving the port, D is the port diameter, dp is the density difference between the effluent and ambient fluid, ρ is the ambient fluid density, and g is the gravitational acceleration.

When momentum is neglected, the principal dilution equation for a single, round-port discharge into an unstratified ambient is:

$$S = 0.089 \, (g \, \Delta\rho/\rho)^{1/3} \; y^{5/3} \, /Q^{2/3} \qquad\qquad (4.25)$$

in which S is the plume centerline dilution at elevation y above the discharge point, and Q is the discharge rate.

When there is a linear ambient density gradient $d\rho_a/dy$, that is, a constant rate of change of seawater density with respect to height, y, the plume will rise to a maximum elevation, y_{max} above the outlet:

$$y_{max} = 3.98 \, (Qg \, \Delta\rho/\rho)^{\,1/4} \, (-g/\rho \; d\rho_a/dy)^{-3/8} \qquad\qquad (4.26)$$

The centerline dilution at the height, y_{max}, the elevation below which the plume spreads out, is:

$$S = 0.071 \ (g \ \Delta\rho/\rho)^{1/3} \ y^{5/3}/Q^{2/3} \tag{4.27}$$

a result identical in form to Equation 4.25, but whose constant coefficient is 20 percent smaller, due to the fact that for the last part of its buoyant rise, the plume passes up through its own submerged cloud, and entrains itself rather than clean ambient seawater.

These conservative equations are for the rare case of zero ambient current and for linear density gradient (or zero gradient). In a real-world current, the dilution, S, attained by the time the plume has reached a given height, y, will be greater than with no current. Agg (16) has proposed the following correction:

$$S_{(moving \ water)} = S_{(still \ water)} \ exp \ (1.107) + 0.938 \ log \ (U/V_p) \tag{4.28}$$

where U is the ambient velocity and V_p is the jet velocity at the port.

With stratification, the presence of a current reduces y_{max}. Results obtained by assuming a linear density gradient can be also applied in the more common case of a non-linear density gradient, as discussed in Section 4.6.4.

4.7.3 Initial Dilution for Plumes from a Line Source

The merging of individual round-port plumes issuing from many ports evenly spaced along a diffuser can be approximated by a "line plume" or a "two-dimensional plume". Equations for the dilution and rise height of line plumes are analogous to those for round plumes in a still ambient (no current), given in the previous section.

When the ambient is of uniform density:

$$S = 0.38 \ (g\Delta\rho/\rho)^{1/3} \ y \ /q^{2/3} \tag{4.29}$$

in which q is the effluent discharge per unit length of the line source. When the ambient has a linear density gradient dp/dy, the maximum height of plume rise is

$$y_{max} = 2.84 \ (g \ \Delta\rho/\rho \ q)^{1/3}/ \ (- \ g/\rho \ d\rho_a/dy)^{1/2} \tag{4.30}$$

and the dilution ratio at the height of rise is:

$$S = 0.31 \ (g \ \Delta\rho/\rho)^{1/3} \ y_{max} / \ q^{2/3} \tag{4.31}$$

a result similar to that of Equation 4.29, but with dilution reduced owing to the plume's entraining its own spent cloud, as in the case of Equation 4.27.

These results are for a line source infinitely long, in an ambient with no current. Following earlier work by Fan (5) and Wright (15), Roberts (13) studied the case of a line source of finite length in a current of uniform density, a case of great practical importance in outfall diffuser design. Roberts' research on line plumes perpendicular to, parallel to, and at 45 degrees from the current direction showed that the ratio of line-length to water depth is not important. His ratios varied between 3.7 and 30.

For $\mathbf{F} = u^3/ (g \, \Delta\rho/\rho)$ less than about 0.1, the effect of current appears to be negligible, and for all current directions:

$$Sq/ud = 0.27 \, \mathbf{F}^{-1/3} \qquad (4.32)$$

This reduces to

$$S = 0.27 \, (g \, \Delta\rho/\rho)^{1/3} \, y_{max}/ \, q^{2/3} \qquad (4.33)$$

which is essentially identical to Equation 4.31 when d is set equal to y_{max}. The 15 percent difference between the coefficients is well within the margin of uncertainty for these analyses.

For $\mathbf{F}$ greater than 0.1, the effect of current direction begins to be evident, and with increasing $\mathbf{F}$, the ratio Sq/ud tends to the following constant limits:

a)	Line perpendicular to current:	Sq/ud	tends to 0.58
b)	Line 45% to current:	Sq/ud	tends to 0.37
c)	Line parallel to current:	Sq/ud	tends to 0.15

(Note that for the ideal case of effluent mixed uniformly from surface to bottom, and uniformly across a swath just equal to the length of a line source perpendicular to the current, the dilution would be $S = Ud/q$; or $Sq/Ud = 1$. This "ideal" value may be compared with the three cases above.)

4.7.4 Initial Dilution from a Line of Port Clusters

Generalized means are not yet available to predict the initial dilution for a line diffuser in which the ports are clustered at rose diffuser caps. Provisionally, however, it is suggested that one compute the initial dilution for the case where the ports are distributed uniformly along the diffuser alignment, and assume that the cluster configuration will perform equally, as long as the spacing between adjacent clusters does not exceed the plume rise height, or in any case the water depth.

4.7.5   Quick Estimate of Likelihood of Plume Submergence

Equations 4.26 and 4.30 above are used to predict the rise height, y_{max}, of a plume rising in a linearly stratified environment. If y_{max} is less than the water depth, d, the plume is submerged.

Far more usual than a linearly stratified profile is the case of irregular and changing stratification profiles. Sometimes the information available is limited to surface, mid-depth, and bottom values of temperature and salinity, hence density.

To come to an adequate understanding of the receiving water stratification regime, it may be necessary to compare many seasonal profiles, construct several composite design profiles, and make linear approximations to them. Alternatively one may use numerical models to analyze nonlinear profiles, as described in the next section.

However, the following procedure is relatively easy to apply, and makes good and complete use of available data. It takes advantage of two features. Linearization of a density profile between the height of rise, y_{max}, and the discharge elevation, $y = 0$, introduces only modest error into computations. In essence, the height of rise of a plume depends much more on the ambient density values at $y = 0$ and $y = y_{max}$ than on the details of the density distribution in between. For outfall planning purposes, it is just as useful to ask whether a plume will be kept submerged beneath the surface and whether it will be kept submerged beneath mid-depth as it is to ask precisely what y_{max} will be, for a given plume and ambient.

The procedure illustrated in Figure 4.13 for a 400-meter multiport diffuser section in water 40 m deep shows the conditions under which the plumes will (i) rise the full 40 m to the sea surface ($y_{max} > 40$ m), (ii) rise to y_{max} more than 30 m but less than 40 m, or (iii) rise to y_{max} less than 30 m.

The left part of the figure contains a plot of data from all available density profiles for the example study area, expressed as the ambient water density difference between the surface and 40 m depth (solid circles) and between 10 m and 40 m depth (open squares). These density differences are plotted against calendar date to show seasonal trends. Inner envelope curves are fitted, such that few solid circles fall beneath the "10 m to 40 m" curve. The plot shows that from April through August there is a fair guarantee of up to 0.4 kg/m^3 of density difference to keep a plume submerged beneath the sea surface.

On the right side of Figure 4.13, the required density difference for plume submergence is plotted as a function of discharge rate, Q from a point source using Equation 4.26, for $y_{max} = 30$ m and 40 m; and as a function of discharge per unit length, q, from a line source, using Equation 4.30 for $y_{max} = 30$ m and 40 m. The values of $g\Delta p/p$ and of p are noted on the figure.

Now, the right side of Figure 4.13 shows that a line source discharge rate of q = 0.10 m^3/s requires a density difference (between $y = 0$ and $y = y_{max}$) of at least

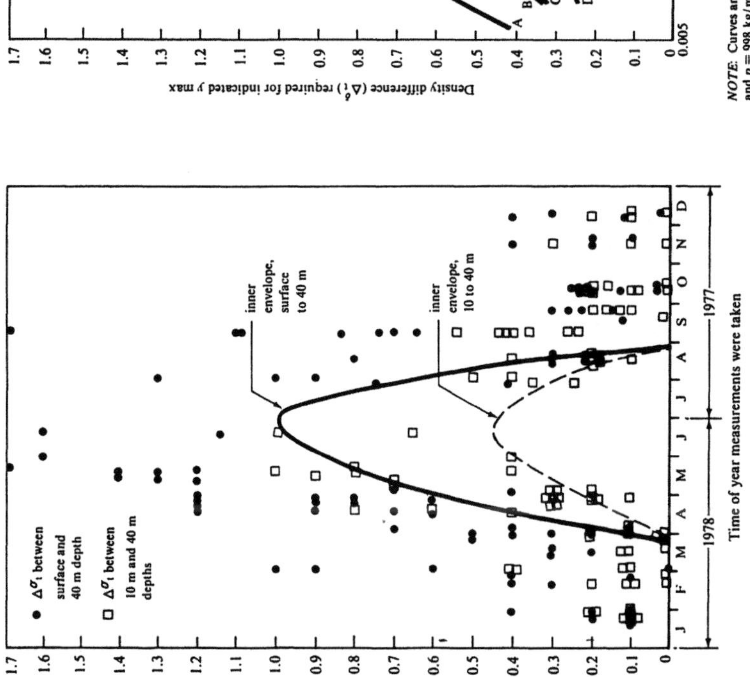

Figure 4.13. Case study of the density difference required and available for plume submergence.

0.32 kg/m^3 consistently from mid-April to mid-August. The density difference 40 m to 10 m consistently exceeds 0.42 kg/m^3 during parts of June and July.

The point-source curves can be used in the same way, either for simple open-end-of-pipe outfalls, or for the individual plumes from each of the many ports of a diffuser. Brooks (3) advises that in some situations y_{max} may be attained before the component port plumes have risen and spread to merge into what is effectively a line plume. Therefore, it is well to compute the rise heights both of an idealized line plume from a diffuser, and from a typical single component port, and use the more conservative of the two results.

In summary, when this method is used to survey the likelihood of seasonal plume submergence, the following steps should be taken:

1. Collect density profile data, and plot density differences versus month or season as shown on the left side of Figure 4.13. (Note that the data need not be full profiles; simple pairs of, say, surface-bottom or mid-depth-bottom density values can be used.)
2. Using equations 4.26 and 4.30, plot required density difference versus Q or q, for chosen values of y_{max}. In the equations, replace dp_a/dy with the linearized form, (density difference)/y_{max}.
3. For any q or Q, examine the seasonal prospects for submergence for each y_{max} considered.

4.7.6 Diffuser Design

Orientation. Often it is possible to extend an outfall pipeline from the shore into the sea, with the diffuser along the last few tens or hundreds of meters with no significant bends required in the construction. Usually the major current flow is parallel to the coastline, so that a diffuser perpendicular to the coastline is also perpendicular to the currents, an orientation that provides greater initial dilution than any other, when current speeds are significant.

If the alignment can be selected to be level throughout the diffuser section, the diffuser can be easily designed to provide uniform distribution at all flow rates. Where the bottom profile is steep, the diffuser section can be laid on a depth contour for uniform distribution of discharge at all rates of flow.

Where there is significant sediment bedload movement past an unburied diffuser, Sediment may block the diffuser ports. In contrast, sediment may be scoured from beneath the diffuser pipe.pipe. Such problems will be avoided by having the diffuser turned parallel to the direction of bedload travel in the direction of current.

In these cases, the initial dilution can be achieved as well with as with a perpendicular alignment by using a somewhat longer diffuser section. In any event, the diffuser orientation does not affect initial dilution when current speeds are small. If there is not a well-defined current direction, design the diffuser to get adequate initial dilution in a parallel current, and good performance for any current direction will be assured.

Diffuser Length. The next step is to choose either a simple open-end or a multiport diffuser. For the latter determine the appropriate diffuser orientation and select a diffuser length, b, that yields q = Q/b small enough to meet near-field water quality standards and to achieve (or avoid) plume submergence

Port Size and Spacing for Uniform Distribution. It is desirable to have a uniform discharge per unit length issuing from the ports along the diffuser length. This is achieved largely by (i) keeping the diffuser section at a constant depth, and (ii) ensuring that the sum of the cross-sectional areas of the ports is less than about 60 percent of the cross-sectional area of the diffuser pipe.

If effluent is less dense than the ambient water and the diffuser alignment is sloped, the effluent at low flows will favor the higher ports, with no flow (or even inflow) through the lower ports.

Ideally, the pipe cross-sectional area should taper continuously and linearly from its greatest size, at its connection with the outfall pipe, down to zero at its end. This, with uniform discharge per unit length, will provide constant velocity and nearly uniform hydraulic conditions within the pipe for scour velocities to be maintained for effluent settleable solids.. However, it is neither practicable nor necessary to provide a continuous taper of pipe area. Large outfalls have been designed with one or more decrements in diffuser diameter. Many smaller outfalls operate well with constant-diameter diffusers.

Detailed computation of discharge distribution in a multiport diffuser follow Brooks (1,2) and starts at the offshore, or downstream end where the pipe flow velocity is close to zero

1. Note that the port area criterion (guideline [ii] above) dictates that the end of the diffuser be closed.
2. Assume a discharge rate through the farthest downstream port in the diffuser.
    3. Using a stage-discharge relationship involving port area and shape, calculate the excess energy head, E, within the diffuser pipe at its downstream end required to achieve the end port discharge rate. Laboratory studies (3, 11) of the function $C = C(V^2/2gE)$, where C is the discharge coefficient for the port, have shown that

$$p = C \, (\pi/4) \, a^2 \, (2gE)^{0.5} \tag{4.35}$$

in which p is the discharge rate for a port, a is the port diameter, V is the average pipe flow velocity, and E is the energy head in the diffuser pipe, relative to the ambient. For smooth bell-mouth ports, flowing full:

3. $C = 0.975 \, (1 - V^2/(2gE))^{0.375}$ $\tag{4.36}$

For sharp-edged ports, flowing full:

$$C = 0.63 - 0.58 \, V^2/(2gE) \tag{4.37}$$

Other port designs have their own functional relationships for C.

4. Calculate the average pipe flow velocity, V, in the section of diffuser pipe feeding the last port.
5. Move up the pipe to the next port. Compute the energy head there by adding the friction head lost between ports, to the energy head at the last port.
6. Compute the discharge through the next port. Compared to the last port, the flow at the next port may be greater, because of increased energy head required to overcome friction losses in the pipe; or less, because the increased velocity in the pipe effectively reduces the discharge coefficient in the total head-discharge equation.
7. Compute the increased pipe velocity in the pipe section leading to the next port upstream, etc., etc.

This analysis proceeds iteratively, and can accommodate any changes in port size, port spacing, pipe diameter, pipe internal friction, or pipe elevation. By the time the discharge for the upstream port has been calculated, the discharge through each port and the total head (relative to the ambient pressure outside the diffuser) required to drive the diffuser are known. The absolute value of these quantities depends entirely on the discharge rate assumed for the first port, so several computations are usually necessary to get a satisfactory result for a particular design Q. This diffuser analysis procedure is easily handled using standard spreadsheet techniques.

For well-screened or well-treated effluents, some designers have selected port diameters may be as small as 5 cm. Larger diameters have the advantage that they are less likely to be plugged by objects trapped in the pipe during process interruptions or breakdowns, although they can be removed by pigging. If the effluent is only given coarse screening, the port diameter should be no less than about twice the screen bar spacing. Of course, variable port sizes can be used.

Buried Headers Using Riser Pipes for Ports. The diffuser pipe may be laid either exposed on the seabed with a properly prepared bed and with revetment; or it may be buried in a backfilled trench. In the former case, ports are simply holes in the sidewall of the pipe. In the latter case, the ports consist of small pipes rising from the diffuser pipe through the backfill and extending to 0.5 m or more above the finished grade of the backfilled pipe trench. There are pros and cons to each type of port, and selection is largely a function of local conditions. If the diffuser pipe is laid upon the seabed, the cost of riser pipes is avoided; there are no riser pipes to break off if snagged by trawls or anchors; and the diffuser pipe is relatively accessible for cleaning or repair. However, the pipe itself is subject to damage from anchors, trawls, or waves. If there is sediment bed load transport at the site, the pipe may cause sediment to be deposited on one side, and may possibly allow sediment to enter the pipe via the ports. On the other hand, sediment may be eroded out from under the pipe; but damage to the pipe from this may be avoided by bedding the pipe adequately on crushed stone.

A buried diffuser with riser pipes costs more to build, clean, and repair, but may be the desired alternative when there is significant shipping, anchorage, or sediment transport due to the longshore drift or (orbital) wave movement. Sediment bedload can pass the riser pipes without difficulty. The riser pipe should be designed long enough to accommodate any seasonal aggradation and degradation of the seabed.

It is well to give the riser the simplest exterior form possible, that is without a crown or T-head of any sort, so that an anchor or trawl caught on a riser can be freed by hauling it straight up.

An important design criterion for diffusers with riser pipes is to ensure that the pressure loss through the risers, that is between the buried diffuser pipe and the outside ambient fluid, exceeds the product of riser pipe height, gravitational acceleration, and the density difference between the sewage effluent and the ambient seawater. This condition must be met if the effluent is to successfully purge seawater from all the risers.

4.8   Other Hydraulic Design Considerations

4.8.1   Pipe Diameter

Selection of an outfall pipe diameter involves striking a balance among a number of considerations. Decreasing the pipe diameter usually reduces construction costs, reduces incident wave forces, and increases the flow velocity in the pipe, which is important if settleable solids in the effluent are to be kept in suspension or to be resuspended. However, decreasing the pipe diameter also increases the required driving head for a given discharge rate. Often, the balance struck is one in which the average flow velocity is not less than 0.6 m/sec in the early design life of a project, yet the system head is not excessive at peak flow in the mature years of the project. Minimum velocities are particularly important in purging saline wedges that develop during low flow periods (24). (see also Section 12.2)

4.8.2   Thrust at Bends

Should there be a bend in the pipeline to provide uniform diffuser depth, it must be provided with a resistance against a pipe motion due to the change in momentum of the flow within the pipe. For example, at a 90-degree bend, the thrust in each direction is equal to pQV; the resultant thrust is $\sqrt{2}$ (pQV).

4.8.3 Hydraulic Transients

Hydraulic transients must be considered for a pumped discharge to a long outfall. If there is a sudden shutdown of the pumps, a means for gradual deceleration of the flow within the pipe must be provided, to avoid damaging downsurges, subsequent reflective upsurges, and water hammer within the pipeline. In some cases, column separation may occur. Where pumping heads are not excessive, an open surge chamber just downstream of the pumps is appropriate. The chamber freeboard should be high enough to accommodate the hydraulic grade line at the greatest peak flow projected for the outfall. The chamber volume should be great enough to continuously feed water into the pipeline from the moment of pump shutdown (when the head is at its level pump-rated for discharge, Q) to the moment that flow in the pipe comes to rest. It may be advisable to throttle the flow to and from the surge chamber to dampen the downsurge somewhat. For design details on surge tanks and other control devices such as air chambers and air relief valves, see a standard text on surge analysis, such as that by Parmakian (10). Modeling hydraulic transient scenarios is both feasible and advisable for pumped discharges for which onvenient software.is available   As always, analysis should be by, or under the close supervision of, a properly trained engineer. A different kind of transient occurs during periods when the flow begins to increase, often quite rapidly, from daily low flows

Non-steady flow. Besides the problems of maintaining flushing velocities inside the diffuser section and risers and jet stirring velocities outside in the ambient water, effects of changing flows and velocities occur elsewhere in an outfall. For example, when early morning low flows in the Los Angeles County outfall tunnel coincide with the lower-low water of a spring tide, water levels drop well below the soffit. of the outlet inlet chamber for tens of minutes. Rising flows entrap air and are associated with a downstream hydraulic jump. Similar forces are at work in the San Diego outfall in the area where the seafloor slope increases rapidly at the edge of the continental shelf and where air pockets can form and gradually accumulate hydrogen sulfide and sulfuric acid (J.A. Foxworthy, 1995, personal communication). Hydraulic model studies can reveal details of these events and, in the case of an sharp increase in bottom gradient, support the choice of a tunnel on a constant grade over laying an outfall on the seafloor.

4.8.4 Excess Hydrostatic Head

Excess hydrostatic head must be considered when discharging a freshwater effluent, with a density of 998 kg/m^3, to a sea with a significantly different density of about 1025 kg/m^3. Suppose an outfall discharges to a saltwater ocean at a depth of 10 m. At the outlet, the hydrostatic pressure imposed by the seawater outside the pipe is equal to the 10 m seawater depth, times g, times the density of the seawater. This is balanced by the freshwater effluent inside the pipe. Because it is less dense, the depth of internal freshwater required to balance the saltwater pressure is the seawater depth of 10 m, adjusted upwards by the ratio of density of the two waters, (1025 / 998), i.e. 10.27 m. With no flow in the outfall filled with freshwater effluent, the water level found in an outfall manhole at the shoreline would be 0.27 m above sea level. The excess hydrostatic head, 0.27 m in this example, must be considered when designing manholes and surge tanks.

4.8.5 Drop Structures

It is desirable to keep all but very small quantities of air from being carried through the outfall. Air would impede the flow of effluent, and might cause unsightly bubble boils when released through the outlet. Sometimes the treatment plant discharging through the outfall is on a cliff or bluff, or there may be a high overland force main between the treatment plant and the outfall. In either case, there is potential for high-velocity flow in a partially full pipe to go through a hydraulic jump in the pipe and entrain large quantities of air into full-pipe flow in the outfall (3) This will reduce the outfall's hydraulic capacity, may introduce flow unsteadiness (e.g. a periodic burping), and create unsightly bubbling at the surface above the discharge (9).

There are three design approaches to limit the entrainment of air into the outfall:

1. Allow the hydraulic jump or waterfall to occur, but provide a stilling section downstream to allow the entrained air bubbles to rise to the crown of the section for venting before the flow proceeds into the outfall pipe; or
2. Provide a vortex drop structure in which the water swirls in a helix as it falls, in which the water is centrifuged to the walls and air is collected at the core of the shaft, and purged up the shaft counter to the water flow direction, or
3. Provide a pressure-reducing valve in the line at the bottom of the descent, at or below sea level.

The second and third approaches have been used successfully on very large outfall systems (3); but the first approach, with good design, will probably be adequate for most small outfall systems. The rise velocity of bubbles larger than a millimeter in diameter is 0.1 m/sec or greater (9).

The stilling section following a section where air entrainment occurs in a jump or cascade should have a length:

$$L = \frac{V, \ m/s}{0.1 \ m/s} \times flow \ depth \tag{4.38}$$

in which V is the flow velocity in the section.   This will allow any bubble larger than a millimeter to rise from the invert to the water surface and escape, before being swept out of the stilling section. Obviously, the stilling section should be designed with appropriate crown venting.

4.8.6   Provision for Pigging

With advancing age, pipelines can lose hydraulic capacity because of grit and/or grease, sediments, and air pockets clinging to the wall, which both increase wall roughness and reduce the pipe flow cross-sectional area. Hydraulic capacity can be restored at least partially by "pigging", an operation in which large cylinders of a sufficiently soft material, and of a diameter essentially equal to the pipe internal diameter, are driven through the pipe by hydraulic pressure. The technique is feasible for pipelines up to 1.5 or 2 m diameter.(A typical pigging operation is described in detail by French (19).

A common means to insert the pig is to provide a tee in the discharge line just downstream of the headworks to the outfall, with the stem of the tee of the same diameter as the discharge line, and extending vertically sufficiently far that it can be surcharged to provide the required head to drive the pigs through the outfall, yet not so high that pigs cannot be lifted into it by a crane. The pig-entry tee can be installed in the discharge line, with a very short stem capped with a blind flange, at relatively small incremental cost during original construction. Many years later when the pipe must be pigged, there will be great cost savings if this tee is present and accessible.

Likewise, at the discharge end of the outfall, it is good design practice to provide a means for a pig of the same diameter as the pipeline to be released from the pipeline, either through a tee at the start of the diffuser section, or by keeping the entire diffuser header diameter equal to that of the main pipeline, and with a removable blind flange at the farthest downstream end of the diffuser.

4.9   Appendix.   A Note on Post-Audits

Post-audits of wastewater management programs and projects for coastal cities are defined in general terms in Section 10.7. Most of them are in response to wastewater demand projections that are not linked to water supply projections, underestimated construction costs including those of excess capacity (Sec. 12.4), and reliance on elegantly calibrated but incompletely verified state-of-the-art numerical models (24).

Sensible laboratory and field studies of initial dilution and diffusion from outfalls began with the classic 1930 work of Rawn and Palmer (21) from the Los Angeles County Sanitation Districts. These efforts gathered momentum after about 1950 that is continuing in industrial countries (16, 22, 23).

Measuring initial dilution in a sewage field is conceptually simple but operationally difficult, particularly when the designer has met the goal of keeping the wastewater below the surface. Dilutions are best estimated from temperatures, salinities, and calculated densities of mixtures of intrinsic fresh (waste) water in sea water rather than by measuring concentrations of extrinsic tracers that are often non-conservative Radioisotopes whose decay functions are precisely known might work better, but are rarely environmentally or politically acceptable.

Post-audits of outfall operations based on measuring dilution of a chemically stable fluorescent dye tracer such as Rhodamine B have been used for some 30 years because of low cost and simplicity; they work well in clear offshore waters. Attempts to use them for wastewaters don't work because being very good dyes, they are adsorbed onto suspended solids so that fluorometry readings are reduced. Calculated (virtual) dilutions are increased accordingly. Other errors include false increases in fluorescence due to plankton. Fluorometer calibrations could be done concurrently and in the same water where the dilution studies are being made (17, 20), but simplicity and low costs would be lost.

In any event, post-audits include distinguishing between calibration and verification of numerical models. This is essential because the conditions under which calibration studies were carried out have been changed as a result of those studies (a reflection of the Heizenberg principle) and because officials and their consultants responsible for subsequently financing and operating the changed systems intuitively need the verification to maintain their own credibility.

4.10 References

1. Brooks, N.H. 1959. Diffusion of sewage in an ocean current. In Pearson, E.A., Editor. Waste Disposal in the Marine Environment. Pergamon, New York.
2. Brooks, N.H. 1972. Dispersion in Hydrologic and Coastal Environments.. W.M. Keck Laboratory of Hydraulics and Water Resources Report KH-R-29. California Institute of Technology, Pasadena.
3. Brooks, N.H. 1962. Report on Model studies for the San Diego Outfall Drop Structure. W.M. Keck Laboratory of Hydraulics and Water Resources, California Institute of Technology, Pasadena.
4. Camp, Dresser, and McKee, Inc. 1978. Alexandria Wastewater Master Plan Study, Volume 4, Marine Studies. Boston.
5. Fan, L.N. 1967. Turbulent Buoyant Jets into Stratified or Flowing Ambient Fluids. W.M. Keck Laboratory of Hydraulics and Water Resources Report KH-R-29. California Institute of Technology, Pasadena.

6. Fischer, H.B., List, E.J., Koh, R.C.Y., Imberger, J., and Brooks, N.H. 1979. Mixing in Inland and Coastal Waters. Academic Press, New York.

7. French, J.Z. 1972. Internal hydraulics of multiport diffusers. Jour. Water Pollution Control Federation, Vol 44, 782-801.

8. Grace, R.A. 1978. Marine Outfall Systems: Planning, Design, and Construction. Prentice-Hall, Edgewood Cliffs, NJ.

9. Haberman, J. and Morton, R.H. An experrimantal study of bubbles moving in liquids. Trans. Am. Soc. Civil Engrs., Vol. 121, 277-252.

10. Paarmakian, J. 11963. Waterhammer Analysis. Dover, New York.

11. Rawn, AM, Bowerman, F.R., and Brooks, N.H. 1961. Diffusers for disposal of sewage in sea water. .Trans. Am. Soc. Civ Engrs, Vol 126, 344-388.

12. Mullendorf, W.P., Soldate, A.M., Baumgaartner, D.J., Schuldt, M.D., Davis, L.R., and Frick, W.R. 1985. Initial Mixing Characteristics of Municipal Occean Discharges, Vol 1, Procedures and Applications. Publ EPA/600/3/-85-073a. USEPA, Washington.

13. Roberts, P.JH.W. 1979. Line plume and ocean outfall dispersion, Jour. Hyddr. Div, ASCE, Vol 105, 313-333.

14. Tetra Tech, 1982. Rev ised Sec. 301(h) Technical Support Document. Report to USEPA. Bellevue, WA.

15. Wright, S.J. 1977. Effects of Ambient Crossflows and Density Stratification on the Characteristic Behavior of Round, Turbulent Buoyant Jets. W.M. Keck Laboratory of Hydraulics and Water Resources Report KH-R-36. California Institute of Technology, Pasadena.

16. Agg, A.R. 1978. Application of Coastal Pollution Research 3 - Initial Dilution, Water Research Centre, SStevenage, England.

17. Botes, W.A.M., and Kapp, J.F. 1994. Diution studies on large offshore pipelines. Preprints, Conference on Marine Outfall Systems, Istanbul, November 1994, International Association on Water Quality, London

18. CERC, 1984. Shore Protection Manual, 2 vols. Coastal Engineering Research Center, US Army Corps of Engineers, Vicksburg, MS.

19. French, J.A. 1995 Pigging submarine outfalls. Journal, Environmental Engineering Division, Amer. Soc Civil Engrs, 121, 5, 396-401.

20 Gunnerson, C.G., qne McCullough, C.A. 1964. Limitations of Rhodamine-B and Pontacyl Brillian Pink as tracers in estuarine waters. In Ichiye, T.N., editor. Proceedings, Symposium on Diffusion in Oceans and Fresh Waters. Lamont Geological Observatory, Columbia University. New York City.

21. Rawn, AM, and Palmer/ 193. Predetermining the extyent of a sewage field in sea water. Transactions, Amer. Soc. Civil Engrs. Vol 24, 1036-1081.

22. Rogerts, P.J.W. Ssnyder, W.H., and Baumgartner,D.J. 1989. Ocean outfalls; submerged wastefield formation, evolution, and diffuser design. Jour. Hydraulic engineering, 115, 1, 1-70.

23. Roberts, P.J.W. 1995. Mamala Bay (Hawaii) Study. Plume Modelling. Project MG-4., Scchool of Civil and Environmental Engineering, Georgia Institute of Technology, Atlanta

24. Wilkinson, D.L. 1984. Purging of saline wedges from ocean outfalls. Jour. of Hydraulic Engg, 110 (12).

5 Construction Materials

Outfall construction materials and methods are closely interdependent. Materials include those for coating and lining as well as the pipe itself. They establish the construction options, mobilization requirements, and resistance to failure during construction and operation. These properties are more important than their costs. In any event, the costs presented in Chapter 12 are site-specific.

5.1   Pipe Materials

Early outfalls used either cast iron or concrete pipe, and occasionally wrought iron. Recent outfalls have been made of steel or plastic to reduce construction costs. The oil and gas industry has used steel exclusively for offshore construction for over a quarter of a century.   Plastic pipe or lining has excellent hydraulic characteristics, but has mostly been selected for its corrosion resistance, a practice that is increasingly unnecessary because of environmental concerns. This chapter presents information on cast iron, wrought iron, asbestos-bonded corrugated metal (still in service but increasingly difficult to obtain because of occupational hazards in its manufacture), reinforced concrete, and coated steel. Standards for pipe dimensions and materials are promulgated by trade associations and governments.

5.1.1   Cast Iron

Historically, cast iron has been one of the most popular of all pipe materials. Its strength approaches that of steel pipe and more corrosion resistant so that long service lives are common. It is much less flexible with less impact resistance, and more expensive. Cast iron pipe is available in diameters up to 135 cm. Standard lengths vary, and joints are flanged, bell-and-spigot, or ball-and-socket (Figure 5.1).
Older outfalls used bell-and-spigot joints that were less suitable for offshore pipelines because they required accurate line and grade, and in many cases the driving of pile bents. More recent pipelines use ball-and-socket joints that permit up to 15 degree seal-tight deflections between adjacent lengths of pipe, without shear stress on the bolts. Joint flexibility compensates for soil settlement and minor undercutting and will allow the pipe to follow some variations in topography.

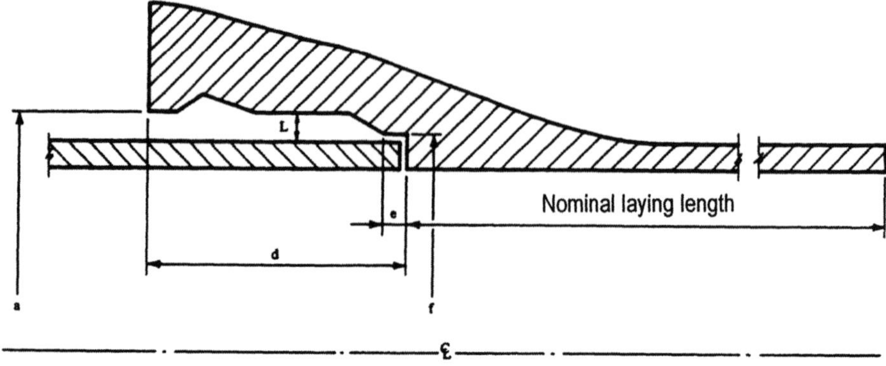

Bell and spigot joint

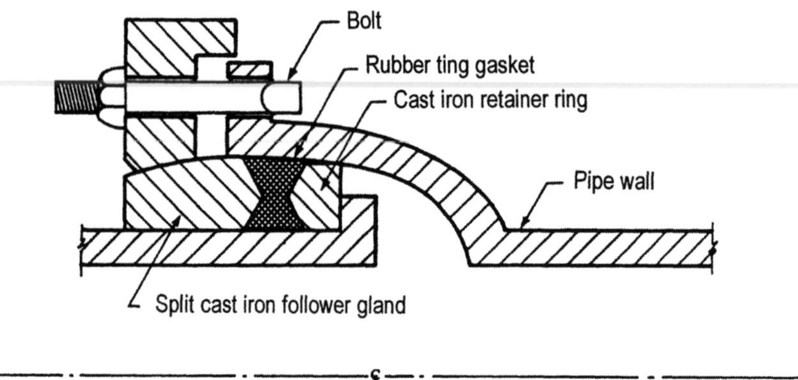

Figure 5.1   Standard Pipeline Joint Connections

The relatively weak joint designs and short standard lengths limit the installation of cast iron pipe to joint-by-joint operations that make them an expensive option.

5.1.2   Wrought Iron

Wrought iron pipe welded joints and mechanical couplings has been used for ocean outfalls on the U.S. west coast. Wrought iron lends itself to construction by either short or long lengths, but it is relatively expensive. Manufacturers claim it is more resistant to marine corrosion than cast iron. The maximum diameter in the U.S. is about 100 cm.

5.1.3   Plastic

Plastic pipes are now available in diameters up to 365 cm (12 ft) and for some materials, larger upon request. Depending upon the material and manufacturing technology, the practical limits for outfalls may be somewhat smaller. Smaller diameters may be formed in an extrusion machine on the coast and floated into place or laid from reel barges. There are two main classifications of plastic pipes:

Thermosetting resin pipes include glass fiber reinforced polyester pipe and polyvinyl chloride (PVC) pipe. These pipes retain their shape and strength throughout a wide range of temperatures. In fiberglass reinforced pipe (FRP) the resin is reinforced with glass fiber by filament winding, contact hand-lay-up molding, or centrifugal casting. Standard diameters are up to 365 cm with pressures up to 1000 kPa at temperatures up to $90°C$ and a specific gravity of about 1.73. Cast FRP is suitable for pipe-jacking. PVC pipe is available in smaller diameters. Both kinds can be designed for high axial loading (see Section 9.3.13)

Thermoplastic pipes are made from resins that are heated to a critical temperature and then formed. The greater flexibility of the smaller diameters that allows them to follow bottom contours may be seen as an advantage during construction and during operation in responding to intermittent scour. This is more than offset by the potential for air pockets to form and reduce capacity. Physical characteristics can be improved for specific applications, but plastic materials tend to be more susceptible to abrasion or puncturing by rock backfill. Some representative ones are in Table 5.1.

Initial costs of plastic pipe are competitive with coated steel; it is 1.65 to 1.5 times the cost of RCP. It floats, even when filled with water, and is usually installed by the surface pull method (Section 9.3.10). where it is joined on shore, weighted when in place, and sunk by slow flooding into the empty pipe. Anchors or weights may be required for on-bottom stability. Plastic is seldom suitable in areas with strong currents.

Joints are critical factors. HDPE sections are usually joined by butt-fusion. Pipe ends are squarely machined by a special planing tool, softened by heating, and joined. Curing occurs during a few minutes of cooling. Mechanical joints of

Table 5.1. Selected characteristics of plastic pipes.

Pipe material	Maximum diameter for acceptable joints, centimeters	Standard lengths, meters	Specific gravity	ABrasion resistance
Fiberglass reinforced epoxy (FRP)	260	various	1.73	higher
High density poly ethylene (HDPE)	160	12	0.95	lower
Polypropylene (PP)	120	various	- - -	- - -
Polybutylene (PB)	60	various	- - -	- - -
Polyvinyl chloride	60	To 10 m	1.38	lower

steel or alloy flanges can be fitted onto special molded pieces of HDPE pipe. Polybutylene requires the same type of butt joins as HDPE, except that from 1 to 7 days are required for curing.

FRP joint connections include low-profile bell and spigot, filament-wound sleeve, and the built-up type. Pipe ends are butted, the exterior surfaces are ground down, and a coupling is made on shore with alternate layers of epoxy and fiberglass, a slow process. FRP bell and spigot connections are made with serrated rubber ring gaskets or O-ring seals. PVC pipe sections are joined with flange, bell-and-spigot, or slip-on connectors. It is primarily a small diameter pipe for pressures up to 150 psi (1000 kPa) at temperatures to 150° F (65° C). It is available only in smaller sizes and is reported to be susceptible to attack by marine life.

5.1.4 Reinforced Concrete

Most submarine outfalls with diameters over 200 cm have been built of reinforced concrete pipe (RCP). The special care needed for the joints has led to the use of cast or wrought iron pipes in the past and steel for smaller pipes. Concrete is highly resistant to abrasion and corrosion, and generally to attack by marine organisms. The materials are less expensive, but installation costs for RCP are quite high. Figure 5.2 shows RCP sections being lowered into place.

RCP is commercially available in sizes of about 60 to 400 cm and weights of 10 to 500 kg/m. Prestressed RCP up to 250 cm with its advantages of lighter weight and greater density. Smaller sizes are available in 5 m lengths and larger sizes in 10 m lengths. Steel bell-and-spigot joints are used for higher pressures. At moderate pressures, concrete bell-and-spigot joints with rectangular grooves for rubber gaskets on the spigot ends are used. At low pressures, joints with double grooves for rubber gaskets on spigot ends are covered by double-bell rings of steel or reinforced plastic.

Long-radius curves can be made with joint openings on straight pipe or with beveled-end pipes or with bevel adapters. Elbows for short-radius deflections, reducers, wyes, tees, and closures are standard items. Special fittings can be supplied or cast in place.

Figure 5.2, Laying reinforced concrete pipe. Source: Flexifloat Systems Catalog. Robishaw Engineering, Houstonb, Texas

5.1.5.   Coated Steel

Steel pipe has been used extensively by the oil and gas industry and for outfalls because it allows for a great deal of flexibility in construction. Steel pipes are generally lighter, less expensive, easier to transport, and easier to install than cast iron. Standard diameters up to 250 cm are available; larger spiral-welded diameters can be obtained. Standard lengths range from 6 to 12 m.

Steel pipes for marine use are usually joined by welding, although flanged connections may be used at the end for future extensions. Welded joints are usually checked by X-ray radiography.

Steel corrodes in seawater. Measures for controlling corrosion or described in Chapter 8. Initial costs of steel pipe and cathodic protection may be higher than for concrete or plastic pipe. The advantages of steel pipe include its adaptability to rapid fabrication and installation, its joint tightness, inherent structural integrity, and its higher strength that allows for future change from gravity to pumped flows.

5.2   Recommended Reading

1.   American Society of Civil Engineers. 1982. Gravity Sanitary Sewer Design and Construction. Manual of Practice No. 60. New York.
2.   Fresenius, W., Schneider, W., Böhnke, and Pöppinghaus. 1989. Waste Water Technology. Springer-Verlag, Berlin. pp 383-409.
3.   Imhoff, Karl. 1989. Novotny, V., Imhoff, Klaus R., Olthoff, M., and Krenkel, P.A., editors. Handbook of Urban Drainage and Wastewater Disposal. Wiley-Interscience, New York, pp 3-66.
4.   Kramer, S.R., McDonald, W.J., and Thomson, J.C. 1992. An Introduction to Trenchless Technology. Chapman and Hall, New York

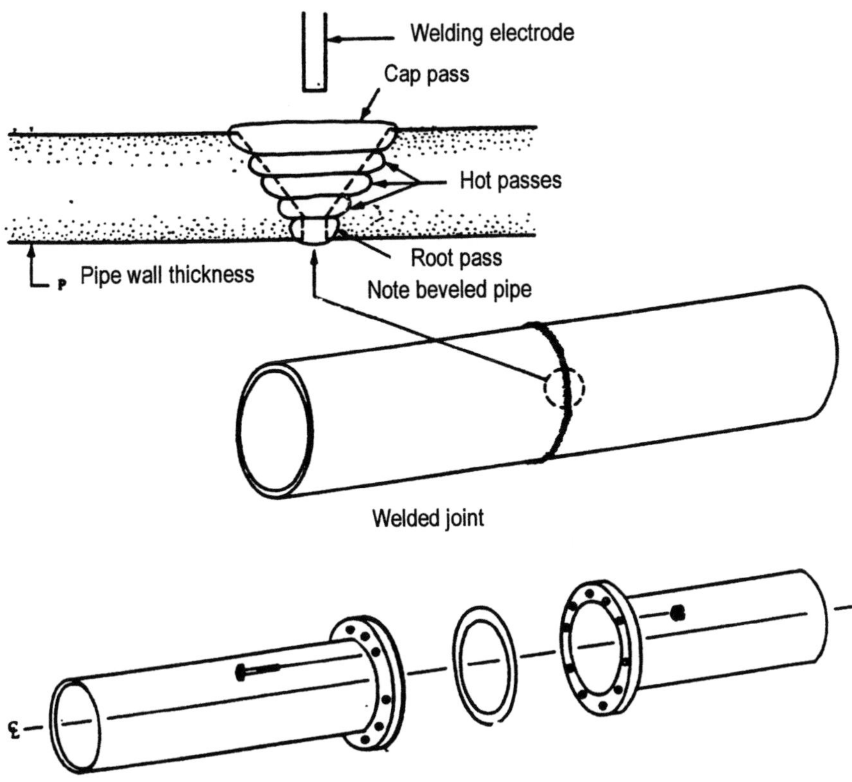

Figure 5.3. Steel pipeline joint connections

6   On - Bottom Stability

On-bottom stability is one of the most critical elements in the design of submarine pipelines. Empirical research into vertical and horizontal stability during installation and operation of both buried and unburied structures began with ancient harbor structures. This entered into a new phases with the mid-19th Century transatlantic cable crossings and ocean outfalls of the 1890s. While sanitary and environmental engineers have looked at the behavior of effluents in the sea, the increasing development of offshore petroleum production since about 1950 has sparked most of the theoretical and empirical studies of on-bottom stability that are identified in this chapter and that constitute the operational extensions to Terzaghi's classic work (24).

Pipelines are generally empty during installation and filled with water during operation. The pipeline weight in both conditions must be calculated and then used in various stability calculations. Most outfall pipelines are buried near the shoreline and unburied at the discharge end. In this case both buried and unburied stability must be calculated.

Recent construction of outfall tunnels provide for either the outfall terminus or a series of risers to diffusers that extend above the sea floor. Here, the weight of the tunnel boring machine can be an additional design factor.

6.1   Forces

The external forces of importance to on-bottom stability include soil forces and hydrodynamic forces.

6.1.1   Soil Forces

Soil forces depend upon soil type. Generally, the vertical soil forces are known as bearing capacity. Horizontal forces are the sum of frictional forces and passive soil resistance (3,5,21,25). Here, the possibility for liquefaction requires special attention.

For horizontal stability, the frictional force is a function of the effective vertical loading. Passive soil resistance is a function of the embedment of the pipeline into the soil. For rocky Seabees, hard clay, and dense sand the frictional forces are dominant with little or no contribution from the passive soil resistance that dominates in soft clay and loose sand. In conventional pipeline design, only frictional soil forces were considered, occasionally with higher friction coefficients that were considered to account for some pipeline embedment into the seabed. Investigations carried out during the 1980s provide better descriptions of the two parts of soil resistance for sand and clay (5,21)

6.1.2   Hydrodynamic Forces

Traditional pipeline design considered only steady current flow and corresponding force coefficients (22). Recent investigations have clearly demonstrated that the loading caused by waves, sometimes combined with that from steady flow, is substantially larger (6,7,13,23). In steady flow, water passing over a pipeline creates a decrease in pressure both on the upper and downstream sides of the pipeline. This pressure field is integrated from two force components, the horizontal drag force, F_D, and the vertical lift force, F_L, where

$$F_D = \frac{1}{2}\rho \, D \, C_D \, U^2 \tag{6.1}$$

and

$$F_L = \frac{1}{2}\rho \, D \, C_L \, U^2 \tag{6.2}$$

where C_D and C_L are experimentally defined coefficients.

Wave current velocities vary over time and are reflected in water particle accelerations (see Table 2.2) that create a pressure field around the pipeline. The integration of this pressure field yields a force in the horizontal direction, the inertial force:

$$F_I = \frac{\pi}{4} \, \rho \, D^2 \, C_M \cdot a \tag{6.3}$$

where a is the water particle acceleration at the seabed, and C_M is the inertia coefficient. The total horizontal hydrodynamic force is then described as

$$F_H = F_D + F_I = \frac{1}{2}\rho \, D \, C_D \, U(t) \, |(Ut)| + \frac{\pi}{4} \, \rho \, D^2 \, C_M \cdot a(t) \tag{6.4}$$

whereas the lift force remains:

$$F_L = \frac{1}{2}\rho \, D \, C_L \, U^2(t) \tag{6.5}$$

Recent literature provides more refined force descriptions (20, 23).

The vector sum of the hydrodynamic driving forces and the soil resistance forces determines the stability of the pipeline. Special consideration on loading needs to given in the case of suspended pipeline sections. The vortex shedding and associated dynamic loading can cause severe vibrations and associated fatigue (see Chapter 7)

6.2 Vertical Stability of Unburied Pipelines

When an outfall pipeline is laid on the seabed it will (i float to the surface because of buoyancy, (ii) lie on top of the seabed, (iii) sink slightly into the bottom until the bearing area of the pipe is sufficient to provide support, or (iv) sink substantially into the soft bottom sediment because of overestimated soil bearing capacity.   Figure 6.1 illustrates these conditions.

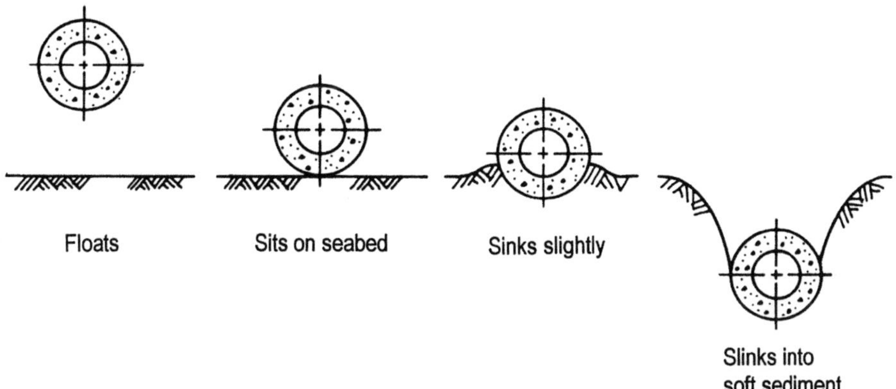

Figure 6.1.   Possible results of laying an outfall on the seabed.

Vertical instability of an unburied outfall pipeline causes it to float to the surface or to sink into the sediments until it forms a catenary that exceeds design stress. Here, the total weight should be controlled to within ±7.5 kg/m. A pipeline may be filled with air to make it lighter and easier to handle in the water while it is being installed. This can be tricky.

In one U.S. more precisely engineered case, a minimum negative buoyancy of 3.6 to 5.4 kg/m was specified. However, it is difficult to maintain pipe joint weights within this tolerance. As a result, the pipeline weight was less than the design weight, the line floated to the surface, was caught in cross currents, buckled, and broke into pieces that either floated to the shore or sank and were lost forever. It cost less to replace the pipe than to locate and repair it. As a result there were high cost overruns, delays, and legal actions brought by the owner, constructor, and engineer against each against each other.

The most common design consideration is the sinking of the pipeline into the bottom sediments. If a short segment of the outfall settles or sags, a low spot is formed there solids tend to settle and reduce the flow capacity. If this is at the discharge end, sediment blockage or even structural failure of the pipe will occur. Soil survey data are used to predict the depth to which a pipeline will sink.

If the seabed is mostly sand, adequate bearing capacity may be assumed. Sands normally support the anticipated loads except where it liquefies under upward flow of water, seismic shock, or wave action.

Most of the methods used to analyze the settling problem are based on infinitely long strip footing theories. Two common methods for determining bearing capacity are those of Reese and Casbarian (4) and Terzaghi (5). Figure 6.2 shows results of Terzaghi's method where B is the buoyancy, R is the soil bearing capacity, D is the outside diameter, W is unit weight of the pipe, and C is the cohesive shear strength of the soils. If the intersection of the pipeline specific gravity and pipeline diameter falls well below the applicable C curve, there is little need to make an in-depth analysis of the soil bearing capacity. If the intersection is near or above the C curve, an in-depth analysis is required.

Another method of estimating the bearing capacity of cohesive soils is that of Reese and Casbarian (4):

$$Qu = kcd, \quad \text{where} \tag{6.6}$$

Qu = ultimate soil bearing capacity per unit length of pipeline , c is the soil cohesive shear strength, d the outside diameter of pipeline, and k the bearing capacity coefficient

Suggested values for k are a linear function between 5.7 at the soil / water interface and 11.0 at a depth of 4d, below which k remains at 11.0. When calculating the pipeline weight that the soil must support, note that the maximum weight is after installation and the pipe is filled with water.

The vertical stability also depends upon the strength of currents flowing across ore beneath the line as shown in Figure 6.3. Scour is difficult to predict accurately, although ongoing research is beginning to provide means for assessing the problem (2,4,17,18). Scour may be minimized by armoring the pipe with coarse stone on a triangular or trapezoidal cross-section or by laying an articulated concrete blanket (3) over it. The blanket is made up of thick, pre-cast concrete, rectangular blocks laced together by cables similar to those used for stream revetments. Anchoring with large saddle-type weights or collar weights tends to increase scour between anchors.

In soft sediments, it is advisable to provide anti-scour measures at the offshore end. This can be as simple as digging a large hole at the terminus and filling it with stone 0.3 to 1 m in diameter extending to at least 1 pipe diameter on either side. If an annual or semiannual inspection detects scour or settlement, the problem can be solved by installing heavy-duty inflatable plastic liners under the pipe span and inflating them with mortar until the pipeline is supported. Alternatively, sandbags or fabricated supports may be placed by divers.

6.3   Vertical Stability of Buried Pipelines

A buried pipeline becomes unstable when it (i) floats to the soil-water interface and is exposed to wave and current forces that is was designed to resist, or (ii) continues to sink into the trench bottom due to soil bearing failure.

It is not uncommon to find buried pipelines "floating" out of their trenches because designers tend to select a pipe that is as light as possible. The assumption is the overburden or weight of backfill in the trench will keep the pipe

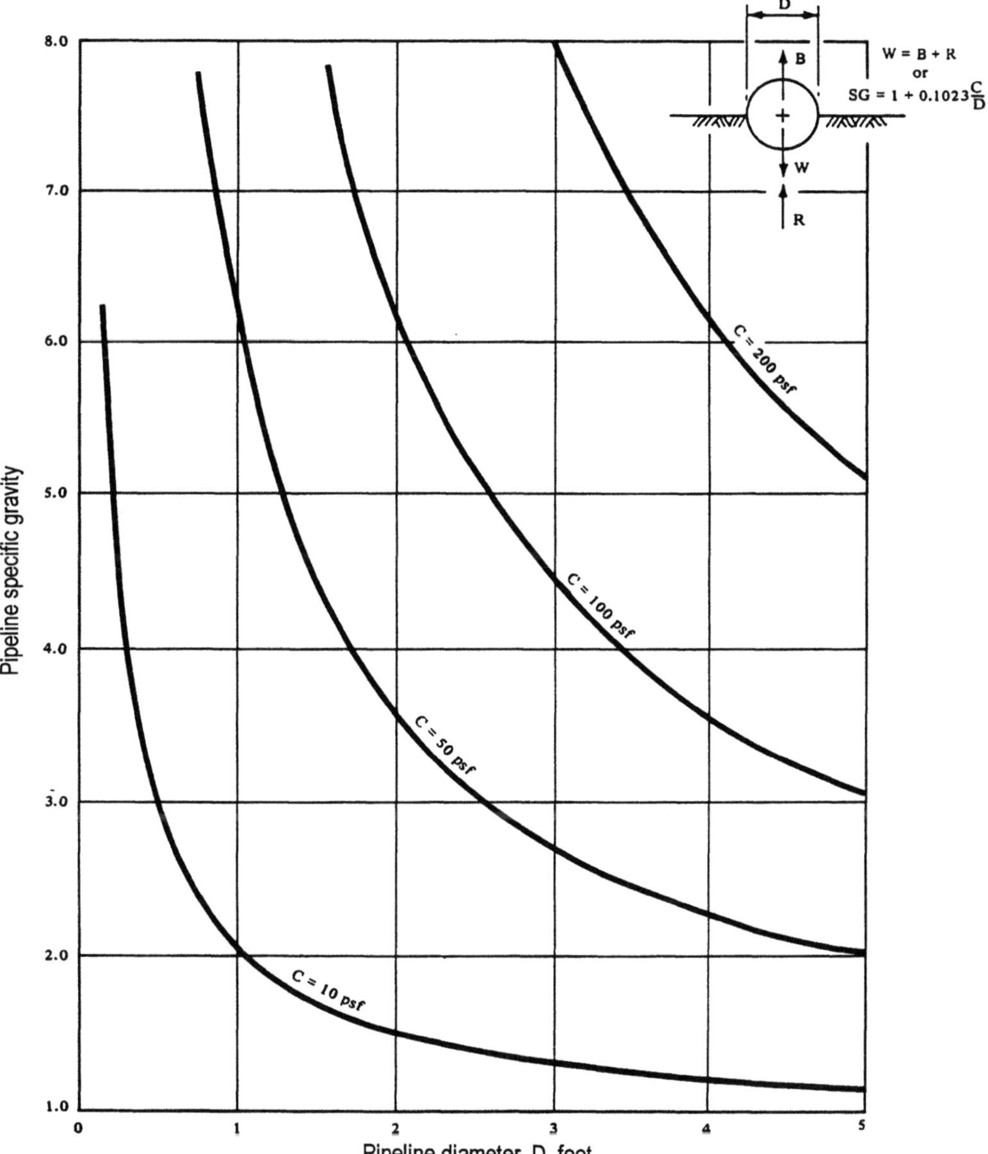

Legend (n foot-pound-second units): Weight, W, = buoyancy, B, + soil resistance, R, C = cohesive strength, and Specific gravity = 1..1023 C/D

Figure 6.2. Specific gravity to cause bearing capacity failure of an outfall

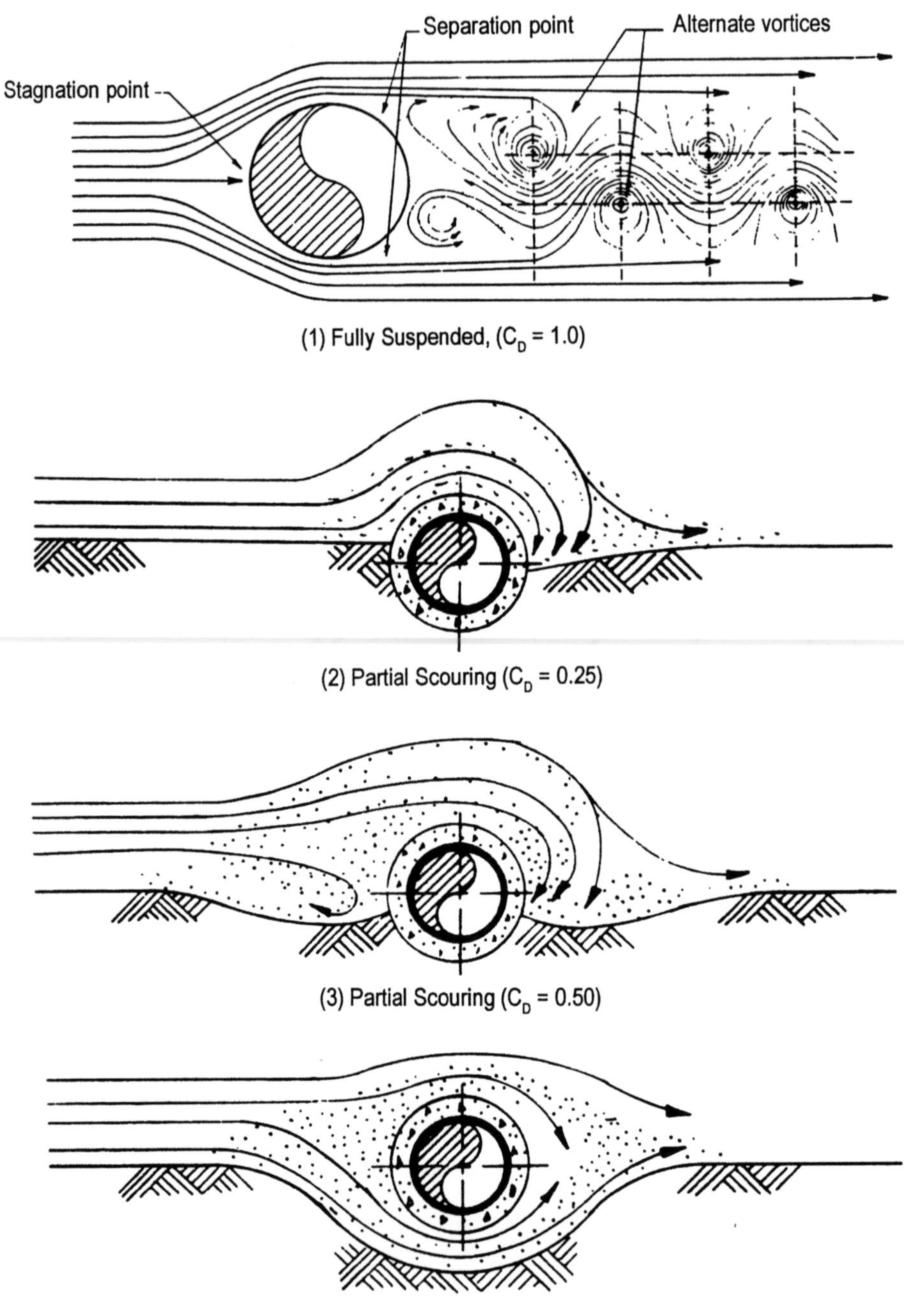

(1) Fully Suspended, $(C_D = 1.0)$

(2) Partial Scouring $(C_D = 0.25)$

(3) Partial Scouring $(C_D = 0.50)$

(4) Complete Scouring of Sediment from under Pipeline $(CD = 1.0)$

Figure 6.3. Vortices .

buried. Whether this happens depends on the soil conditions, backfill material, and method of burial. For example, some designers choose a pipeline weight that will allow the line to be floated into place and filled with water so that it can be sunk in marshes or periodically flooded areas. Often the line is covered or backfilled with the material that had been excavated from the trench. Generally it is a heavy (specific gravity ~ 1.3) liquid-mud with little or no shear strength because it has been handled twice and replaced in the ditch. Alternatively, the trench may not have been backfilled at all under the assumption that natural sedimentation will fill the trench. Also delayed liquefaction may be due to earthquake shock or unusually heavy surf. Pipelines that are light enough to be floated into place should be analyzed for these conditions.

All trenching methods that produce slurries contribute to vertical instability, particularly the "jet-sled" method that liquefies soils under the pipeline by means of eductors, the pipeline sinks, and the often heavier excavated material flows back into the trench and lifts the pipe. This happens when the designer doesn't answer the question of why didn't the pipeline sink into the mud in the first place?

In soft clay sediments, the problem is to select a design weight that is heavy enough to remain buried in the liquefied soil but not so heavy as to cause bearing failure. One approach clay and silt is to determine the liquid limit (ASTM specification D423) by filling an Atterberg Cup with soil at specified water contents, the sample is divided with a groove, and the cup subjected to a series of blows or taps. The number of blows at which the two partitions become one is recorded for each water content. This number (usually in the range of 10 to 50) is plotted and extrapolated as shown in Figure 6.4. Water content in percent is the weight of the water divided by the weight of the soil in a specific volume.

The water content at 0.01 blows is estimated from Figure 6.4. It has been empirically determined that at this water content, the soil and liquid mixture acts as a viscous, dense fluid that produces maximum buoyancy on the pipeline.

Scour is difficult to predict accurately. Figure 6.5 shows the effect of moisture content on soil density values used on buoyancy calculations. The rationale for this procedure is presented in a 1961 ASCE report, "Rational Design for Pipelines Across Inundated Areas" (1). Nearshore sands present special problems. A striking example of pipeline failure in Australia when a pipeline with a specific gravity of 1.3 was trenched across a sand shore approach and backfilled with the same sand. After a severe storm, the pipeline was lying exposed on the beach. An analysis indicated that the basic cause was in inertial or cyclic shock caused by breaking waves. This force liquefies the soil and imparts a cyclic vertical force to some depth under the sea bed. The pipe tends to rise slightly each time a wave breaks and sand particles move under the pipe. Designers can avoid this by specifying deeper burial, a heavier pipe, or by constructing the shore approach as a groin.

For outfall lines buried on an active coast, both short-term and long-term changes in coastal profiles influence the burial depth. Seasonal storms can erode considerable of material, and the long-term effects of coastal currents , particularly downstream from river basin development projects, can remove shoreline materials (10).

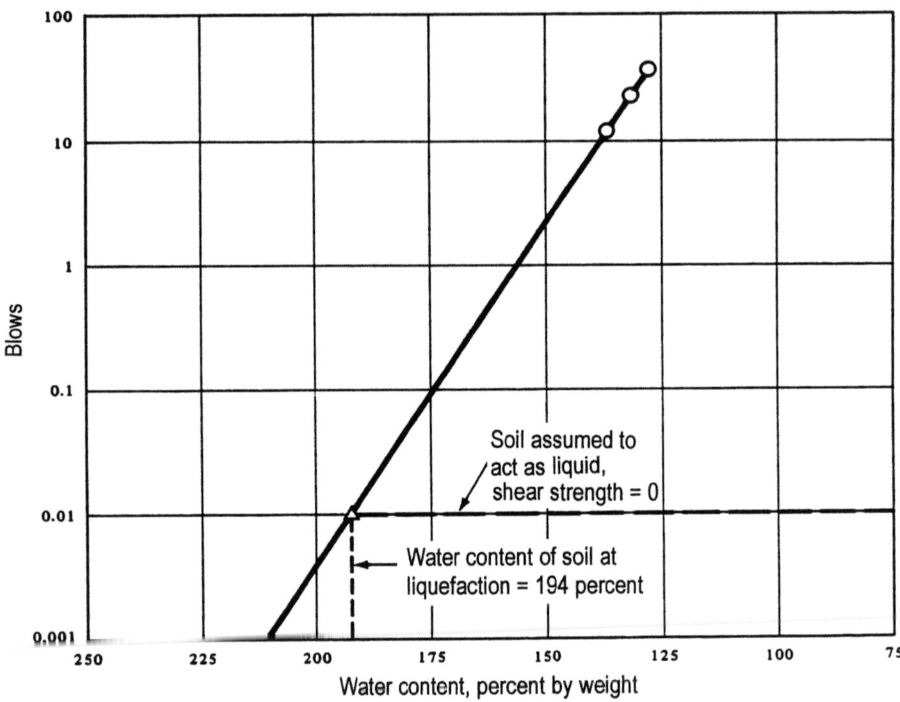

Figure 6.4. Liquid limit curve (ASTM D 423)) for determining water content
Circles represent laboratory test data. Triangle point obtained by extrapolation.

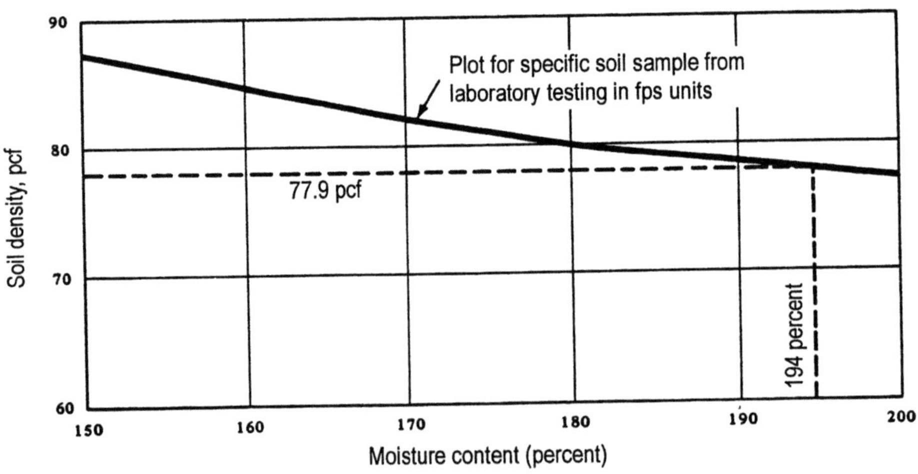

Figure 6.5.   Typical moisture content and soil density for marine sediments

6.4   Lateral Stability of Unburied Pipelines

When a pipeline lies on and is not anchored to the seabed, it must weigh enough to resist lift, drag, and inertial forces. The design wave forces and current velocities used in the design are based on a statistical analysis of historical weather records and on-site measurements. Two sets of data are collected, one for installation and the other for operation. Because installation is of relatively short duration, a set of design values for maximum wave heights and current velocities not to be exceeded more than 5% during any one year is sufficient. Operations design values are usually based on a yearly probability of exceedence of 1%, or a 100-year storm.

Drag, inertial, lift, and soil resistance forces are the principal factors in horizontal stability. Buoyancy, submerged weight, and lift forces contribute indirectly since friction force is generally taken to be proportional to net downward force, an assumption within the accuracy of the other variables.

Inertial forces are much less than drag forces for non-breaking shallow-water waves and are of the same order for deep water waves (see table 2.2) Detailed analyses of wave and current forces, including those from breaking waves forces are necessary for stability design.

The stability requirement is for the horizontal driving force to be less than the total soil resistance. For the simple case of steady current flow and soil resistance being only the frictional force (the case usually considered), the stability requirement can be expressed as

$$W_s/D \geq (C_L + C_D / f)\frac{1}{2} \rho U^2 \tag{6.7}$$

where W_s is submerged weight of pipeline (including content) per unit length, D is the outside diameter, C_L and C_D are the lift and drag coefficients, f the friction coefficient, ρ the density of the sea water, and U is the current velocity. More refined procedures are given in references 9, 15, and 16.

Coefficients for lift, drag, and friction have been the subject of much research, particularly by the offshore oil and gas pipeline industry. For a steady current, lift and drag coefficients as a function of the Reynolds number (UD/ny< ny being the kinematic viscosity of water) and surface roughness have been published (22). Figure 6.6 shows these coefficients for a smooth pipeline surface. See references 7 and 19 for hydrodynamic force coefficients for a wide range of conditions, including partial embedment of the pipeline.

Friction coefficients of 0.65 to 0.75 are used for sand, depending on the designer's confidence in the sampling and laboratory testing procedures. Those used for estimating axial pulling friction are usually less than those for lateral stability.

The effective friction coefficient for clay is generally higher than the laboratory value and varies with pipe weight and soil shear strength. As its submerged weight increases, the pipe settles into the clay, more pipe surface is covered and a soil wedge is formed on each side of the pipe, both of which increase the effective resistance to sliding. Conservative values for friction coefficients in clay vary between 0.5 and 1.0. Note that in more refined designs, the combined effects of frictional and passive soil resistance can be included, where only the frictional part is influenced by the lift force.

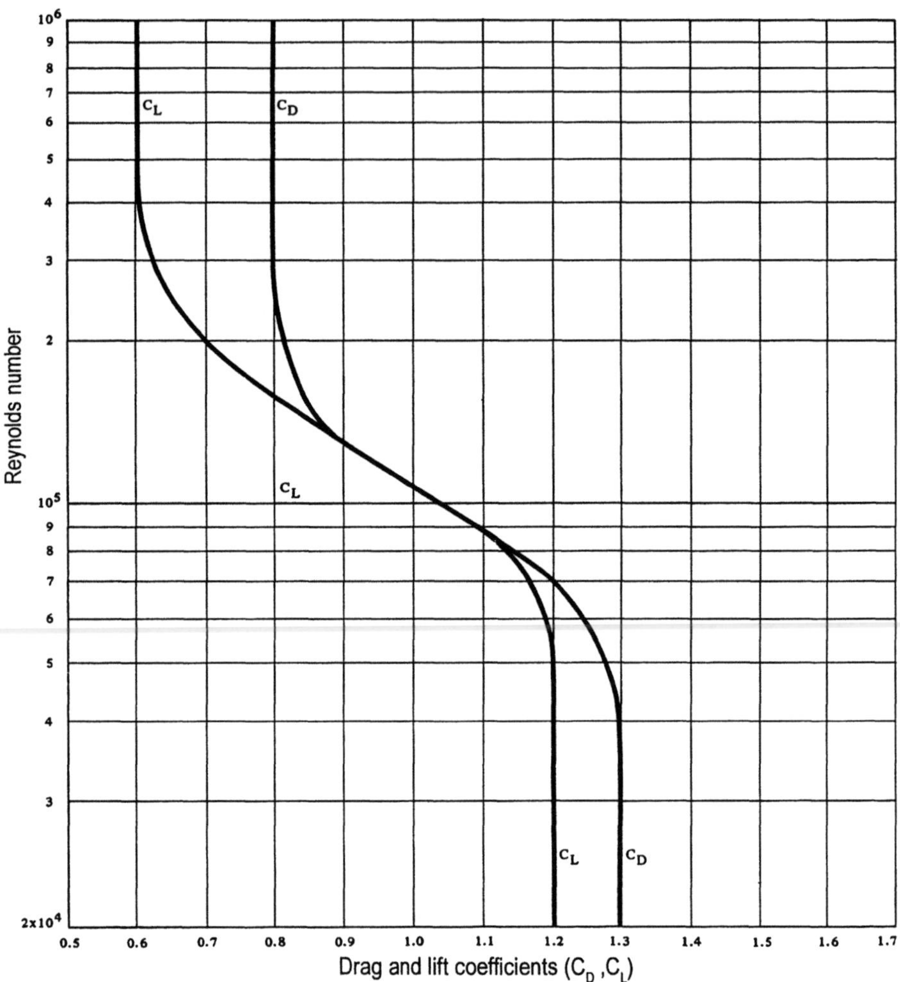

Figure 6.6. Drag and lift coefficients for pipe laying on ocean floor

6.4.1 Density Anchors

Density anchors add weight to a pipeline to increase its negative buoyancy to a level that will be stable under the design criteria. Continuous reinforced concrete weight coatings are commonly obtained by substituting heavy aggregate such as iron ore for the conventional aggregate. A maximum thickness of 14 to 15 cm is recommended to avoid the concrete spalling that accompanies greater thickness.

Concrete weight coatings may be applied by machine or, for small quantities of pipe, by pouring into a form around the pipe. Forms are normally thin sheet metal. Small spacers are used to hold the form. A slot is cut midway into the spacer and wire mesh (usually 15 x 15 cm, 10-gauge reinforcing mesh commonly used in concrete foundations on grade) is fixed into the slot. This holds the reinforcing steel midway between the form and the pipe. Concrete with just enough water to be poured is fed into a slot left open on top of the form which is then vibrated to ensure uniform density.

Machine-applied concrete coating is very dry an can therefore achieve a higher strength. During application the pipe is continuously rotated and moved axially past the applicator. The applicator is either a belt or set of brushes that rotate at high speed throwing the concrete onto the rotating pipe, where it is dry enough to stay in place without forming.

Reinforcing 6x6 cm, 17 gauge wire mesh, known as "chicken wire" is automatically embedded into the concrete. If the coating is more than 6 cm, two layers of mesh are used.

Set-on or bolted-on reinforced concrete anchors made with high densities concrete (Figure 6.7) are semicircular sections that are placed onto the pipeline Bolt-on weights are made by pouring concrete into molds with reinforcing bar in place. Each half has lifting hooks. Rock shield is sometimes attached to the inside of the weights to protect the pipe coating from damage during construction. Bolt-on weights have bevels on each end to prevent snagging on obstacles if they are used on a pull section.

U-shaped set-on weights are set over the pipeline after it is in the trench (Figure 6.7). They should have as low a center of gravity as possible. The legs are 5 to 7.5 cm longer than the pipe diameter to prevent the weight from rolling off the pipe and to enable the ditch bottom to take the load. Large-diameter thin-wall pipelines can be overstressed if the pipeline has to support the set-on weight in addition to resisting the stresses due to pressure and bending. Rock shield is sometimes attached to prevent damage to the pipe coating. Set-on weights are usually poured on the job site. They are reinforced with steel and have lifting hooks. Both the bolt-on and the set-on weights are installed with a side-boom, dragline, or other machine capable of lifting the weight. Set-on weights are the least expensive density anchors, but they tend to be moved together along the pipe by wave and current forces with consequent losses in stability.

6.4.2 Mechanical Anchors

Mechanical anchors are usually steel and are not designed to add weight. They maintain a minimum hold-down force on the pipe when properly installed. The

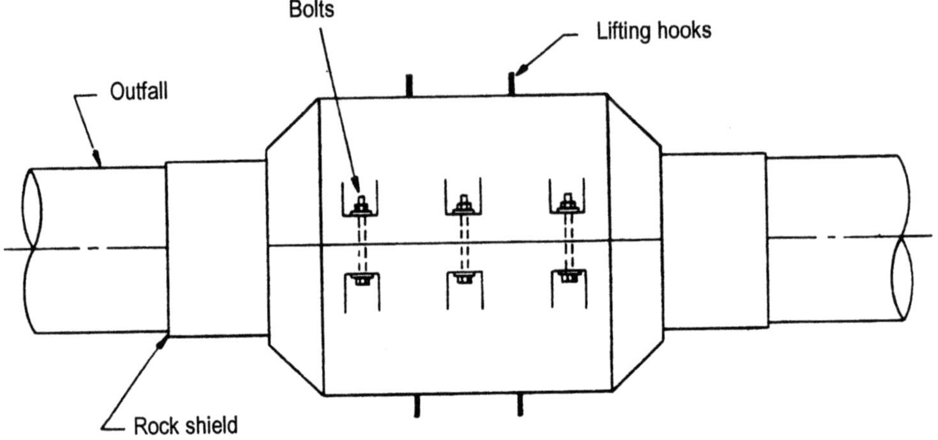

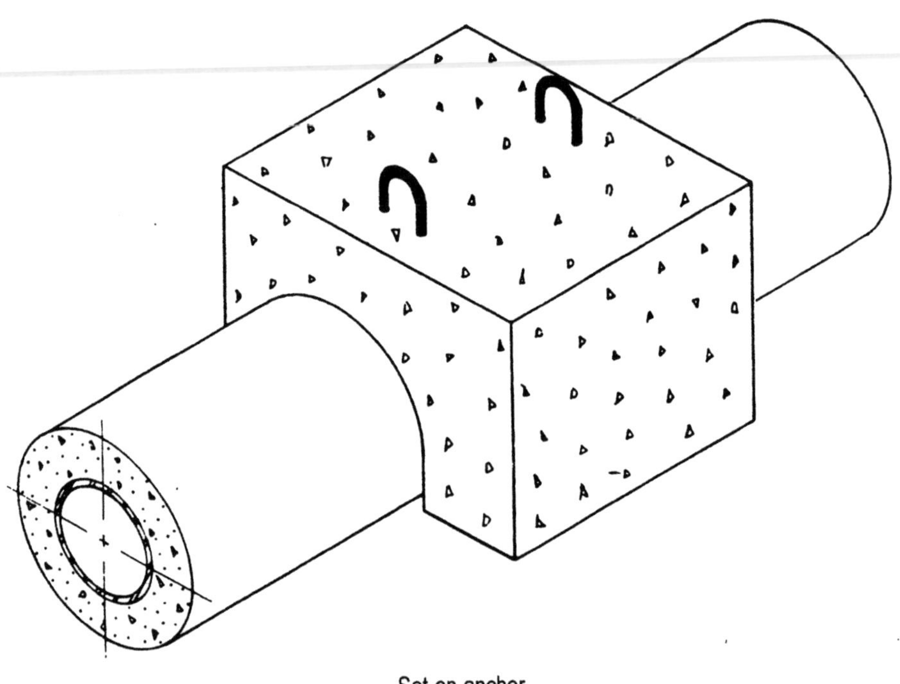

Figure 6.7. Density anchors

holding power of mechanical anchors is considered to be more efficient than that for density anchors, particularly on large-diameter lines.

The most commonly used mechanical anchor is the auger-type (Figure 6.8) This anchor consists of a round steel plate shaped like an auger that is attached to the end of a long steel rod. The other end is threaded so it can be attached to a clamp. This system consists of two augers and a strap shaped to fit the pipeline.

An anchor is installed on each side of the pipeline and the strap is attached snugly over the pipe. The strap is usually padded to protect the pipe coating. The anchors and strap are hot-dipped galvanized to extend their service life. Small magnesium anodes can be attached to each anchor to provide corrosion protection (see Chapter 8). In most applications, the anchor will last as long as necessary for the backfill to compact and gain sufficient shear strength to hold the pipeline in place.

Expanding mechanical anchors (Figure 6.8) are used in the same manner as the auger type. The anchor rod has hinged flukes on one end that expand outward by turning the threaded anchor rod which is run through a nut in the center of the flukes. The anchor is either expanded in a previously bored hole or forced to the desired depth and expanded. The expanding increase the effective area and corresponding shear strength.

6.5   Lateral Stability of Buried Pipelines

Lateral stability is a problem for buried lines only when the pipeline cover is scoured or removed under normal or storm conditions. High nearshore currents during storms can cause pipelines to be exposed to current and wave forces. Both current and wave forces can suspend or liquefy sediments in a trench. Long-term records as well as site-specific data collected over two storm seasons should be used to determine the required burial depth.

6.6   References

1. ASCE Committee on Pipeline Installation. 1961. Rational design for pipelines across inundated areas. Proceedings, Pipeline Division, Amer. Soc. Civil Engrs, New York.
2. Bernetti,R., Bruschi.R., Valentini, V.,and Venturi, H. 1990. Pipelines placed on erodible seabeds. Proc. 9th International Conference on Offshore Mechanics and Arctic Engineering, Vol. V, 155-163. Houston.
3. Brennodden, H., Sotberg,T., Leing, J., and Verley, R.1989. An energy-based pipe-soil interaction model. Paper 6057. Proc. 21st Offshore Technology Conference, Houston.
4   Bijker, R., Staub, C., Fredsøe, and Sumer, B.M. 1991. Scour-induced free spans. 23rd Offshore Technology Conference. Houston.
5   Brennoden, H., Sueggen, D., Wagner, D.A., and Murff, J.D. 1986. Full-scale pipe-soil interaction tests. Paper 5378. Proc. 18th Offshore Technology Conference, Houston.

152

6. Bryndum, B., Jacobson, V., and Brand, M.B. 1983. Hydrodynamic forces from wave and current loads on marine pipelines. Paper 4454. Proc. 16th Offshore Technology Conference. Houston.
7. Bryndum, B., Jacobson, V., and Tsahalis, D.T. 1988. Hydrodynamic forces on pipelines: model tests. Proc. 7th Offshore Mechanics and Arctic Engineering Conference. Houston, 9-21.
8. CERC. 1984. Shore Protection Manual, 2 vols. Coastal Engineering Research Center, Vicksburg, MS.
9. Det Norske Ventas. On-bottom stability design of submarine pipelines. Recommended Practice E 30. Oslo.
10 Foster, T., Jacobson, V., Brøker, I., and Johnson, H. 1994. A deterministic approach to burial depth requirements for pipeline shore approach design on an eroding coast. Proc. 10th Conference and Exhibition. Offshore East Asia.
11. French, J.A., Gustafson, P., Mark, J.S., and Burn-Lucht., K.M. 1994. Offshore rescue. Civil Engineering, October 1994.
12. Grace, R.A. 1978. Marine Outfall Systems Planning, Design, and Construction. Prentice Hall. Englewood Cliffs.
13. Grace, R.A., Nicinski, R.A. 1976 Wave force coefficients from pipeline research in the ocean. Paper 2676. Proc. 8th Offshore Technology Conference. Houston.
14. Graveson, H., and Fredsøe, E. 1983. Modeling of liquefaction, scour, and natural backfilling processes ii relation to marine pipelines. Proc., Offshore Oil and Gas Pipeline Technology European Seminar, Copenhagen , February 1983.
15 Hale, J.R., Jackson, V., and Lammert, W.F. 1989. Improved basis for static stability analysis and design of marine pipelines. Paper 6059. Proc., 21st Offshore Technology Conference. Houston.
16. Hale, J.R., and Lammert, W.F. 1988. Submarine pipeline on-bottom stability. Pipeline Research Committee report for projects PR-178-516 and PR-178-717. American Gas Association.
17. Hansen, E.A. 1982. Scour below pipelines and cables: a simple model. Proc. 11th International Conference, Offshore Mechanics and Arctic Engineering, Calgary, Alberta, Canada Vol. V-A, 133-138.
18. Hansen, E.A., Staub, C., Fredsøe, J., and Suymer, B.J.. 1991. Time-development of scour-induced free spans of pipelines. Proc. 10th Offshore Mechanics and Arctic Engineering,
19. Jocobsen, V., Bryndum, M.B., and Bonde, C.L. 1989. Fluid loads on pipelines - sheltered or sliding. Paper 6056. Proc, 21st Offshore Technology Conference, Houston.
20. Jackson,, V., Bryndum, M.B., Tsahalis, D.T. 191988. Prediction of irregular wave forces on submarine pipelines. Proc. 7th Offshore Mechanics and Arctic Engineering Conference, Houston. 23-32.
21. Jacobson, V., Palmer, A.C., Steenfelt, J.S., and Steensen-Bach, J.O. 1988. Lateral resistance of marine pipelines on sand. Paper 2598. Proc. 20th Offshore Technology Conference, Houston.
22. Jones, W.T. 1976. On-bottom pipeline stability in steady water currents. Paper 5338. Proc. 8th Offshore Technology Conference, Houston.

23. Lambrakos, K.F., Chao, J.C.,, Beckman,, H., and Brannon, H.B. 1987. Wake model of hydrodynamic forces on pipelines. Ocean Engineering, 14, 2, 117-136.
24. Terzaghi, Karl. 1971. Theoretical Soil Mechanics, John Wiley and Sons, New York.
25. Verley, R.L.P., Sothberg, T., and Brennodden, H. 1990. Breakout resistance for a pipeline partially buried in sand. Proc, Offshore Mechanics and Arctic Engineering Symposium, Houston. 121-126.

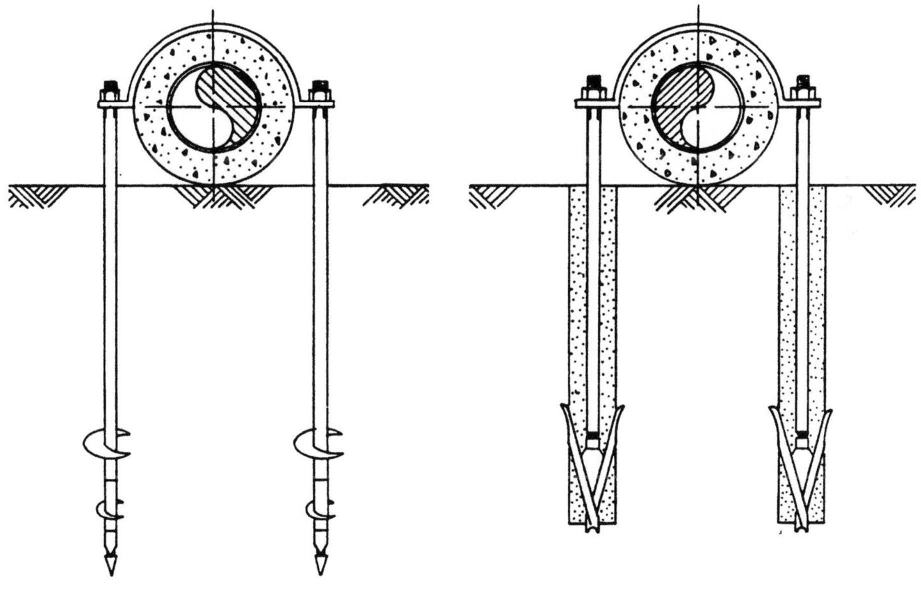

Figure 6.9. Mechanical anchors

7 Stress Analysis

The design of marine outfall pipelines must take into account four structural factors: (1) installation stress, (2) stress induced by anchoring if applicable, (3) collapse / buckling analysis, and (4) unsupported pipe spans. Installation stress can be analyzed by calculating the estimated magnitude of forces acting on the pipeline induced by the selection construction method and equipment. For instance, if the pipeline is to be pulled into place, the maximum pulling force can be divided by the pipe wall thickness area to determine stress. Operational stresses are due to environmental factors such as slumping of deltaic soils, scouring of supporting sediments by wave or current action, or even flotation due to removal of overburden, all of which create supported lengths.

7.1 Identifying and Analyzing Stress

7.1.1 Stress from Anchors

When anchors are use to provide on-bottom stability to an outfall pipeline, they become supports for uniformly loaded beams when the loading is the drag force caused by currents. Figure 7.1 illustrates the principle involved.

7.1.2 Unsupported Span Analysis

When an offshore pipeline is suspended on seabed obstructions or over excess scour, unsupported pipe spans have water flowing above and below them. This flow creates vortices that are shed alternatively from with side of the pipeline, which induces a periodic dynamic loading. A sound design of the maximum allowable span of unsupported pipe includes both static and dynamic analyses.

In static pipeline analysis all static forces acting on the pipeline must be considered when calculating the maximum span allowable without exceeding the design stress. Span ends can be assumed to be fixed or pinned, depending on soil characteristics and external supports.

Vortex shedding produces a large transverse force acting on the pipeline that leads to vibration unless there is adequate restraint. The frequency of the vortex

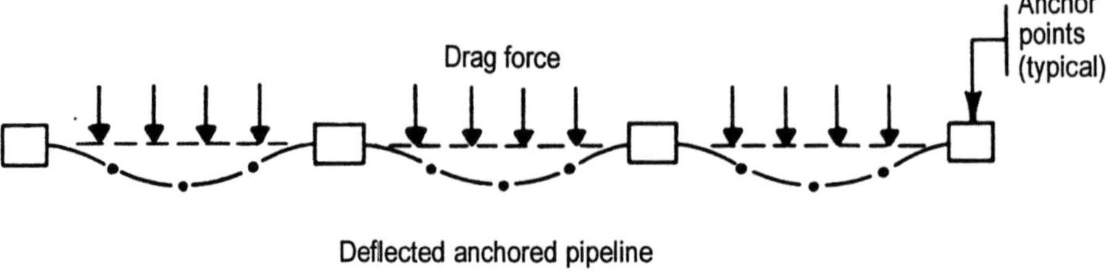

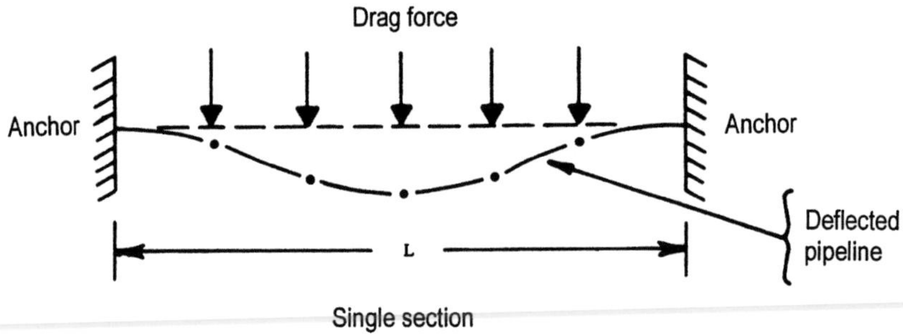

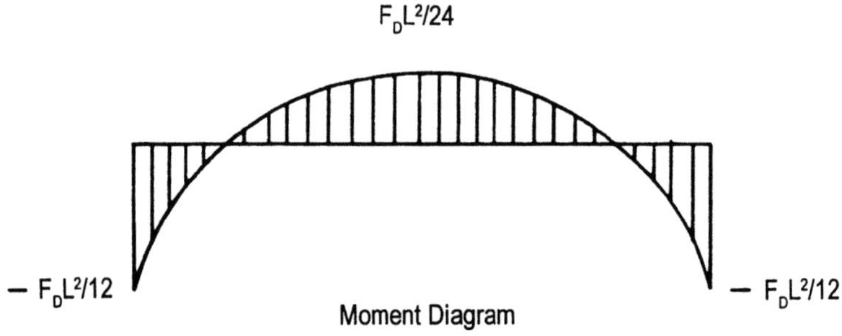

Figure 7.1. Deflection of an anchored submarine pipeline

shedding depends upon pipe diameter, flow velocity, and a non-dimensional parameter called the Strouhal number. The Strouhal number depends upon the Reynolds number and the pipeline surface roughness (1,2,10,11) For rough surfaces, the Strouhal number is close to 0.2, regardless of the Reynolds number. The Strouhal number, S, is defined as

$$S = FD / U \tag{7.2}$$

where F = excitation frequency of vortex shedding, cycles per second, D = pipe diameter, and U = flow velocity .

The excitation frequency of vortex shedding can be determined by calculating the Reynolds number, estimating the Strouhal number from Figure 7.2, and solving equation 7.2 for F.

When the frequency of vortex shedding coincides with the natural mode of oscillation for the pipeline, resonance and amplitude amplification occurs (57.8.13) To avoid this, the natural frequency of the pipeline should be made higher than the highest anticipated vortex shedding frequency. This can be done by increasing the stiffness of the pipe and its restraints on the pipe. During construction, spoilers can be attached to the unsupported span. If resonance vibration develops, the pipe can be dropped quickly to shorten the span. Analyses for vortex-induced vibrations of pipelines have been published (10,11,12,13)

During installation, the unsupported spans may be limited by leveling the pipeline route or by providing intermediate supports. Construction specifications should specify the allowable unsupported span length and acceptable methods of correcting excessive spans. If the span occurs in an area or trench that is to be covered or backfilled with soil, the supports should be installed before covering. Consideration should be given to the impact load of placing the cover plus the material weight so as not to overstress the pipe across unsupported lengths.

7.1.3 Collapse / Buckling Analysis

Special precautions are required for pipes filled with air during construction.   Air-filled, smaller-diameter pipes installed in deep water or larger lines in moderate depths are subject to hydrostatic pressure. If the pipeline is to be installed by a method where it open-ended, such as with joint-by-joint bottom assembly, the external and internal pressures are equalize.

Failure by buckling occurs when there is excess bending of the pipe. Buckling and collapse can occur at the same time under conditions of excess loading and external pressure. In the case of an offshore pipeline, two conditions should be investigated. First, the vertical position (or water depth) of maximum bending during installation should be located and quantified. The combination of maximum bending and external pressure should be checked for buckling failure. If the pipe is filled with air during construction, a second check should be made for maximum moment near the seabed under maximum hydrostatic pressure.

A structure's buckling resistance depends on both its material and its shape. The physical properties of a steel pipe are characterized by the D/t ratio (outside

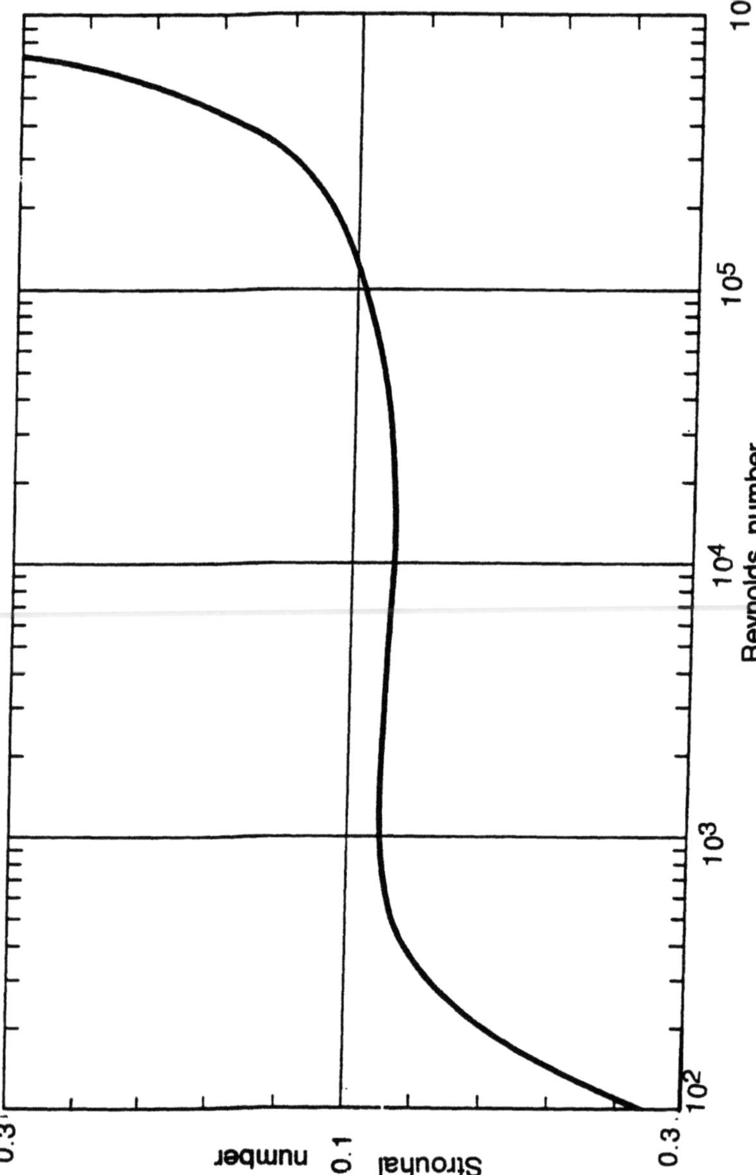

Figure 7.2   Variation of Strouhal number, $N\,D/V_0$, as a function of Reynolds number, $V_0 D/\nu$.   Source: Steinman, (3)

diameter divided by wall thickness) and the properties of the material defined by the stress-strain curve. The elastic limit for buckling in steel pipe occurs at a D/t ratio of approximately 250 or more, depending on the material. Below this level, buckling occurs in the plastic range of the material and is accompanied by a flattening of the pipe's cross-section. The resulting ellipse lowers the bending resistance by reducing the section modulus of the pipe. For steel with a D/t ratio over approximately 60, buckling occurs before the pipe develops its full resisting moment. For a D/t ratio less than 60, the resisting moment tends to drop off before buckling. An alternative approach for determining the critical collapse pressure, P_c, is given by Reynolds (9):

$$P_c = (2E / (1 - \lambda^2)) (t^3 / D) \quad \text{where} \tag{7.1}$$

E is Young's modulus, λ is Poisson's ratio, t is thickness, and D the average diameter.

Another type of failure is the propagating buckle in which the geometry of an air-filled pipeline forms a transverse dent or crease to a longitudinal collapse which then propagates along the pipeline. The external pressure required to maintain the propagation of the collapse is less than that required to initiate it. Initiating pressure, propagating pressure, and theoretical collapse pressure should be computed and plotted for different D/t ratios for pipe materials under consideration. While the equations to analyze collapse and buckling are usually solved by computer, a rule of thumb for conservative design is:

Maximum depth, meters	D/t
300	≤ 25
150	35
75	45
38	55

7.2   References

1. Achenbach, E.. 1971. Influence of surface roughness on the cross-flow around a circular cylinder. J. Fluid Mechanics 43,2,321-335.
2. Achenbach, E., and Heinecke, F. 1981. On vortex shedding from smooth and rough cylinders in the range of Reynolds numbers 6x103 to 5x106. J. Fluid Mechanics 109, 239-251
3. Bruschi, R., Simantiras, P., Vitali, L., and Jacobsen,V. 1994. Vortex shedding-induced oscillations on pipelines resting on uneven seabeds. predictions and counter-measures. Int. Conf. on Hydroelasticity in Marine Technology, Trondheim, Norway, : (also available in Italian).
4. Bruschi, R., et al. 1991 Free-spanning pipelines: a review. ISOPE Conference. Edinburgh.
5. Bryndum, M.B., Bonde, E., Smitt, L.W., Tura, F., and Montesi, M. 1989. Long free spans exposed to currents and waves: model tests. 21st Offshore Technology Conference, Houston.

6. Bryndum, M.B., Nielsen, K.G., and Jacobsen, V. Free-spanning pipelines. Lecture notes. Design and analysis of slender marine structures: risers and pipelines. WEHGEMT Graduate School. Trondheim.

7. Bryndum, M.B., Vested, H.J., Nielsen, K.G., and Gravesen, H. 1989. Improved design methods for spanning of pipelines. Offshore Pipe Technology, 1989 European Seminar.

8. Jacobsen, V.J., Bryndum, M.B., Nielsen, RF., and Fines, S. 1984. Cross-flow vibrations of a pipe close to a rigid boundary. J. Energy Resources Technology, Dec. 1984, v.108.

9. Reynolds, J.M. 1981. Design and construction of sea outfalls. Chapter 13, Coastal Discharges. Thomas Telford, Ltd. London. 119-126.

10. Rolf, Th. von, and Ruback, H. 1924. The frequency of the eddies generated by the motion of circular cylinders through a fluid. British Advisory Committee for Aeronautics, R.&M. 917. Summarized in Ref (3).

11. Steinman, D.B. 1946. Problems of aerodynamic and hydrodynamic stability. Proc., 3rd Hydraulics Conference. Studies in Engineering, Bull. No. 31, University of Iowa, Iowa City.

12. Tura, F., Dminrescu, A., Bryndum, M.B., and Smed, P.F. 1944. OMAE 1994, Houston, Vol. V, 247-256.

13. Tsahalis, D.T. 1984. Vortex-induced vibrations of a flexible cylinder near a plane boundary exposed to steady and wave-induced currents. Proc. 3rd OMAE Conference, New Orleans.

8 Corrosion Control

Concrete pipe, cast iron pipe, plastic pipe, and wrought iron pipe are generally considered corrosion resistant. Corrosion of steel is difficult to control, however, on steel outfalls and is usually accomplished by a combination of pipe coatings and cathodic protection.

8.1 Corrosion Protection for Steel in Seawater

Galvanic corrosion (the term includes oxidation and rust formation) occurs when a small electric current flows through an electrolyte between dissimilar metals or, in a single structure or object, between small local cells with small differences in composition or structure. It is also caused by stray electric currents. Various factors that affect the corrosion of steel in seawater include dissolved oxygen (most common factor), surface film (rust or corrosion product), pH (depends on the dissolved carbon dioxide and/or hydrogen sulfide), salinity, and temperature (corrosion rates approximately double with each 10° C increase in temperature). Marine organisms can influence corrosion rates by penetrating the protective coating. Sulfide corrosion attack is also common beneath barnacle encrustation.

Much of the potential corrosion is normally controlled with pipe coatings. Care must be taken to protect these coatings during pipe handling and installation.

Galvanic-type corrosion is induced by the flow of electric current from the more positive areas on a metal surface (anodes) through the electrolyte (soil, seawater) to the less positive (cathode). This electrochemical action removes metal at anode areas where metallic ions leave the metal and enter the electrolyte.

The best practice for corrosion control is to use both pipe coating and cathodic protection.

8.1.1 Cathodic Protection

Cathodic protection imposes an electric current that reverses the natural electrolytic action of corrosion. The basic principle is that the entire metallic surface to be protected is made a cathode in an electric circuit by supplying an external anode. The external anode must have sufficient driving potential (electromotive force) to deliver the required current to the metallic surface through the electrolyte (seawater) so that corrosion ceases.

There are two common sources of protective current, both of which are used on offshore pipelines. One is direct current (DC) obtained from rectification of alternating current (AC). The rectified current is distributed to the electrolyte, usually by means of semi-inert, long-life anodes such as graphite, high-silicon cast-iron, or lead-silver alloys. The other common source of current is from sacrificial anodes distributed in the electrolyte and connected to the pipeline. These anodes are consumed in the process at rates depending upon the choice of anode material, the integrity of the pipe coating material, and environmental factors. Cathodic protection systems commonly have design lives of up to ten years.

Anodes for rectifier/groundbed installations (called impressed current systems in European practice) are normally arranged in a long "groundbed" parallel to the pipeline but seldom closer than 60 meters to the pipeline. (Scrap steel rails been used for this purpose.) A 60 m spacing between pipeline and groundbed provides for current distribution along the pipeline. A groundbed may consist of two or three anodes to as many as forty, depending on soil resistivity and local conditions. Anode spacing may vary from 3 to 6 m. Each anode is connected to the rectifier by an electrical cable, often one of a pair that ensures redundancy. Soil resistivity governs the depth to which anodes are buried. Anode are buried in the lowest resistance soil available. Soil resistance can be measured on location with special instruments or from test samples sent to a soils laboratory.

Figures 8.1 and 8.2 illustrate sacrificial anode and impressed current (rectifier) cathodic protection systems, respectively. System selection is based on soil resistivity, local costs of materials and labor, available energy sources, and maintenance requirements. The present trend is toward the use of long-life, low-maintenance sacrificial anodes to protect offshore pipelines. Suitable sacrificial anodes are made from magnesium, zinc, or aluminum alloys. One measure of the degree of protection is measured in ampere-hours of protective current per kilogram of metal consumed.

Magnesium was widely used for marine sacrificial anode installations during the 1950s and 1960s. Magnesium has a high (driving) potential (-1,5 volts open circuit to silver/silver chloride half-cell reference) and a high operational current availability. However, the high current output causes rapid depletion of the anodes which must be frequently replaced at considerable expense.

Zinc also has a long history as a sacrificial anode and has an open circuit potential of -1.05 to -0.5 volts (Ag/AsCl). The relatively low driving potential of zinc against steel, as compared with magnesium, has been proven advantageous in extending the life of sacrificial anode systems.

According to theoretical electrochemical performance and cost criteria, aluminum is the most attractive material for a long-life galvanic anode. However, unalloyed aluminum has little use because the normally passive surface film limits current output. Alloyed aluminum overcomes this limitation and has an open circuit potential of -1.0 to -1.05 volts (Ag/AgCl).

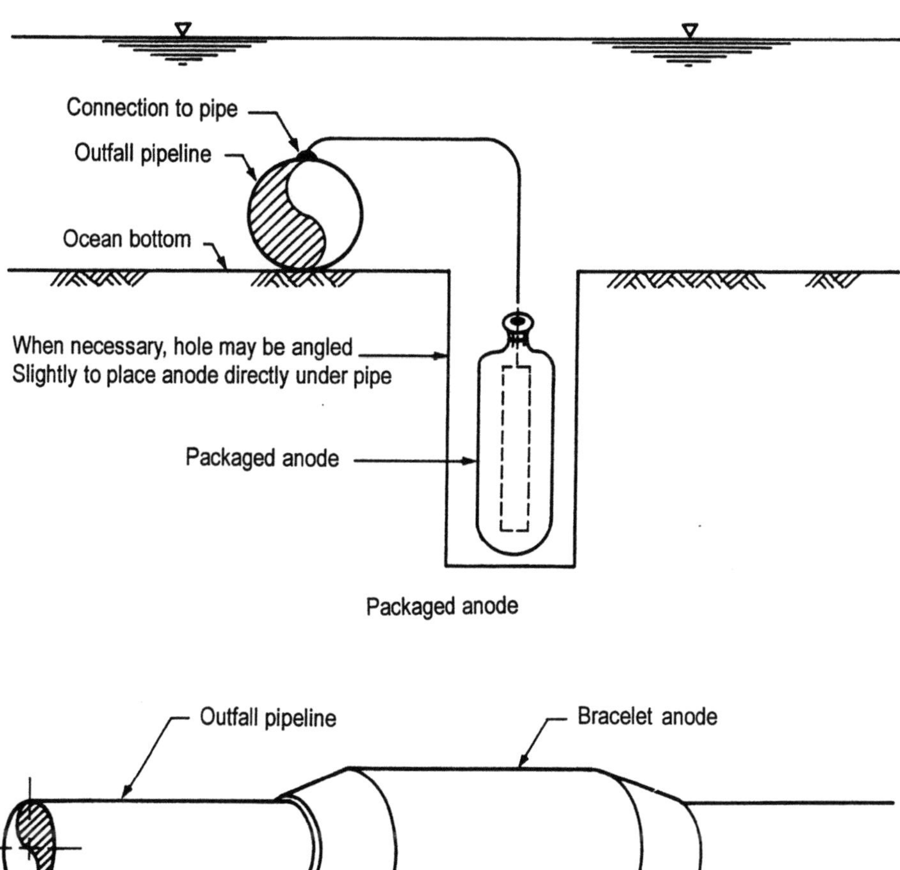

Figure 8.1   Sacrificial anode cathodic protection system for a marine pipeline

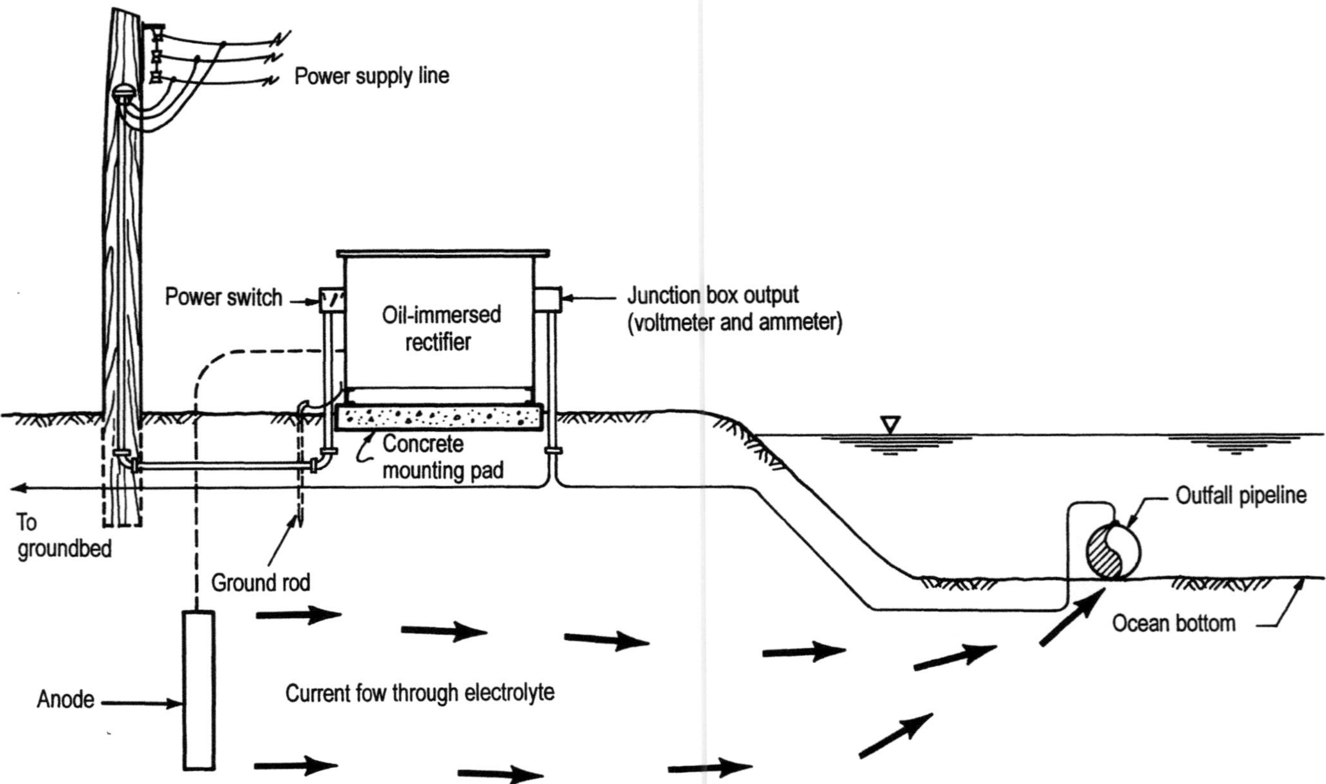

Figure 8.2. Rectifier cathodic protection system for a marine pipeline

8.1.2   External Coatings

External coatings are not required for concrete or plastic pipe outfalls, or normally for cast iron outfalls. Steel outfalls almost always have external and internal protective coatings in addition to cathodic protection systems.

Two types of external coatings are included in the design of most steel pipelines; a corrosion prevention coating and a weight coating. Weight coating provides stability and protects the external corrosion prevention coating from damage. Pipelines with diameters less than 30 cm are often installed with only a corrosion prevention coating. When larger diameter steel lines are installed with no concrete weight coating, the corrosion prevention coating is often damaged.

Most types of corrosion preventive coatings are suitable for marine pipeline applications. Tape coatings are normally not used offshore or in areas where the pipe is submerged. Here, they are easily damaged by the weight coating and tend to detach from the pipe when submerged.

Thin film epoxy coatings are occasionally used but they are expensive. The most popular coating materials are coal tar enamel, asphalt enamel, polyethylene, and polypropylene.

Coal tar enamel and asphalt enamel coatings are known to withstand the marine environment. They are normally 2 to 3 mm thick and are overlaid with felt.

The American Water Works Association (AWWA) publishes the C203-73 AWWA Standard for Coal Tar Enamel Protective Coatings for Steel Water Pipelines. Surface preparation includes sand or shot blasting to provide a clean surface. A high-quality primer is used to ensure a strong bond between the pipe metal and coating. Manufacturers' recommendations on the primer and coating material should be followed exactly with respect to shelf life, surface preparation, application temperature, and ambient conditions during application. An inspector representing the owner should be present during all coating operations.

Polyethylene and polypropylene are normally used as thin films around the pipe and the annulus between film and steel filled with mastic. An extrusion process is used to apply the coating. The type of coating is generally limited to 25 cm or smaller pipelines that require no weight coating. Concrete coatings are not applied over such coatings. Field joints are usually coated with heat-sensitive tapes or heat shrinkable sleeves.

Factors that govern the selection of the proper corrosion coating materials for marine pipeline projects are availability of materials and proximity of coating yards in the area, temperature conditions, and potentially corrosive conditions.

Some materials should be carefully evaluated for their costs and service life before imported materials are called for. A survey of coating yard capabilities is also necessary to ensure that the specified coating can be properly applied. Normally the corrosion prevention coating and weight coating are applied at the same yard to prevent damage to the corrosion prevention coating during shipment.

High ambient temperatures can cause some coatings to soften and become susceptible to damage. If it is very cold, coatings become brittle and crack during installation where the pipe is stressed by bending

Temperatures of chemical process or thermal wastewaters must be known. If a maximum rated service temperature is exceeded, disbonding and deterioration will occur. An alternate coating, e.g. heat-cured epoxy, may be selected.

Unusually corrosive conditions occur at the shore approach of a marine pipeline. The soil may be soaked with previous discharges of oil or other materials that are detrimental to a particular coating.

Coatings are normally applied in a permanent coating yard well in advance of construction to protect the pipe from oxidation-type corrosion in storage and to prevent delays during construction owing to coating yard breakdowns. Because steel pipe is usually welded, about 15 to 20 cm at each end of the pipe is left uncoated. After welding, this uncoated area must be cleaned with a wire brush, primed, and coated with a material compatible with the other coating. It is essential that an inspector be present during joint coating operations.

Coating integrity can be inspected with a portable, battery-operated electrical resistance monitor, commonly called a "jeep" or "holiday detector." If a hole in the coating ("holiday") is found, an alarm will indicate the exact locations where repairs are needed. This device should be used to inspect both yard and field coating operations. The machine is readjusted for different coatings and thickness.

8.1.3 Internal Coatings for Marine Outfalls

Non-corrosive pipe materials such as concrete, plastic, and cast iron do not require linings. Steel outfalls should always be lined. A wide variety of coating materials, including baked-on epoxy, or other resins and paints can be used. However, the most commonly used linings for water and wastewater service are coal tar enamel, coal tar epoxy, and cement mortar because of their low costs. Cement mortar lining is not recommended because of its deterioration in seawater. It is also more brittle than coal tar. Most lined outfalls use coal tar.

Other internal lining materials that have been used for water, oil, and gas include zinc silicate epoxy (potable water), polyamide-cured epoxy (oil and gas), epoxy-phenolic (oil and gas), and polyurethane (abrasive slurries). Whether the extra expense of these coatings is warranted will depend upon the corrosive or abrasive nature of the wastes and possibly upon improvements to pipe flow characteristics.

During welding operations, the lining will be "burned back" or charred for five centimeters or more. As with exterior coatings, these areas must be cleaned and recoated with a compatible primer and coating. On large pipeline this presents few problems since a man can crawl into the pipe and coat the joints. On smaller lines, an automatic cleaning and coating machine is sometimes used, but the results have not been consistent.

8.1.4   Recommended Reading

Degrémont. 1991. Water Treatment Handb ook. Sixth edition. Lavoisier
    Publishing. Paris. Chapter 7. The effects of water on materials

9 Construction

Construction usually accounts for about three-fourths of the costs of an ocean outfall. Both the costs and methods are site specific. The best solutions depend on local environmental requirements, successful and failed construction experiences elsewhere, and rigorous comparisons of established and state-of-the-art alternatives. Some of the latter are identified in the following paragraphs.

9.1 Practical Limits to Current Construction Practices

The methods listed in this chapter for constructing ocean outfalls have been successful. Failures have occurred when they have been misapplied. The following paragraphs summarize some of the more commonly used methods. Section 9.2 describes these methods in more detail

9.1.1 Selecting a Construction Method

The principal factors to be considered when selecting a construction method include pipe size, pipe material, length of outfall, on-shore space for pipe storage and assembly, tidal and storm currents, and construction equipment availability.

Outfalls with diameters greater than 150 cm are normally reinforced concrete pipe (RCP). Smaller pipes can be of RCP, steel, cast iron, or plastic. Those less than 30 cm diameter are lighter and more flexible and provide for a wider range of construction options. Lines larger than 60 cm in diameter are increasingly rigid and require more attention to pipe-lifting capacity and pipe-bending stress analysis.

RCP sections are usually installed by connecting 3 to 10 m sections on the sea bed with a crane barge. Joints are usually a bell-and-spigot design with gaskets. Above-water assembly is not practical for RCP as the joint is weaker than the pipe. Steel and plastic pipes have joint designs that arc as strong as the pipe itself so that more construction techniques are feasible. Cast iron is no longer widely used because of availability only in smaller diameters, joint weakness, brittle nature, and higher costs.

Construction methods for short outfall lengths of , say, 100 to 400 m depend upon local contractor experience and availability of equipment. Longer outfalls over 600 m have either been pulled into place from on-shore fabrication sites, or installed from a lay barge. Recent developments in construction materials and technologies make other methods practical, such as reel barges for small diameter lines, microtunneling (including directional drilling and boring), and tunneling.

A level space at or near the shore end of the outfall is ideal for conventional construction methods. Figure 9.1 shows a representative on-shore assembly site for a bottom-pull operation. Where on-shore space is limited, or where large-diameter sections are being assembled, the contractor can use a barge or other floating platform as shown on Figure 9.2.

The availability and location of construction equipment is a significant factor in selecting pipe materials and construction methods. This is particularly true for small projects where mobilization and demobilization costs characteristically exceed 30 percent of the construction costs (see Chapter 12).

Strong tidal and storm currents my be an asset in dispersing wastewater in the sea, but they are a liability during construction. Strong tidal currents usually preclude use of floating sections or surface pull methods.

9.1.2   Classification of Construction Methods

Construction methods include bottom (seabed) assembly, surface assembly and lowering from an offshore barge, bottom pull from onshore, or opposite shore, surface pull (flotation), remote assembly tunneling, and directional drilling.

Bottom assembly refers to assembling and connecting short lengths, 3-10 m, on the sea bed from crane barges or jack-up barges. Divers make the connections.

With surface assembly, pipes are joined on a floating or jack-up barge that lowers the assembled pipeline to the seabed as it progresses along the pipeline route. Lay barges range from very simple units used for small diameter or shallow-water projects to elaborate floating assembly plants. Lay barges used on offshore oil and gas pipelines and some longer outfalls are capable of fabricating up to 3,000 m of pipeline per day with a crew of 100 to 200. Substantial auxiliary floating equipment is necessary to support the operation of larger barges. Most lay barges are located near offshore oil or gas activities and are feasible for nearby outfalls of 1,000 m or more. Many lay barges can install pipes up to 12 cm diameter and with minor modifications, longer lines can be installed.

In bottom pull, the line is pulled along the final alignment of the pipeline. This method requires a minimum of floating construction equipment and may be used in many geographic locations and sea conditions.

The surface-pull method is essentially the same as the bottom-pull method except that the pipeline is light enough to float either by itself or with auxiliary buoyancy. After it is pulled into place, water is allowed to enter and sink the line into a prepared trench or along the specified route.

An example of remote assembly is the reel barge where lengths of pipe up to 30 cm diameter are welded together at an onshore fabrication site, reeled onto a floating reel barge, and transferred to the job site for laying. It is also possible to assemble long sections or, where there are no tidal currents, the entire length, by floating into place and sinking it along the route. Connections are made on the seabed as necessary.

Microtunneling (including boring and uni-directional drilling, also advertised as trenchless technology) follows the directional drilling technique developed in the

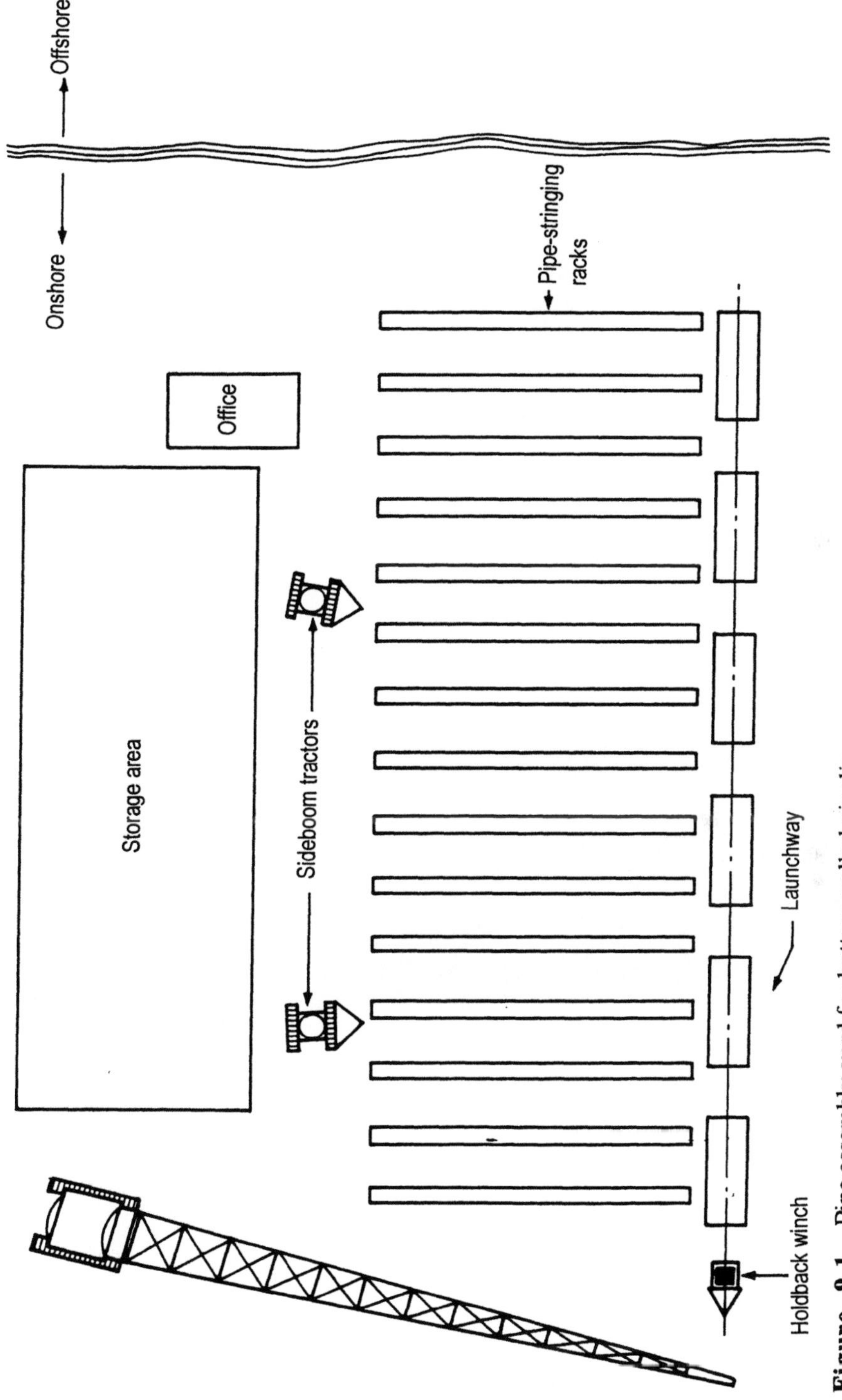

Figure 9.1. Pipe assembly yard for bottom-pulled pipeline.

Figure 9.2. Pipe assembly yard for lay barge operations. Source:Robishaw Engineering, Houston

oil and gas industry (see Section 9.5. An on-site drilling rig is expensive, but it minimizes construction area requirements and eliminates traffic interference. Both the reel barge and directional drilling technologies are for small pipelines carrying oil, water, or other high-value products. The are suited more to water supply than to wastewater disposal.

For outfalls more than 1,000 m long, one construction method may be used for the shore approach and another for the offshore portion. Pre-dredged trenches are often laid in the shore approach zone and the line laid or pulled into the trench. Trenches in surf zones tend to silt in quickly so that planning and scheduling are critical. Trenching and backfilling are discussed in Section 9.3. Meanwhile, there have been a number of attempts to lay pipelines across open shorelines and bury them later; this is a recipe for failure.

9.1.3. State-of-the-Art Constraints

Innovative and improved outfall constructions are constantly being developed. Limitations listed in Table 9.1 are updated from 1988 when the first edition of this book was prepared.

9.2 Construction Methods

Descriptions and design factors for a number of construction methods are presented in the following paragraphs. Construction costs usually account for 60 to 80 percent of the overall installed cost of a submarine pipeline. Costs and methods are site specific. The ability to select the least-cost method from among those described below depends on familiarity with and experience in marine and coastal construction practices.

9.2.1 Bottom Assembly Methods

Final connections of pipe lengths (joints) are made on the seabed. Variations of this concept include laying from a mobile jack-up platform, a trestle, or a crane barge. All the variations require positioning and lifting capacity with sufficient precision to lower pipe lengths into place, alignment and on-bottom connection of pipe lengths by divers. Research on and prototype testing of unmanned, remote-controlled connection systems are being carried out by oil and gas companies on steel pipe in water deeper than 300 m. Meanwhile bottom assembly methods are limited to water depths and site conditions in which divers can perform useful work, which is presently less than 100 m for conventional diving and less than 250 m for saturation diving.

Table 9.1. Practical limitations of outfall construction technologies

	Maximum diameter, cm	Minimum length, m	Maximum length, km	Maximum depth, m	Maximum current, cm/s
Shore approach					
Pre-trenching	---	—	---	---	3
Mechanical dredge					
Dipper	---	—	---	30	3
Dragline	---	—	---	100	3
Clamshell	---	—	---	60	3
Continuous	---	—	---	60	3
Hydraulic					
Hopper	---	—	---	---	3
Dustpan	---	—	---	20	3
Cutterhead	---	—	---	---	3
Post-trenching					
Jet sled	---	—	---	100	3
Cutterhead	---	—	---	---	3
Plow	**NOT**	**RECOMMENDED**			
Cofferdam	---	—	0.3	20	---
Trestle	---	—	0.3	20	---
Offshore installation					
Bottom assembly					
RCP, bell and					
spigot	350	—	not limiting	250	1
Steel, alignment					
frame	250	—	not limiting	250	1
Cast iron	125	—	not limiting	250	1
Lay barge	50	1	not limiting		
	75	1	not limiting	200	3
	120	1	not limiting	100	3
Reel barge	30	10	not limiting		3
Bottom pull,	300	—	0.5	150	
steel pipe	100	—	30	—	2.5
Surface pull,	---	—	0.1	—	2.5
steel pipe	---	—	0.3	—	0.5
Tunneling					
Directional drilling	150	—	1.5	—	---
Pipe jacking	350	—	3.0	—	---
Tunneling	practical minimum 150 cm		---	—	---
	to unlimited maximum		---	—	---

Note: minimum lengths are based on mobilization costs while maximum lengths and other limits are based on technological constraints. Microtunneling applies to diameters less than 1.5 m and runs up to about 200 m between shafts. Descriptive information on selected technologies is found on the following pages.

9.2.2   Pipe Laying from a Mobile Jack-up Platform

Mobile jack-up platforms have been built for specific projects using large diameter reinforced concrete pipe. The platform is floated into place, large anchors (either gravity anchors or drilled-in) are placed to maintain the barge on location, and the legs are lowered to the seabed. Large jacks are used to raise the work platform out of the water free from the effects of normal waves and currents.

After each pipe section is added to the line, the platform is advanced by lowering its buoyant deck until the whole structure is floated, pulling itself on its anchor lines to the next position and jacking the platform out of the water to continue laying operations. Figure 9.3 shows a photograph of the mobile platform used to install the 1957 5-mile (9 km), 12-foot (4.3 m) diameter Hyperion outfall. Figure 9.4 is a simplified drawing showing the platform operation. A design for work in the surf zone is the skid-mounted platform is described by French al (3).

9.2.3 Pipe Laying From a Trestle

The inshore sections of many outfalls have been laid from trestles, or temporary piers. The trestle is used in waters that are too shallow for pipe-laying barges or where wave and surf conditions might rock a floating platform excessively. Although the trestle does not eliminate all wave surge problems in inshore waters, it ensures that the work can be performed from a fixed platform. Trestles are begun by driving H-piles into the sea floor. They are usually not grouted, so that the piles can be retrieved later. A pile bent is made by setting beams across pairs of piles. Bolting rather than welding eases recovery. Longitudinal beams are then laid across the pile bents, and rails for the crane, pile-driver, and pipe-laying equipment are placed to complete the structure.

The crane advances seaward for pile driving and follow-up work, and when the work is completed, returns to shore, taking up the piles as it proceeds. While the trestle is being extended by one crane, another crane can be driving sheet piling further inshore, (preferably with a vibratory hammer) or excavating with a clamshell to prepare the seabed for the pipe laying operations. Either one or two standard cranes or a gantry crane (Figure 9.5) can be used to lower the pipe, supported by two or three slings, into place on the seabed.

9.2.4   Pipe Laying from a Floating Crane Barge

This method requires a large barge and auxiliary pipe transport and supply barges. The crane on the lay barge lowers a length of pipe horizontally to the bottom, where it is joined to the existing line. Pipe lengths can vary from 3 to 100 m. This method can be used for small to very large pipelines.

In protected waters, a gantry or crane barge (Figure 9.5) and a strongback (heavy beam supporting the length of a pipe) can pull a pipe joint into the bell of a previously completed pipe section with the aid of cables. Final seating may be done with a vacuum in the space between the pipe ends, or by using flanged connections and bolts.

Figure 9.3. Mobile pipe-laying platform in Santa Monica Bay, CA. Source: Hyperion Engineers (c/o Daniel, Mann, Johnson, and Mendenhall, Los Angeles)

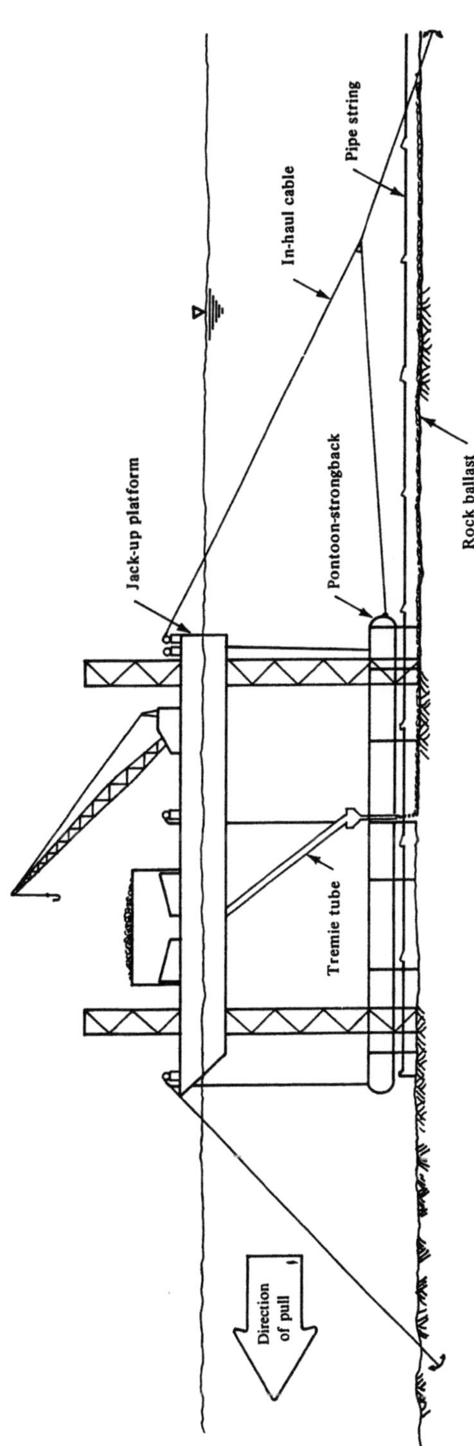

Figure 9.4.   Sketch of mobile platform operation

Figure 9.5.   Gantry operation on articulated platform. Source: Robishaw Engineering, Houston

If, in moderate to heavy seas, there is poor control over the position of the section being laid, pipe ends can be damaged and it is extremely difficult to insert a new section into the completed line. This problem can be overcome by using a pipe handling frame or a pipeline horse. shown in Figures 9.6 and 9.7, respectively. Handling frames such as these can assemble lengths of pipe and install the pre-assembled section underwater. Hydraulic rams provide vertical control at quarter points, transverse adjustment at each end, and longitudinal motion for inserting the new section into a previously laid section. The frame works independently of the crane barge. The frame is supported by the seabed and controlled by a console aboard the barge. The crane is free to prepare and pre-assemble joints of pipe.

Figure 9.6. Pipe handling and aligning frame.

Figure 9.7. Pipe handling horse. Source: Daniel, Mann, Johnson, amd Mendenhall, Los Angeles.

9.2.5 Surface Assembly with an Offshore Lay Barge

This method is used in deep water by the offshore oil and gas industry, but it can also be employed in laying sewer outfalls. Although cast iron pipe has been laid from a lay barge, the pipe usually laid by this method is coated steel pipe.

Lay barges can be single or multiple barge units with fabrication facilities, automatic positioning systems, and living accommodations. Simple barge arrangements are usually suitable for nearshore, shallow water, 6 to 60 m deep.

Lay barges used by the oil and gas industry are large floating pipeline construction facilities, with four to seven welding stations, a radiographic weld inspection station, and a joint coating station. The conventional lay barge method of construction in waters less than 60 m deep is illustrated in Figure 9.8. The pipe configuration consists of a straight portion, supported by rollers on the barge assemblyway, an overbend at the end of the barge, support of the descending pipe by rollers on a buoyant ramp (the stinger), an unsupported sagbend between the end of the stinger and the seabed, and support of the pipe by the seabed.

Pipe lengths are joined together on the lay barge supplied by auxiliary pipe barges. As the pipe is joined, the lay barge pulls itself forward on its anchor lines, which are progressively moved forward by an attending tugboat. The pipe moves down the barge assemblyway and down the stinger, which extends close to the bottom so that large stresses are not built up in the unsupported pipe section between the end of the stinger and the seabed. Stinger length and angle are designed to keep the pipe stress below approximately 85 percent of the minimum specified yield stress. This is done by controlling the radius of curvature of the pipe in the overbend and sagbend regions. At water depths of 60 to 200 m the length of the unsupported pipe between the end of the stinger and the seabed increases. This results in overstressing the pipe unless axial tension is applied to the pipeline. For instance, at 160 m of depth, a 90 cm diameter pipeline could have an unsupported sag bend length of up to 650 m.

A lay barge for use in deep water is shown in Figure 9.9. The pipe configuration is similar to that of Figure 9.8, except that the tensioning device is installed in the assemblyway before the end of the barge, and a curved or articulated stinger is used instead of a straight stinger. Articulated stingers are made from several structural sections, usually 20 to 40 m long, hinged at the joints to allow more flexibility than is possible with straight or fixed curvature stingers.

Straight stingers up to about 150 m in length provide support and protection for the pipe over most of its path from the barge to the ocean floor at 60 m. Longer stingers up to 215 m have been used but are unwieldy, more susceptible to breakage, and more difficult to maintain. They have the advantage that when rough weather approaches and it is necessary to drop the pipe and stinger to the ocean floor to prevent breakage, the operation can be performed quickly and with much less risk than is the case with shorter straight or articulated stingers.

Articulated stingers are shorter and more mobile. Also, the stinger configuration can be altered to meet changing conditions. They are difficult to lower to the bottom with the pipe, especially in deep water; time must first be taken to alter the stinger configuration so that it is straight. Curved stingers offer the same advantages and disadvantages as articulated stingers, except that their radius of curvature is fixed at the beginning of a job and can only be altered by disconnecting it and towing it back to a shore-based facility.

Anchoring of the lay barge is critical to successful pipe laying, since excessive or sudden barge motion due to anchor slippage can cause the pipe to be overstressed or the stinger to fail. Problems occur where anchors are too light or improperly designed for the particular bottom conditions. At present, anchoring is normally accomplished by using from 8 to 12 anchors in deep water and 6 to 8 anchors in calm environments such as estuaries and bays. Anchors are positioned by tugs and attached to the barge winches by steel cables.

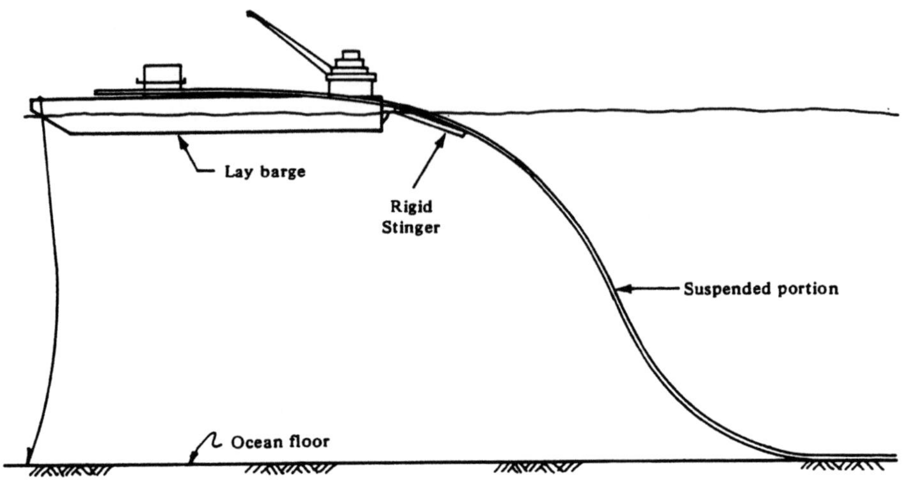

Figure 9.8 Lay barge for steel pipe in shallow water up to 60 m depth

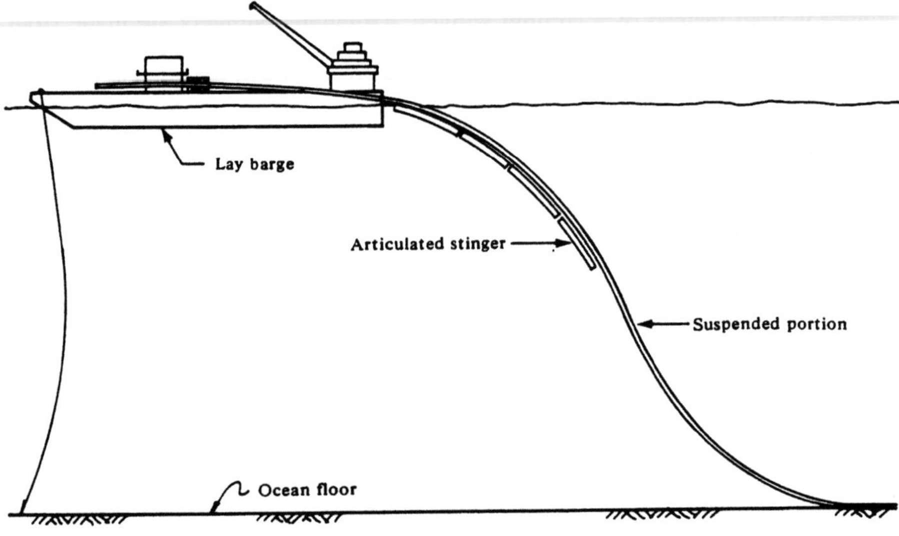

Figure 9.9. Lay barge for steel pipe in deep water up to 200 m depth

Laying pipe by lay barge is fast but costly. Lay barge construction spreads require auxiliary tugs and barges and tugs for moving anchors and supplying pipe, supplies, and personnel. In 1980 the costs for typical lay barge spreads ranged from approximately US $100,000 per day in the calm water areas of the world to more than US $300,000 per day in the North Sea. Apart from cost considerations of the lay barge method, there is considerable difficulty in accurately positioning the barge along a predetermined alignment. Most modern barges are equipped with computerized automatic positioning systems. An outfall less than about 1,000 m is normally too short for the lay barge to be a viable alternate. Many lay barges can install pipe sizes up to 120 cm With minor modifications to rollers and tensioners larger pipe up to 150 cm can also be installed.

9.2.6 Bottom-Pull Method

These methods are suitable for smaller diameter outfalls in both industrial and developing countries and uses a minimum of floating equipment. A configuration for the bottom-pull method is shown in plan view in Figure 9.10   Steel and, under unusual conditions, high-density polyethylene plastic pipes have been installed by this method. Several variations of the bottom-pull method have been successfully used when site conditions have been favorable.

Sections of precoated steel pipe are assembled onshore in 20 to 500-m lengths, welds are nondestructively tested, and the bare pipe at the field joint is given a corrosion-resistant protective coating. Each length is set aside on skids or runners parallel to the route of the pipeline to await the actual installation.

The pulling winch may be on a barge anchored several hundred meters beyond the pipe terminus and directly in line with the route of the outfall line. Alternatively, the winch could be located onshore with a pile cluster and sheave located offshore or on the opposite shore of an estuary.

Before the pull, one end of a cable or wire rope is connected to a pulling head welded to the leading section of pipe to be pulled. An onshore track or roller system may be needed. As the pipe enters the water, buoyancy relieves some of the pipe weight, but it is often necessary to add more buoyancy to the outfall line, particularly if the line is to be pulled over a rough or rocky bottom. If a pipeline weighs more than 15-30 kg/m in the water, the external coating could be damaged by rough or rocky bottoms. Pipes are pulled empty and often fitted with timber floats or buoyancy tanks (pontoons). Such tanks can be permanently sealed or fitted with ports and valves for controlled buoyancy. It may be difficult to release buoyancy devices at depths greater than 60 m. The usual practice is to provide 8-30 kg/m of negative buoyancy for a pipeline when it is being bottom-pulled. Maximum pulling speed is of the order of 6 m per minute.

Under no circumstances should this method be used when the line must be left lying immobile between tides, particularly where the tide might come in laterally to the line.

With too little negative buoyancy, a line being pulled can easily wander off course, and attempts to straighten the line by pulling usually result in breaking the pull cable, the pipe, or the pipe coating material.

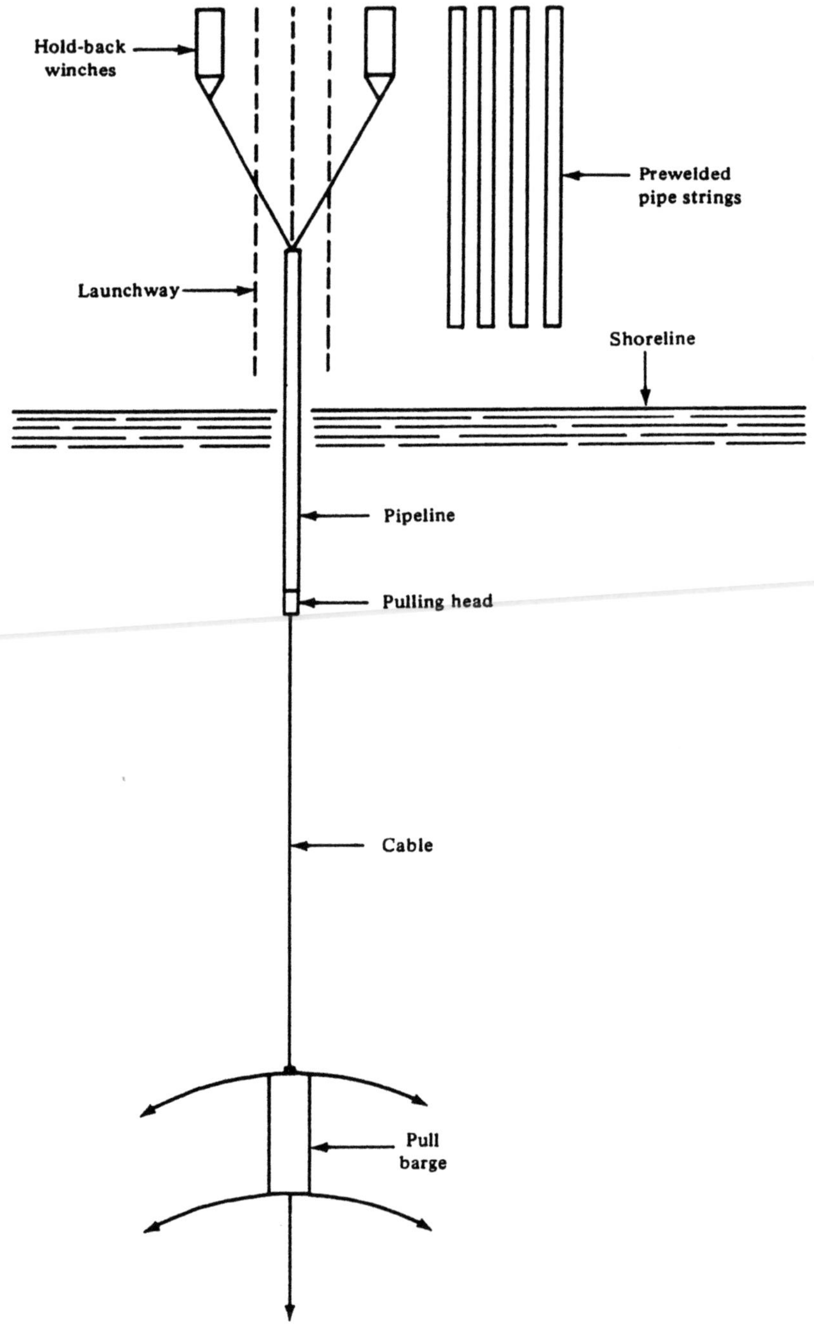

Figure 9.10. Configuration of bottom-pull method

A serious problem that arises from the effects of friction and cohesion during pulling is that on resumption of pulling operations following a pipe section tie-in, the starting force may be of such magnitude that the elastic limit of the pipe may be exceeded, with consequent damage to the pipe and its coating. A series of excessive stress applications could ultimately rupture the pipe. Starting forces from two to five times the continuous pulling load have been encountered.

The bottom-pulling method is not compatible with laying lines around obstacles such as rock outcrops, reefs, wrecks, or isolated deeps. This is not to say that bottom-pulled lines cannot be laid on broad sweep curves. However, a great deal depends on pipe stiffness and bottom conditions. The best practice is to pull the line in straight route alignments.

The method requires a relatively large, level land or barge area near the shoreline and in line with the proposed offshore route. This allows onshore storage and assembly of pipe sections prior to pulling.

9.2.7 Floats-and-Chains Method

This method has been developed for towing long sections of pipe that are fitted with buoyant units and chains. The pipe is towed while being suspended a few meters above the seabed. It requires surface facilities that are less expensive to operate than lay barges. This method has been used to lay very long sections of pipe, over 10 km in length at depths up to 100 m. A diagram of the pipe fitted with its buoyant units and chains is shown in Figure 9.11.

The weight of the raised chains balances the amount of the thrust load of the buoyant units and the weight of the pipe. The pipe is towed empty. Where the sea floor is flat, the chains maintain the pipe at a constant level above the seabed. Where the sea floor is uneven the effect of the chains and buoyant units is similar to that of an elastic mattress placed between the pipe and the seabed. The length of the chains pulled on the seabed is calculated so that pipe stability is ensured in the maximum lateral current likely to be encountered on the towing route. The tensile force required for towing a section of line is directly related to the maximum force exerted by the lateral current on the section of line over a flat area. This method is of great interest, especially in areas with small currents. The method differs from conventional bottom pull in that a retaining force is exerted on the shore end of the pipeline to avoid buckling.

9.2.8 On-Bottom Connection of Short Lengths

This method is a variation of the bottom-pull method. Several long sections of pipe are bottom-pulled and connected underwater by welding or bolted flanges. The first section is pulled into place at the most distant location offshore. Then a second section is pulled by the same method and axially aligned with the first section. Final adjustments are made by an alignment frame that is lowered onto pipe ends from the surface and is operated by divers. Alignment frames also contain a chamber that enables a diver-welder to join the sections. Succeeding

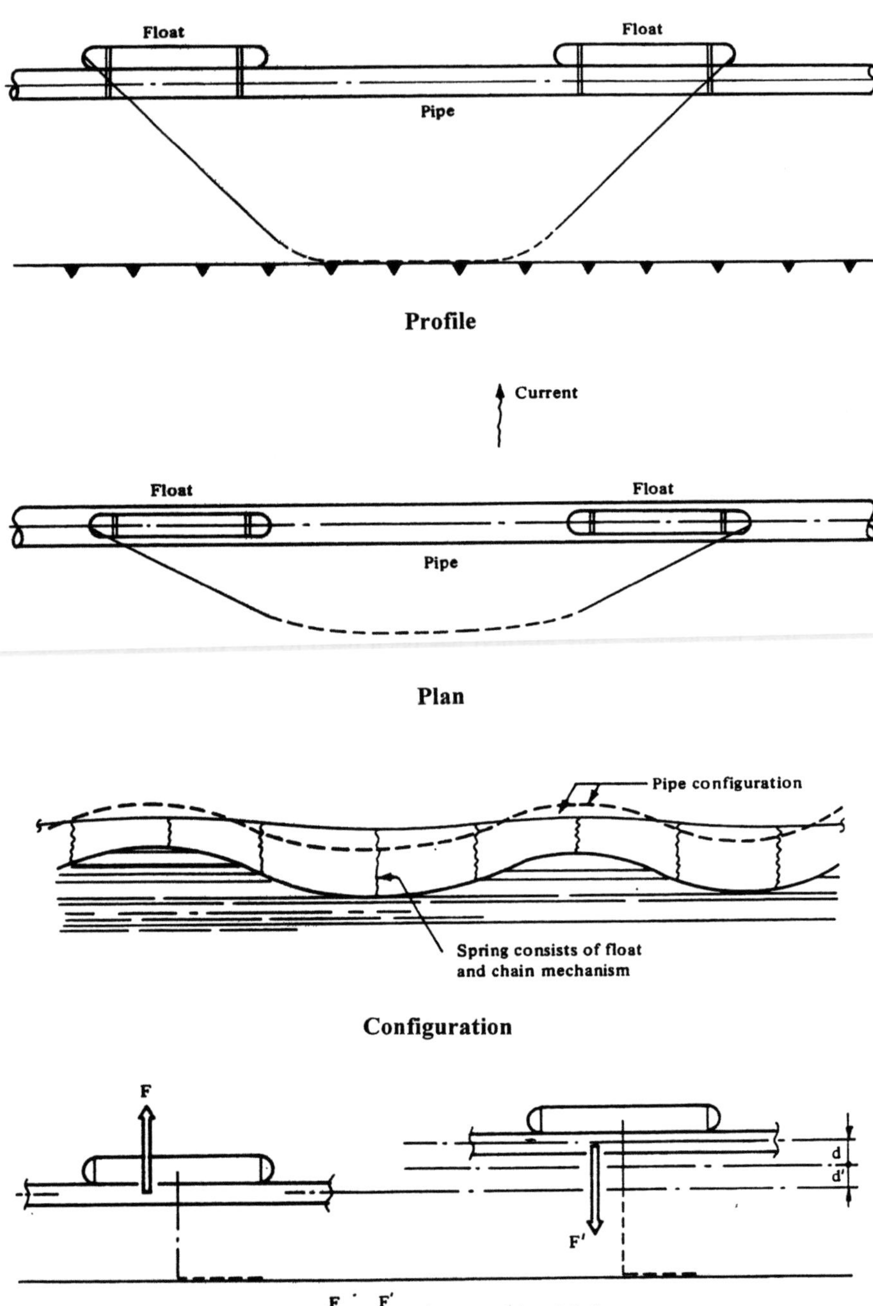

Figure 9.11. Floats and chains method

strings follow the same procedure of pulling, aligning, and joining until the job is completed at the onshore end.

While rough positioning of the pipe is accomplished by conventional pipe pulling procedures, final alignment of the joint is made in the alignment frame with tie-in fixtures. Pipe strings that are made up on shore are fitted with weld-neck anchor flanges on the leading and trailing ends of each string. These flanges are used to attach pulling heads, to cables and anchors for rough alignment, and to secure the alignment-welding fixture. After rough positioning, the alignment frame is lowered to the pipe on taut lines. Divers remove the pulling heads and attach the frame fixtures to the pipe. The hydraulically actuated system of the frame is used to position the pipe in the frame and to pull the pipe ends into a chamber for sealing, cleaning, and final alignment. Welding, coating, and inspection take place in a dry, controlled atmosphere. The welding and alignment fixture is also able to cut pipe and install valves. Figure 9.12 shows an arrangement for such a fixture.

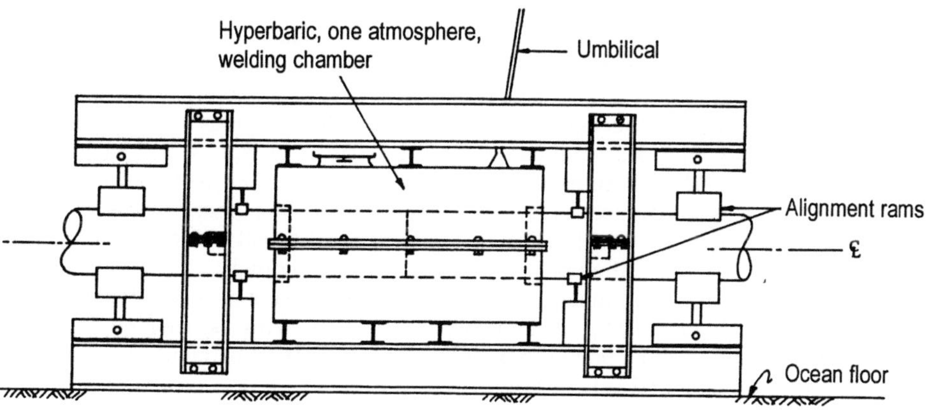

Figure 9.12. Underwater welding and alignment fixture.

9.2.9   Bottom Pull from Floating Work Platforms

This configuration of the bottom-pull method is used where the entire offshore operation is conducted from three floating platforms or barges. One barge, the pulling barge, is used to pull the pipeline offshore. A second platform is made up of one or more barge units designed to operate near the shoreline in shallow water. This work platform is used for blasting, excavating, and making up the pipe joints. A third utility barge is used in mounting a crane for offshore excavation and for pipe handling.

With the work platform and utility barge in position, the utility barge crane supplies pipe lengths to the work platform for welding and field jointing. The fabricated pipe length on the work platform is pulled into the sea by the bottom-pull method. Another pipe joint is set into position on the work platform by the crane. It is welded, field joint coated, and the pull cycle repeated.   Figures 9.13 and 9.14 show various barge unit configurations used with this method.

9.2.10   Surface Pull (Flotation) Method

Steel pipe. Steel pipe is assembled in long sections and buoyancy pontoons are attached on a launchway parallel to the direction of the assembly (Figures 9.15 and 9.16) Each string is pulled into the water and towed into position as a floating unit. The tie-in barge, which holds the offshore end of a previous string, makes the connection. Then the pontoons are released except for those near the end of the completed pipe string. Another string is pulled into the water, floated to the site, and connected. The process continues with the remaining pipe strings. By substituting weights for flotation, the same method can be used for high-density polyethylene.

This method can be hazardous even in moderate seas because the pipe can oscillate even under small wave conditions. Currents tend to push the pipe off-line. so this method is generally used only in protected waters. Since the pipe hangs between pontoons, large pipe stresses can be built up if this spacing is too large, especially during lowering operations. In a variation on the above method, the sections are joined onshore so that the need for a tie-in barge is eliminated.

Plastic pipe. A successful method for polyethylene pipes is a variant of the above. The pipeline is assembled onshore in strings of 200 - 300 m by butt-welding the 10 -15 m lengths together. Each string is then ballasted by adding concrete weights at distances of 2 - 3 m, depending on the pipeline diameter. Each end of the string is fitted with a flange. The ballast is usually 90 to 95 percent of the buoyancy of the air-filled pipe which will float accordingly.

The pipe strings are launched into the water by means of a crane and floated into position (Figure 9.16). When in position, the string is lowered carefully onto the bottom or into a prepared trench. Final ballast adjustments can be made with concrete weights added after have been verified.

Laying starts from shore which means that the first string is connected to an onshore manhole or surge tower. The following strings are towed and lowered into position where the flanges are connected by long tie bolts The method is

Figure 9.13. Articulated platform for offshore assembly in bottom-pull system. Source: Flexifloat Systems, Robishaw Engineering, Houston

Figure 9.14. Articulated platform for near-shore assembly in bottom-pull system. Source: Flexifloat Systems, Robishaw Engineering, Houston

188

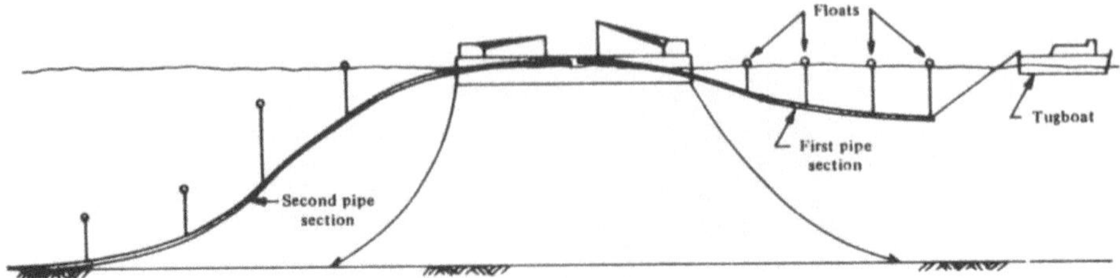

Figure 9.15 Flotation method. Configuration for offshore assembly of outfall

Figure 9.16. Flotation method. 1989 launching of 710 mm HDPE for 650 m outfall at Kerteminde, D. Sources: Mogens Pedersen Nyborg Constructors, Ltd., and KWH Pipe Company, Denmark.

sensitive to rough weather, although where strong currents are mostly in one direction, their effects can be controlled by driving piles into the bottom. The entire operation requires a few experienced divers and a powerful boat whose size is determined by the size of the pipeline. This eliminates the need for the barges and auxiliary equipment needed for installing steel, and accounts for the total costs per meter of pipeline being from 10 to 50 percent of comparable steel listed in Table 12.1 (personal communication, J.F. Knudsen, 1995)

9.2.11   Remote Assembly Methods

Remote assembly refers to pipe sections that are joined together in a location away from the job site. In some cases this may involve joining the entire line together at a remote work site, floating it into place over the route and sinking the line to a prepared seabed. This is applicable to short outfalls of less than 300 m (1,000 ft). It requires calm weather and currents of less than one knot.

Reel barge method. Figure 9.17 shows the reel barge method, that may be suited to small lines generally more suited to water supply than to wastewater disposal. Polyethylene pipe, sometimes reinforced by steel bands for high pressure service, with diameters up to about 30 cm is feasible for lengths greater than 10 km. The pipe is plastically deformed for storage on a large reel or spool. The remaining steps are the same as those used with a lay barge except that the pipe is unwound from the reel in a continuous, uninterrupted fashion. Holding tension on the pipeline is applied to the reel mechanism as required. This approach is attractive because :(i1 Work stoppages due to weather are practically eliminated onshore and greatly minimized offshore.(ii) Capital investment can be reduced.(iii) Welds and protective coating of long lengths of pipeline can be fully tested iv) Large volumes of pipe can be transported and handled with ease.(5v) Very high laying speed can be achieved (1,500 to 3,000 m per hour) with small-diameter steel pipe.

Steel pipe laid with the reel-barge method is normally coated with either a polyethylene or epoxy coating. Heavy wall pipe is normally used to attain sufficient negative pipe buoyancy with this technique.

9.3   Trenching and Backfilling

9.3.1 Controlling Factors

Submarine outfalls may be trenched and buried for almost their entire length or only throughout the shore approach, or they may be unburied for their entire length. The length of outfall to be in a trench depends on
(1) Local regulations for controlling and protecting the shoreline.
(2) Aesthetic considerations (for instance, a scenic beach that would be adversely affected by an exposed pipeline).
(3) Fishing activities that might be adversely affected (for instance, bottom trawls might hang or snag on an exposed pipeline).

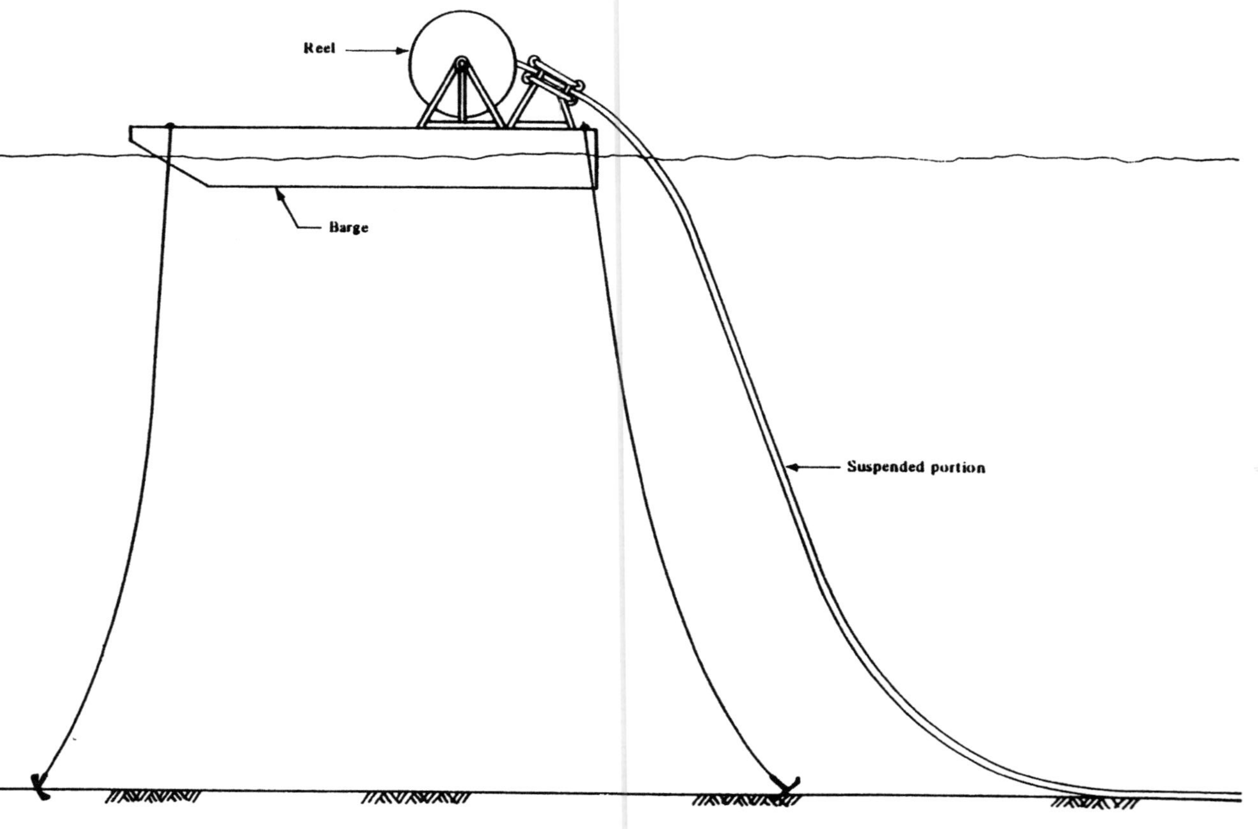

Figure 9.17. The reel barge method.

(4) Heavy breaking surf that may cause bottom instability and damage to an exposed outfall line.

(5) Danger of damage to an exposed pipeline by wave-borne debris or potential damage from anchors of ships or barges during severe storms.

(6) Uneven terrain that requires trenching to maintain a uniform alignment and gradient in the pipeline.

(7) Construction method and pipe material selection constraints.

The best practice is to bury the outfall through the surf zone to protect it from damage and to restore the shoreline to its original condition and use.

At the discharge, trenched and buried outfalls emerge from the bottom just upstream from the end of the diffuser section. The exposed end must be designed to be stable under the design storm-current conditions.

9.3.2   Trenching Methods

Trenching can be done either before or after the pipe is laid. The installation procedure, joint connections, and the pipe material all have an effect on the choice of trenching and pipeline burial methods.

Pre-installation trenches can be dug by explosives, dredges, excavation between two parallel rows of sheet piling, and plows. It is usually necessary to blast through rock or coral, either by drill hole explosives or by using shaped explosive charges. Shaped charges require no drill holes and direct most of the energy into the rock. After the initial explosives have fractured the rock along the ditch line, a string of explosives or "Bangalore torpedoes" is used to clean out the ditch.

Bangalore torpedo strings also have been used to excavate trenches in silts, clays, and sands. A problem with blasting or with trenches excavated prior to pipe laying is that the sediment is disturbed, so that if the sea is turbulent (as in a surf zone), much of the sediment settles back into the trench before pipe installation is finished.

Dredges used offshore include bucket-type dredges (dippers), draglines (Figure 9.18), clamshell dredges and continuous mechanical dredges such as the bucketline dredge. Continuous-type dredges are seldom suitable for outfall line trenching because of the set-up time required and the fact that they cannot work nearshore. establish the original cross sections.

Hydraulic or suction dredges include hopper dredges and cutterhead dredges; their efficiency is significantly reduced if large rocks are present. Suction dredges operate best on loose material, while bucket dredges are more effective in consolidated soils. Unless the trench is being excavated between parallel sheet piling, side slopes on most excavated trenches vary from a slope of 1:5 to 1:20 so that large quantities of soil must be moved and deposited elsewhere.

Large towed plows have been used offshore in Australia and in the North Sea for post-installation trenching of oil and gas lines. These reportedly worked well in both sands and clays (1). Alignment control may be a significant problem because the plow will follow the path of least resistance. If sufficiently hard pockets of soil are encountered, the plow will shift laterally.

Figure 9.18. Dragline dredge on articulated floating sections. Source: Flexifloat Systems, Robishaw Engineering, Houston

Post-installation trenching works best when the pipe material is steel with welded joints. It is not used when rock is present. A previously common post-installation device is a "jet sled" with high velocity jets that loosen the soil immediately under the pipeline . Air lifts or water eductors lift and remove the soil and spread it on each side of the ditch. Jets and eductors are mounted on the sled that is pulled along the pipeline by a surface barge that supports the large pumps and prime movers required to operate the water jets. As the sled is pulled along the pipeline, the pipe settles into the trench only to later emerge because the trench is backfilled with the same material, a dubious practice (see Section 6.3)

Rollers guide the sled along the outfall during operation. Sensors on the rollers relay signals to the barge to monitor the alignment of the sled and correct it when necessary to avoid undue stresses on the pipe and coating.

Jet-sleds work best in silts and medium clays in favorable soil conditions. They can excavate as much as 2 m deep on a single pass. In hardened formations or for deep burial, additional passes are necessary. Because of pressure losses in the hose and water back pressure, this method is less efficient at depths greater than 100 m, although there are reports of eductor systems working at 300 m.

Another post-installation trenching machine uses dual hydraulic powered cutterhead dredges mounted on a sled that travels along the pipe similar to the jet-sled. This machine requires much less energy to obtain the same depth. Additional advantages include a cleaner ditch, less soil disturbance, ability to handle a wider range of soil conditions, less support equipment, and unlimited water depth. Some cutterhead sleds have been equipped with self-propulsion devices which reduce the possibility of damage to the pipeline.

Section 6.3 describes pipeline failure due to jacking by wave or surf action and liquefaction of the bottom sediments. This can be avoided if the proper pipe weight, depth, backfilling, and armoring are selected.

9.3.3   Backfilling Methods

Backfilling requirements are site-specific. Where there are persistent longshore currents, the trench will backfill itself. In most cases, it is best to include backfill in the construction contract. High current velocities require gravel, rock, or riprap. For example, one outfall in an area with design currents of 3 m/s was backfilled with 5.1 cm crushed stone to a height of 0.6 m, then the remainder of the ditch, 0.3 to 1.0 m, was filled with 40 to 60 cm riprap to prevent the lower backfill from being swept away. The placement method should be selected before construction starts and should be done so as not to damage the pipe or coating.

If current velocities are less than 1 knot (0.5 m/sec) on the seabed, the excavated material may be flushed back into the ditch. In such cases the excavated spoil will have to be stockpiled on the seabed or on barges. Owing to the high costs of stockpiling, it is seldom practiced except in very shallow waters with no current.

9.4 Shore Approach

A substantial part of the cost of most outfall pipeline projects lies in the construction of the shore approach. Special attention must be given to protecting the outfall from mechanical damage, hydrodynamic forces, stability in a soil that liquefies in heavy surf, methods of construction, and effects on the usefulness and aesthetics of the beach area or shoreline. These factors are identified below. For an authoritative treatise on coastal engineering, the reader is referred to the Corps of Engineers' Shore Protection Manual (1).

9.4.1   Design Considerations

Governmental regulatory requirements must be considered early in the project identification phase. Designers must coordinate project design periods with local planning and permitting agencies and property owners to identify potential impacts of the outfall on future coastal zone development and particularly shoreline protection works. Failure to identify the latter has resulted in some spectacular pipeline failures. In some parts of the southern United States, all pipelines must be at least 4 m below mean low water (MLW) level through the shallow water shore approach, at the shoreline, and for minimum distance of 300 m on the land side of the shoreline.

Shorelines are unstable for reasons described in Chapter 2, which explain why they are where they are. They are either receding inland or advancing seaward, after at accelerated rates because of shore construction or changing sediment discharges from drainage basins undergoing development. Seasonal removal of several meters of nearshore sediment surface during winter storms and its redeposition during spring and summer is common. If a design specifies one meter of cover over the pipeline in such areas, it must also specify the reason to which this applies. Estimates of nearshore scour and fill may be obtained from historical charts and records, local pilots and harbor masters, nearby university departments, ocean-related government agencies or, better yet, seasonal bathymetric surveys over two or more winters along or near the proposed outfall alignment.

Designers and constructors must know quantities and types of sedimentary and rock materials to be excavated before they can select construction methods and equipment. Unfortunately, sub-bottom data are more difficult to obtain in shallow water than further offshore. Vessels used offshore are seldom suitable for shallow water work. Shore-based operations using divers are complicated by breakers and surf in those areas where bathymetric and sediment data are most important.

One method of obtaining samples is to load a small, 10-20 ton crane on a flat barge, ground the barge in shallow water, and use it a work base to either core or dig a test pit in the bottom. Divers can work during calm, slack tide periods and obtain useful date data with a hand-held jet and a sharp probe.

A sandy shoreline is particularly subject to continuous long- and short-term change. Many pipelines buried across sandy shorelines have gradually become exposed even when littoral drift was insignificant because breaking waves liquefied the soil and the pipeline gradually jacked its way upward through the sand. On way to deal with this action is to increase the empty specific gravity of the pipeline to 2.0. Another is to bury the line beneath the zone of liquefaction.

In sum, factors that have to be established during the design phase include:

(1) The depth of pipeline burial through the shore approach necessary to protect the line from potential shoreline movement and sea level change throughout the design life. of the project.
(2) The depth of burial necessary to protect the pipeline from the particle motion of breaking waves.
(3) The weight of the pipe needed for vertical stability which will almost certainly be different from that required offshore.
(4) The compatibility of the proposed outfull with future plans for the shoreline.

9.4.2   Construction Considerations

The construction methods suitable for the shore approach is often different from that for the offshore, particularly for outfalls over 1,000 m long. Pre-dredged trenches are often prepared through the shore approach zone and the pipeline laid or pulled into the trench before it fills with sediment moving with the littoral drift. Placing riprap over pipe laid in a cofferdam will provide less stability than a mass concrete cover. Attempts to lay the pipe across the winter time surf zone and bury it later work as well as but no better than local weather and sea state predictions even though progress is being made in methods for laying a blanket (linked revetment over exposed pipelines in the surf zone (3). Chapters 2 and 6 describe the forces that remedial measures must deal with. It is better to tunnel or otherwise bury the pipe beneath the zone of liquefaction in the first place.

Several factors to be weighed during the selection of construction materials and methods include the availability of essential skills in the local labor market, availability of construction equipment from local sources or distant ones with higher mobilization costs (see Chapter 12 for information on the relative costs of mobilization), availability of work space on shore, access for heavy construction equipment, impacts of excavation and construction on nearby structures and residents, effects of weather upon construction schedules, impacts of the completed structure upon the shoreline, consequences of temporary increases in water turbidity during construction, and the requirements for information and technology transfer between temporary and permanent staffs.

9.5   Tunneling

9.5.1 Horizontal Directional Drilling and Microtunneling

Tunneling (trenchless) technologies and terminology, particularly for horizontal directional drilling (HDD) and pipe-jacking for trunk, interceptor, and outfall sewers have evolved rapidly, particularly in the U.S. where European and Japanese systems have come onto the market (5). Technologies developed by the oil and gas industry have been adapted to include small diameter utilities distribution systems as well as large diameter gravity sewers. Microtunnels have working faces too small for men to work efficiently and comfortably so that their drilling or boring systems are remotely controlled (Figure 9.20).

Applications of HDD and pipe jacking include river crossings where line and grade can be controlled. To ensure joint integrity on vertical orzontal curves, a rule of thumb for minimum working radius is 1200 times the diameter of the of drill stem or product pipe (Tom Iseley, personal communi-cation, 1995).

Technological improvements are leading to increasing use of tunnels for ocean outfalls of from 3 to 8 m diameter (see Table 12.1). Pipe-jacking of outfalls up to a length of 1,000 m and diameters of over 3 m are technically feasible Diffuser sections consist of risers from the spring-line or soffit of the tunnel to marine bottom waters (see Chapter 4).

Materials for pipe-jacking have included steel, glass-reinforced polyester, asbestos-cement in Europe, vitrified clay pipe in Australia, Europe, and Japan, and more commonly reinforced concrete (5).

In shield tunneling, men work behind shield machines that are driven under coastal areas or river beds. Current practice (see Section 11.2) uses tunneling shields that are normally "earth pressure balanced" or "slurry pressurized" and involve advancing by short pipe sections 0.8 to 1.0 m length. that are bolted across neoprene or rubber gaskets (Zhou Yucheng, personal communication, 1995)

Median unit costs for pipe-jacking are on generally similar to those for lay-barge systems (see Table 12.1), but are much more variable because of both small and large-scale differences in soil and rock characteristics that require changing heads on tunnel boring machines as they proceed along the tunnel alignment.

Construction begins with a specifically designed drill rig that is disassembled into several components for highway transport and reassembled on the site. The heart of the rig is a self-contained hydraulic power unit that travels up and down a ramp the slope of which is dfixed by the pipeline profile or that is is assembled in a construction pit Intermediate pits are used for long sections.

The first stage in construction is the drilling of a small-diameter pilot hole beneath the outfall route, following the pre-planned course as closely as possible. The drill bit is powered by an in-hole hydraulic motor attached to the end of a non rotatingpipe string. The drill string is composed of 10 m lengths of lightweight threaded drill pipe. The in-hole hydraulic motor is attached to a curved section of pipe called a bent sub. This unique feature of bit rotation without pipe rotation and the bent sub makes it possible to achieve the directional control required to produce a curved hole. The drill bit progresses downhole and curves in the direction of the bend. The drilling rate is a function of the composition of the material

A pipe spread is set up on the shore site and the pipeline is welded together. If space permits, the entire line is preassembled into one continuous unit; more often, it is made up into several long sections. In either case, preassembly facilitates pressure testing normally conducted while the pilot hole is being drilled.

After the pilot hole has be completed, it is enlarged by a full-size bit running just in front of the leading edge of the pipeline. The entire assembly is actually jacked or pushed from the shore site. Several roller stands are set up along the centerline of the outfall route to support the pipeline as it is being pushed into the hole. This reduces friction and protects the pipe from being damaged.

(j) **Microtunnel boring machine (MTBM) –**
   (1) Remotely controlled
   (2) Automated soil transportation and rate of
       excavation controls
   (3) Pipe jacking equipment suitable for the
       direct installation of the product pipe
   (4) Active direct control
   (5) Earth or slurry pressure balance face control

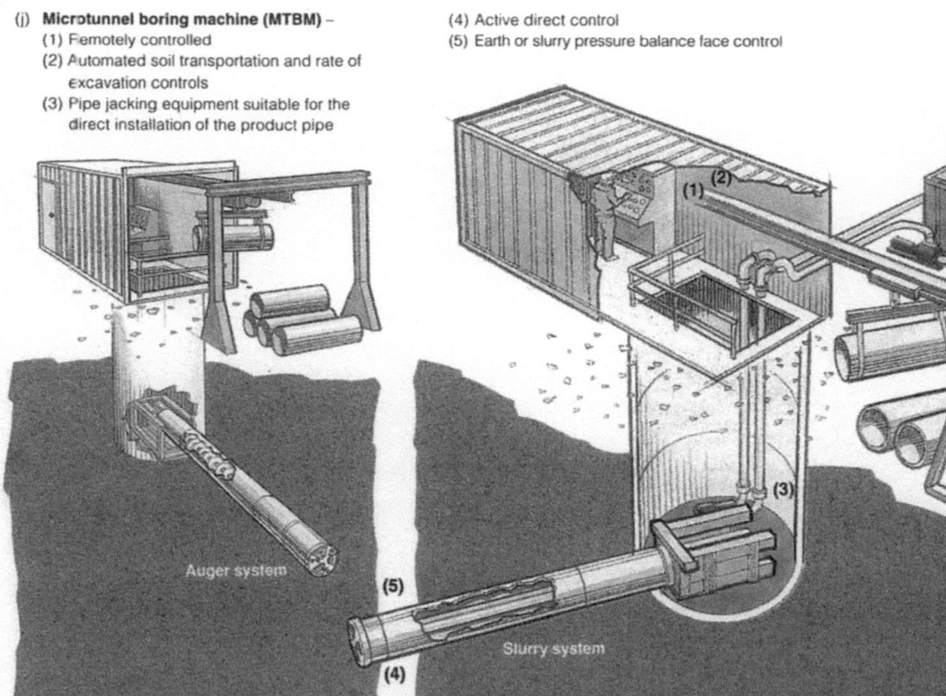

Figure 9.19. Micrtotunnel boring machiine. (MTBM)).)1) Remote control. (2) Excavation and Spoil Removal contyrols. (3) Pipe-jacking equipment. (4) Active direct control. Earth or slurry pressure balance face control. Source: Pipe Jacking Association (6).

PIPE JACKING EXCAVATION METHODS

A range of pipe jacking excavation methods is illustrated. In most cases, the choice of method will also depend on the selection of the appropriate ground support technique:

(c) **Open hand shield** – an open face shield in which manual excavation takes place.

Figure 9.20. Open hand shield wityh manual removal of poing and pipe jacking. Source: Pipe Jacking Association (6)

(d) **Full face tunnel boring machine (TBM)** –
a shield having a rotating cutting head in which
the face may be separated from the rest of the
shield by a bulkhead. Various cutting heads are
available to suit a broad range of ground
conditions.

(e) **Cutter boom or road header shield** –
an open face shield in which a cutter boom is
mounted for excavation purposes.

(f) **Excavator shield** – an open face shield in
which a mechanical excavator is mounted for
excavation purposes.

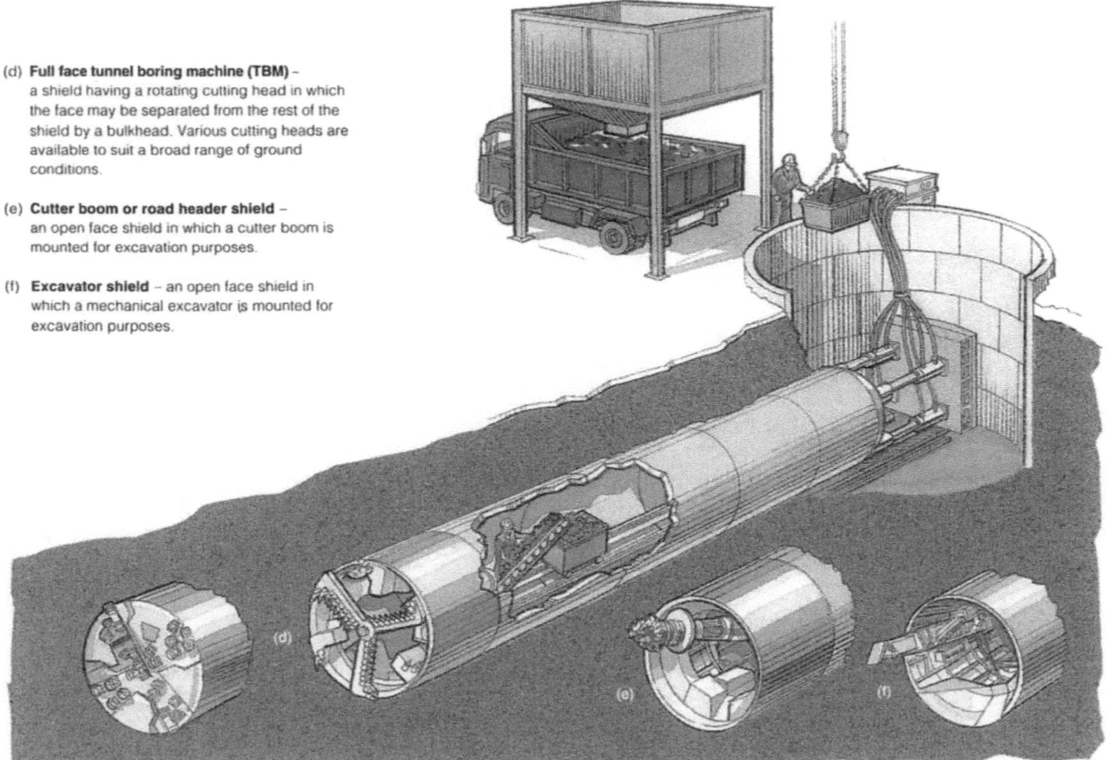

Figure 9.21. Full-face tunnel boring machine in soft ground with slurry removal, slurry pressure that controls earth pressure balance machine ((EPBM) and compressed air pressurized shield. Source: Pipe Jacking Association (6).

9.5.2 Large-diameter Tunnels

Although the technologies for boring microtunnels and the larger tunnels are the same in principle, they are much different in scale and practice, particularly with respect to equipment and mobilization, institutional requirements, wastewater reclamation or disposal, separate or combined systems for municipal sewerage and drainage, and comparative costing described in Chapter 12.

We define large-diameter tunnels as having diameters greater than 1.5 m which is about the minimum for men to work efficiently at the tunnel. Figures 9.21 and 9.22 show both manual and mechanical systems used at the tunnel face and for handling the spoil.

9.6 Construction Monitoring and Inspection

The objective of construction monitoring and inspection is to ensure that the outfall is installed in accordance with the engineer's specifications and with good workmanship. This means that all of the construction contractor's activities must be inspected by the owner's project management team, often furnished by the consultant who designed the outfall. If the project is small, the project management team may only consist of a project manager/engineer and a craft inspector. Except for outfall sections constructed alongside a trestle or in a cofferdam, the project management team requires an inspection boat and access to qualified divers to inspect the completed outfall.

During the bidding phase, contractors should be placed on notice that the successful bidder will submit detailed construction (safety, welding, installation, etc.) procedures before beginning.. The project management team is responsible for either approving or rejecting these procedures before installation begins.

Pipe coating or other corrosion prevention measures are usually the first field activities requiring inspection. Inspectors are assigned to the yard to ensure that the surface is properly prepared for coating, and that the coating is installed in accordance with the engineering specifications and the coating materials manufacturer's recommendations. "Holiday" detectors that measure the electrical resistance of the coating detect holes in the coating ensure coating integrity.

Concrete weight coating is also applied and monitored in advance of pipe installation. Application of the weight, including concrete thickness and pipeline weight per unit length, can be critical. Construction specifications spell out elements to be monitored during weight-coating application.

During trenching and other preparation of the offshore route, it is necessary to monitor the route cross sections and profiles. A small boat with a fathometer can

Construction management and inspection includes inspecting the welding (or other jointing), installation of pipe, backfilling, and cleanup. The number and specialties of the construction management team members depend on the pipe material, installation method, and magnitude of the project. In general, these include a project manager; project engineer to oversee the marine survey and trenching operations; craft inspectors to inspect coating, onshore preparation, and welding; and divers to make a trench inspection and a final inspection.

Certain specialized equipment is necessary, such as a holiday detector for coating inspection, fathometer, survey boat, tender boat for divers, and specialized equipment for radiographic joint inspection.

There is a tendency to rely on the contractor's divers and marine surveyors to reduce inspection costs. This is poor practice. Contractor staff and subcontractors are loyal to the contractor and can almost never be depended on to represent the best interests of the owner. Since the area of inspection is submarine, the only information that the project manager has for decision making comes from diver reports and instrument interpretations. It follows that this information should be gathered by individuals representing and obligated to the outfall owner.

Other inspection techniques such as video tape records and inspection with a submersible are available for large deep water outfall projects. Factors in outfall inspection that affect diving time include water visibility, support equipment required for the submersible, mobilization time, and comparative costs.

Although both owners and contractors benefit by minimizing construction time, there are unique incentives. In one celebrated case that occurred when the storm season was at hand, friction during a bottom pull approached the theoretical limit of the winch and cable system. The contractor elected to push on the shore end of the pipe with a bulldozer. It didn't work. The pipe buckled and broke.

9.7   References

1. Anon. 1977. 7200-foot loading line plowed in at Statford, Petroleum Engineer, September 1977, 10.
2. CERC. 1984. Shore Protection Manual. 2 vols. Coastal Engineering Research Center, U.S. Army Corps of Engineers, Vicksburg, MS.
3. French, J.A., Gustafson, P., Murk, J.S., and Burn-Lecht, K.M. 1994. Offshore rescue. Civil Engineering, October 1994, 42-44.
4. Titan Contractor Corporation, Houston, 1979. Personal communication to J.T Powers.
5. Kramer, S.R., McDonald, W.J., and Thomson, J.C. 1992. An Introduction to   Trenchless Technology. Chapman and Hall, New York
6. Pipe Jacking Association. 1994. A Guide to Microtunneling and Pipe Jacking Design. 56 Britton St., London EC1M 5NA.

10 Performance Monitoring

Wastewater management for coastal cities occupies a geographically small but economically important part of the ocean as a global commons whose stewardship includes monitoring. Here, we define three scales of monitoring consistent with the World Bank's linked hierarchical approach to monitoring environmental progress (91). Performance monitoring is a microscale activity for day-to-day operations and short-term (up to five years) planning for changes in waste management practices. Data collection focuses on that whose use is known.

Mesoscale monitoring looks at near- and far-field effects of urbanization and, in the United States, looks to compliance with laws and litigation intended to provide financial and technological equity among wastewater dischargers regardless of their location (75). Where external resources for data analysis are available, mesoscale monitoring is useful for post-audits and ten-year planning horizons.

Megascale monitoring and analysis is becoming increasingly important for interregional and international research, negotiations, treaties, interceding in inadvertent (90) or experimental large-scale ecosystem changes (48) that cannot be repeated or verified, and protocols leading to global environmental and economic sustainability. For example, large marine ecosystems are described in terms of commercial fisheries while catches decline due to overfishing (Section 2.3.7) Resources used in microscale and macroscale monitoring can, with adaptive and cooperative institutions and, with attention to reaction time-constants (Section 10.9), become essential in meeting the larger goals of megascale monitoring. Some of the issues to be considered are identified in this chapter.

10.1 A Framework for Sustainability Monitoring

Global scales of economic development and exploitation and their environmental costs have given rise to the idea of sustainable development. Although the term is hard to define and measure in scientific and economic terms, it is politically modern, marketable and attractive in industrial countries and in industrial cities of developing countries. It is a subject of research into its problems of scale-velocity functions (18, 58) and supports arguments for regionalizing compliance and performance monitoring for wastewater management in coastal cities (67,68). Meanwhile, terrestrial models with their better theoretical and empirical information base provide the logical starting point for designing models for better defining, monitoring, and sustaining the poorly known marine ecosystems (69).

Table 10.1.   Space and time scales in sustainability .

	Space, log kilometers	Time, log years
Atmospheric events		
Breeze, microburst	- 2.0 to -0.7	-6.0 to -4.0
Convection storm	0.0 to 0.5	-4.5 to -3.5
Front	1.5 to 2.2	-3.2 to -2.0
Tropical storm	2.5 to 3.5	-2.0 to -1.0
Hurricane	3.5 to 4.3	-1.1 to -0.9
El Niño	4.3 to 4.7	0.3 to 1.0
Marine systems	(Physical)	(Biological)
Oceanic stirring or mixing	- - - -	2.7
Oceanic residence times		
Atlantic deep water	3.7 to 4.3	2.5
Pacific deep water	4.0 to 4.3	2.8
Mediterranean Sea water	2.9	1.85
Elements	- - - -	2 to 3
Small particles	- - - -	0.9
Carbon dioxide	- - - -	5
Iron	- - - -	1.8
Large marine fishery ecosystems (inland seas, boundary current areas)	2 to 3	1 to 2
Ocean outfall systems		
Compliance monitoring	0.0 to 1.3	-0.2 to 1.2
Sewage treatment operations		
Biological system failure	- - - -	-2.0 to -1.3
Mechanical system failure	- - - -	-2.0 to --1.1

Sources: Holling (58),   Sherman (78,79), Open University (70,71,72), World
Resources Institute (89),   National Research Council (64,65), Rapport et al. (77)

C.S. Holling begins within the space and time framework in Table 10.1. He
compares the atmospheric scales with biological scales of North American forest
ecosystems ranging from 1 cm and 1 year for single needles to 1,000 to 10,000
years and 100 to 5,000 km for boreal forest biomes. The latter are transformed
into velocity terms as in Table 10.2. From the continuous spectrum of scales,
succession, and velocities in living ecosystems, he demonstrates that the
conventional linear model of sustainable economic development is an incomplete
equilibrium view.

Large marine ecosystems. Analogs of the forest products exploitation cycle
and forest ecosystems are found in large marine ecosystems. These are defined by
their economic importance as marine fisheries. Again, sustainability is

Table 10.2. Succession in forest ecosystems

Individual function	Ecosystem function	Celerity of succession	Attributes
1. Birth	Exploitation	Slow	r-strategy (opportunist, brittle, invasive, short life span, little stored capital, rapid growth)
2. Growth	Conservation	Very rapid	K strategy (steady state at carrying capacity , consolidation, resilient, much stored capital)
3. Death	Release	Rapid	Fire, storm, pests, resilient, little stored capital
4. Renewal	Reorganization	Rapid	Accessible carbon, nutrients and energy, brittle, much stored capital

Source: adapted from Holling (5) Lalli and Parsons (64)

dynamic, successional, cyclic, vulnerable to economic exploitation, and subject to global changes in primary production species such as the brown tide *Aureococcus anophagefferens* (90) Forty nine large marine ecosystems are identified, including the Weddell Sea, the Bay of Bengal, the California Current, the Northeast U.S. Continental Shelf, and the Norwegian Sea. Monitoring information important to fisheries has been obtained cooperatively from most of the 49 by international agencies including the UNEP and FAO, and national agencies such as the U.S. National Oceanic and Atmospheric Administration (NOAA) whose charters and policies are scaled for international cooperation (77,78).

Performance and compliance monitoring in coastal city wastewater management can be essential small-scale elements in sustainable systems for marine ecosystem protection and economic development in coastal areas. All with due regard for professional and cultural thresholds in cooperation for inter-laboratory calibration, data and information sharing and analysis, and management transparency in regional and global monitoring scales.

Operational aspects. Attributes of performance and sustainability monitoring are revealed by comparing United States Environmental Protection Agency (USEPA) and World Bank policy. Environmental monitoring and assessments of all proposed projects and programs are carried out under prescribed procedures prior to approval and financing (51,52, 83). Compliance monitoring of the effluent and receiving waters is mandated by the USEPA. In contrast, the Bank leaves performance monitoring to the local entity without advancing requirements or guidelines for either performance or compliance monitoring.

At the country level, monitoring systems evolve in response to technological, political, economic, and environmental needs. Table 10.1 reveals that scales (time and space resolution) of compliance monitoring under USEPA National Pollutant Discharge Elimination System is insufficient for operational management, and too much for ecological analysis where they are filtered to make them tractable. The system appropriately serves the legislative mandate of financial equity rather than environmental efficiency in waste management (75) although at an additional cost of frequent contention and litigation (61) This may be affordable in a high-income country, but cannot logically be applied to a middle or low income one. It provides quickly warehoused data, bits of which are later mined for research.

Problems in matching mandated mesoscale scale of compliance monitoring to sustainability monitoring are found in the 2.6×10^5 km^2 Northeast U.S. Continental Shelf Ecosystem. 54 million people living in 4.8×10^5 km^2 discharge about 3.5×10^{10} liters/day of wastewaters. An additional scaling factor depends on the locations of the U.S. District Courts in the tributary watersheds on whom the USEPA relies for enforcing financial equity among all dischargers in the watershed. A first approximation of costs to monitor the Northeast U.S. Continental Shelf based on Southern California Bight (64) compliance monitoring operations (\$20 million, 15 million people, and 1500 km^2) yields 75 to 150 million dollars per year range.

In coastal cities, monitoring needs include baseline research and program or project identification, preparation, financing (usually public), construction, operation, and project cost recovery. National, bilateral, and multilateral agencies bear the costs for larger-scale monitoring and evaluation. There are areas of potential monitoring efficiencies and costs-sharing among all parties, providing incentives can be developed to deal cooperatively rather than competitively with institutional issues (local purpose, national identity, and regional cooperation).

Large data systems are defined as those that are hard to turn off with consequent diseconomies of scale. Their inertia and the job security are directly proportional to the amount of money invested in hardware and facilities, the number of careers involved, and especially the length of record. Adapting compliance monitoring to other countries is not recommended because it has always been embedded in the singularly litigious environment of the United States In contrast, small data systems use monitoring for research where data have a known and scheduled use, and for project identification and preparation .

10.2 Some Performance Monitoring Principles

The operational definition of performance monitoring of an ocean outfall system is that which provides necessary information to be used for (i) operating an existing facility, (ii) managing commercial or recreational fisheries and aesthetic resources in the vicinity of the waste discharge, (iii) modifying or expanding waste disposal facilities. or (iv) evaluating conformance with national or international guidelines, criteria, or standards.

Effective monitoring and data interpretation depend upon the use to which the data will be put and the nature of the variance. Periodic events in tidal estuaries, bays, inlets, and straits are due to semidiurnal (lunar), diurnal (solar), fortnightly (lunar), and annual (solar) events (7). Diurnal land-sea breezes affect surface water movements and properties (24). Filtering of (Eulerian) measurements at intervals of 1.67 to 6.00 minutes in tidal estuaries reveals that (Lagrangian) estimates of advection by semidiurnal tidal currents could have been determined from observations taken at 2- to 3-hour intervals (13). Alternatively, measurements taken during a single, lower-low water slack when tide induced changes are unimportant are used to measure longitudinal distributions of estuarine pollution (8).

Optimal intervals for sampling, digitizing, averaging (which always loses some information), or filtering of a time series for descriptive or predictive models can be based on spectral or cross-spectral analysis (Section 10.9).

Navigational requirements for sampling stations depend upon how the data will be used. Table 10.3 lists optical, microwave, and satellite systems in order of increasing accuracy and cost. Note that costs of satellite systems are decreasing.

Table 10.3.. Advances in nearshore navigation system accuracy.

Horizontal sextant angles	10 m over 5 km
Theodolites/laser	5 m over 10 km
Artemis III	16 m over 30 km
Microwave systems	4 m over 30 km
Hydrodist	1,5 m over 30 km
Decca Main Chain, Loran	100 m over unlimited area
Satellite navigation (GPS)	1 to 10 m over unlimited area

Adapted from Sillis (41). and Tetratech (83).

Optimal sampling station location and replication of biological samples are closely related. Both numbers and associations of individual species vary over short distances. Thus many ecologists, regulatory agency officials and their contractors (38, 79), and attorneys in environmental practice insist on say, five or more, closely spaced replicates from the same water mass or benthic sampling station to attest greater statistical elegance. For example, five samples are used by USEPA to impute parametric significance to the non-parametric Kruskal-Wallis one-way ANOVA test (83). This has been shown by Bascom (2, 3) and Word, et al. (43) to be irrelevant when the objective is to determine the area of discharge impacts. They reported that more information for the same effort was obtained by sampling more stations rather than replicating samples from fewer stations.

Selection of water or sediment quality parameters to be measured at particular sampling station locations should be based upon preliminary field surveys of the area that identify probable characteristics of waste and stream discharges and their potential effects upon receiving water uses. Other sources of information include university engineering, physical science, biology, and geography departments; government agencies including those that issue building or operation permits, ferry pilots; fishermen; residents, and media releases from new or expanding industries Occasionally, a pre-project outfall survey or an academic research project will identify an unanticipated need for monitoring a particular material (e.g., the reported finding in the muds of Manila Bay of increasing concentrations of dieldrin, a pesticide whose use but not its export has been banned in the United States and other industrial countries because of its persistence and toxicity).

Collecting environmental monitoring data is easier than interpreting them. Excess data are ordinarily justified because (i) the marginal cost of collecting them is low, even if the costs of interpreting them is not, (ii) another agency (or country) is collecting them within their jurisdictions, (iii) "We've not sure what to measure, so we measure everything. ," (iv) the data impute a simple quantitative measure of accomplishment, (v) large amounts of data confuse critics by their bulk, a (vi) the data may be useful someday, and (vii) collecting them provides employment.

These uncertain marginal benefits rarely justify the marginal costs of excess data storage, retrieval, and analysis. A series of six regional marine pollution monitoring workshops in the United States revealed (i) little utilization of costly monitoring data, (ii) a need to reallocate resources so that data assessment receives essentially the same funding as data collection and (iii) a need to demonstrate that benefits derived from institutionalized analytical quality control justify its cost in environmental monitoring systems (4, 15, 19, 20, 30, 33, 34, 35).

Care with statistics, with due regard for advances made over the past ten years, can save money and provide insights that cannot be had any other way. Sampling system design, replication, institutionalized analytical quality control, data interpretation and analysis have value. With respect to compliance monitoring, Kamlett (23) has argued from the viewpoint of U.S. environmental law that "while accurate and precise data (in compliance monitoring) are obviously preferable to inaccurate and imprecise data, we do not require scientific certainty (if such is attainable) as a predicate for action. Reasonable trends, projections, and potentials are usually sufficient to allow us to respond in the administrative, legislative, and political spheres."

Careful selection of procedures and their calibration within the particular environment to be monitored are essential. For example, Rhodamine B, a dye measured by fluorometry and used to observe and model wastewater diffusion in coastal areas (28), works well in clear water. However, since it is also a very good dye, it adsorbs onto sediments suspended by tidal or other currents (12), and daily recalibration of the procedure is essential to determine the significance of dye losses to sediments. When adsorption losses are unacceptable, another dye such as Rhodamine WT can be used. Meanwhile, frequent recalibration in waters being tested is required in order to adjust for false positives due to fluorescing algae (12).

Scales of Things in Operational Monitoring. Optimal intervals in space and time for marine and estuarine monitoring depend upon (i) mass emission rates of effluent constituents, (ii) the residence time of water in the area under consideration, (iii) reaction rates of physical processes such as sedimentation of effluent suspended solids, (iv) dilution by stirring and mixing processes, (v) decay rates for enteric bacteria and other microorganisms in the effluent, and (vi) mineralization and utilization rates of nutrient and other non conservative constituents. Even if all of these rates could be expressed in linear, periodic, or first-order terms (and they cannot), it is not possible to optimize for more than one of them at a time Research and project identification require more data points than operations simply because the latter responses are known.

10.2.1   Equilibrium Response Times and Monitoring Design

When a waste discharge to the ocean or an estuary is initiated, the physics, chemistry, and biology of the discharge region will be altered. Monitoring is usually intended to identify and measure these changes over time. In practice, once a discharge has begun, changes will take place over a limited period of time during which the receiving ecosystem adjusts quantitatively to a new dynamic equilibrium. Once this is reached, no further changes can or will take place unless (i) the quantity or nature of the discharge is substantially altered, (ii) the area is subject to the influence of other contaminant inputs which are changed, or (iii) the ocean climate, that is best measured by other than microscale performance monitoring, changes. The equilibrium reached is a dynamic one because, in all marine ecosystems, natural fluctuations in physical, chemical, and biological characteristics occur whose magnitudes are ordinarily comparable to or greater than the magnitudes of any changes caused by sewage inputs.

The time required for an ecosystem to respond to a discharge and reach a new dynamic equilibrium depends upon the biogeochemical nature and energy characteristics of the ecosystem and the specific characteristics of the area monitored. For example, when a waste is discharged into a large water body, physical and chemical changes in the water column caused by the discharge occur rapidly and dynamic equilibrium is reached within a day or two. In an enclosed bay, chemical equilibrium may not be reached for days or weeks. This effect is more pronounced in an estuary where seasonal extremes in temperature and outflow result in qualitative and quantitative differences in equilibrium response times and characteristics. In contrast, chemical changes in sediments near the discharge will proceed slowly in response to inputs integrated over several years or, in areas of low sedimentation of effluent solids, decades. In addition, biological changes will lag behind physical and chemical changes particularly where these changes take place quickly compared to the generation times of the affected biota.

In sum, numbers of parameters and sampling locations and frequencies and their analytical quality control procedures set the costs of monitoring. Timely management responses to the monitoring information determines the benefits. The wide variety of health-related issues for which this is true can be developed from the rigorous concepts of Shuval et al. (36) who showed that the marginal benefit-cost ratio of health investments follow a logistic curve. In contrast, well-planned research projects and feasibility studies that follow standard methods, and that have clear objectives are cost-effective in supporting engineering, operational, and management decisions; long-term monitoring programs adopted for reasons of political legitimacy (48) are both more expensive and difficult to manage.

10.3 Hydraulic and Structural Monitoring

Evidence of actual or imminent structural or hydraulic failures of outfalls is provided by surface boils or predicted by diver or submersible inspection, by excessive current requirements for cathodic protection systems (see Section 9.3), or by high operating pressures resulting from reduced cross sectional area. Most failures are due to storms, ship anchors, or heavy fishing gear that bend, move, or break the pipe (9, 10, 21, 40). Partial failures, which may go unnoticed from the surface, include cracking or spalling of concrete weight coats caused by movement during storms or by anchors, loss of diffuser risers (9, 10, 40), gradual accumulation of grease in the line (21), or loss of anchor blocks near the outlet so that the pipe can move laterally and become difficult to locate for inspection (40). Breakage occurred in three of twenty-nine recently constructed British outfalls (10) two of them were in PVC pipe. The remaining steel or concrete pipes and six other plastic pipes were designed, constructed, and functioned satisfactorily.

European and Californian experiences indicate that at least one visual inspection per year is needed, particularly on those outfalls extended beneath shipping lanes, anchorages, and trawling areas, or over shallow, broad continental shelf areas with heavy surf. Plastic pipe sections, which are weaker than steel sections, require more inspection and maintenance (9, 10).

10.4 Discharge Monitoring

Discharge (end-of-pipe) monitoring is either operational monitoring or compliance monitoring. The latter responds to political and regulatory decisions that mandate sampling frequencies, kinds of data, sanctions, and punitive measures where a single sample may set off an alarm. This discussion focuses on operational monitoring in which the data are used to control and schedule waste collection, treatment, and disposal operations. Monitoring information includes wastewater flows, concentrations, variance, and mass loadings of selected dissolved and suspended constituents. Information needs increase with the size, complexity of the system and with residence time (see Chapter 2). In some areas, seasonal or annual observations of nearby shoreline conditions may suffice to monitor impacts of raw, screened, or comminuted sewage. In contrast, daily end-of-pipe monitoring of chlorinated secondary effluent for BOD, volatile and total suspended solids, nitrogen and phosphorus, ether solubles, detergents, heavy metals, and selected hydrocarbons may be used. This may be supplemented by hourly monitoring of chlorine residuals for operational control during bathing seasons. In all cases, the useful frequency of data collection is fixed by the response time needed for corrective measures (see Section 10.8)

10.5 Ecological Monitoring

End-of-pipe and some near-field monitoring systems are dynamic, and data outputs can be acted upon quickly within the treatment plant to that response time for changing an operation is not much more than twice the sampling interval (see

Section 10.8). In contrast, routine far-field compliance monitoring may provide some information on ecological changes but it would seldom suffice for research or for planning remedial measures. Monitoring systems that provide data for explicit needs in imposing quarantines occupy an intermediate position in time.

The urgency with which routine performance monitoring is adapted to research monitoring and to implementation of remedial measures will depend upon expert scientific, technological, financial, and economic appraisals. Local benefits, such as protection of a bathing beach or shellfishery, can be readily assessed. However, the regional implications of locally caused change, particularly those of multiple discharges and other inputs are more difficult to assess.

10.5.1 Public Health

Most outfall system designs are based on meeting bacteriological criteria in near- or far-field water-contact sports or shellfish areas. Discharges from industrialized cities may also contain both manufacturing and post-consumer toxic or noxious materials that accumulate in seafood. Meanwhile, bacteriological quality of ocean bathing or diving waters has not been demonstrated by credible epidemiological surveys to be a high priority public health concern (26,31), even in high-income countries (see Section 3.2.1).

Cabelli's frequently cited pioneering 1980 epidemiological study (5,6,16,25) concluded that health effects of bathing in polluted ocean waters are restricted to occasional low-level gastrointestinal upsets in children and other non-immune populations and to minor ear, eye, nose, and upper respiratory complaints.

More recent U.S. reports (67,68) have focused on the mere presence rather than the epidemiological consequences of pathogens in sea water and have duly concluded from first principles that there should be risks of infection. These works may not support Gillespie's (54) epidemioloical conjectures from Santa Monica Bay in 1942 but they are certainly inconsistent with Moore's 1959 (66) rigorous epidemiological study. In contrast, aesthetic matters are often a very high-priority public concern, especially in touristic or recreation areas that become more remote as cities expand. In practice, designing and operating municipal waste systems to meet applicable microbiological receiving-water criteria normally result in the discharge meeting aesthetic criteria.

Treatment process interruptions can cause the stranding of grease ball, fecal, plastic, and rubber materials on beaches. Two highly publicized cases in New York and California are typical. Beaches were closed and reopened on purely aesthetic grounds to assure a frightened public. Mandated daily sampling showed that microbiological quality standards were met throughout both incidents (11, 16, 37) and no illnesses were reported   The threat of litigation in the event of an infection could not be removed even by hourly sampling because the treatment location and municipal financial resources, accepted by litigants and their counselors as "deep pocket." It follows that weekly or monthly sampling during critical seasons are sufficient for public health monitoring. The marginal benefits of the additional data during storm or flood periods have never been examined, even in California where there popular proposals to impound and treat all storm water runoff from the coastal plain.

Along with microbiological water quality, public health considerations include exposure time in the water, the general health of the discharging population, and, for tourist areas, the health of the visiting populations. For example, during the 1960s, the United States had one typhoid case per 22,500,000 people while Turkey had one case per 10,000 (14). This means that visitors from the United States swimming in the Sea of Marmara would be afforded protection from typhoid equivalent to that in U.S. waters only under coliform standards that were 1/2250 of U.S. standards. Corresponding ratios could be developed for hepatitis and other waterborne diseases. However, comparison of Cabelli's (6) and Hakim's (17) results on gastroenteritis in the United States and Egypt, using enterococci as an indicator, does not indicate that such an effort would provide a worthwhile improvement in public health protection. Fortunately for swimmers (but not for eaters), gastroenteritis and similar infections of foreign tourists are predictably derived more from food than from swimming.

In any event, microbiological monitoring data are assessed by local guidelines, criteria, or standards. Both numerical standards and how they calculated vary widely among World Health Organization, European Community, France, Brazil, Cuba, Japan, Mexico, the United States, and California (Table 3.3).

Contaminated shellfish, eaten raw or under-cooked, carry typhoid and paratyphoid fevers, infectious hepatitis, cholera, and a variety of other gastrointestinal illnesses. Contributing factors include (i) market forces that cause shellfish to be harvested illegally from quarantined waters; (ii) shellfish concentrate microorganisms from their environment, often by 5- to 10-fold; (iii) people ingest perhaps 150 grams of shellfish at a sitting, with many more pathogens than the salt water they could possibly swallow during a swim; and (iv) contaminated particulates tend to settle to the seafloor where survival rates may be greater (32) so that pathogens are readily available for uptake by the shellfish.

The World Health Organization has suggested that shellfish should be harvested only from those areas where the product is acceptable for human consumption without further treatment (44). Otherwise shellfish should be boiled, steamed, or depurated in clean water for several days after harvesting to allow the shellfish to eliminate any pathogens.

Paralytic shellfish poisoning (PSP) is occasionally caused by blooms or "red tides" of toxic dinoflagellates that result in shellfish contamination. These blooms are limited geographically, and although their seasonal preference is known, the years of their outbreaks are unpredictable. For frequently affected areas, routine assays using mice as indicators are appropriate to warn communities of potential PSP outbreaks and to suspend shellfishing.

Pathogens in marine or estuarine finfish are not routinely monitored. Most of the fish will be cooked and the pathogens destroyed. Where local preferences are for raw fish collected from ponds fertilized by sewage or nightsoil, people will be infected by helminths. Where it is affordable, fish to be eaten raw are selected for firmness and taste and ordinarily come from offshore waters.

In contrast, monitoring for toxic hydrocarbons in the North American Great Lakes and other areas has resulted in 1970s - 1980s closures of waters to fishing as a precautionary measure. This has been supported by findings of PCB's (which were arguably considered as indicators of dioxins and furans) in New Jersey finfish, and Wisconsin human body fluids (80,62), Potential costs of such precautionary measures are indicated by the 1969 United States Food and Drug Administration's mercury standards for fish. Although this standard was a response to public furor in the late 1950s over methyl mercury being the cause of Minamata disease (22), subsequent research revealed that the high mercury levels in tuna that led to its prohibition were natural and that reasonable consumption levels of the fish imposed no danger (29).

Monitoring health, indicators and pathogens. Microbiological monitoring to protect health should respond to changing incidence and prevalence of a large number of infections. Some infections are caused by high concentrations of virulent strains of indicators that are ordinarily benign. This has led to changing lists of microorganisms to be monitored in different jurisdictions. Table 3.3 lists ocean bathing water criteria for combinations of total coliforms, E. coli, fecal coli, fecal streptococcus, enterococci, enterovirus, and salmonella. An example of responding to changing patterns of infection throughout the world is found in the increasing attention is being directed to *Cryptosporidium* because of its higher infectivity in AIDS victims as the pandemic continues to expand. Other differences will be found between countries where traditional, wellness based medicine is practiced and those where high technology, sickness based medicine is practiced.

10.5.2 Ecological Interactions

Research monitoring of ecological interactions can help protect the long-term productivity of marine and estuarine areas from the effects of waste discharges. These effects are greatest in benthic animals that accumulate waste constituents slowly over time. Since ecological changes take place slowly, benchmark data from benthic stations sampled over periods of months are considered synoptic. After a baseline survey, areal sampling may be conducted at 5- to 10-year intervals and supplemented by surveys immediately preceding and one year following major structural or other technological changes.

A number of indices have been designed to reflect environmental degradation. Benthic species diversity is reduced in the affected area during a period of continuous discharge, whereas total numbers of adaptive organisms and biomass are increased (3, 30). Similarly, where fish may have been scarce or absent prior to a discharge, substantial populations of finfish may be attracted to where the discharge is a primary food source. Monitoring for diversity and abundance of biota alone will not reveal subtle ecosystem alterations. This requires research into community structure and of sublethal effects on reproduction, feeding, growth, endocrine functions, and tissue pathology. Research into benthic, sediment, and fish sampling; water column chemistry; and toxic substances in food chain organisms may be warranted.

The three environmental characteristics associated with waste discharges that have been demonstrated to have the greatest influence on the benthic community are (i) the concentration and biological availability of organic particulates in the waste, (ii) the velocities of the bottom currents, and (iii) the settling rates of effluent particulates through the water column (13). Virtually all pollutants found in discharged wastes are attached to particles that eventually settle to the seafloor.

Benthic sampling and monitoring programs can vary greatly in complexity and cost. Generally, samples of benthic organisms larger than 1 mm are taken because analyses of smaller benthic organisms are more costly and their community structure is less well-known. Conditions should be maintained as uniformly as possible when sampling, since variables such as depth and time of year can significantly affect the results The sampling device should sample an effective area of the sediment (0.1 m^2 is suggested) and should penetrate at least 5 cm into the sediment, except for rocky bottoms where this is not possible. The sample should reach the deck of the boat undisturbed, and at least two grab samples (one for biological examination and the other for chemical analysis) should be taken. Samples should always be sorted with the same screen size (1.0 mm or less) and preserved in formalin after relaxation with $MgCl_2$ and later transferred to alcohol

10.5.3   Other Parameters

The size distribution, settling velocities of sediment particles, and associated organic carbon, total volatile solids, and occasionally other variables may be measured as indicators of the presence of sewage organic particles and associated toxic contaminants. Monitoring for toxic materials in or adsorbed onto sediments is costly and requires sophisticated technical facilities. Although there are great expectations, both their linkages with living organisms and predictions are weak.

Fish are mobile and their community structure is complex and dynamic. Natural variations, daily vertical movements through the water column, and seasonal migrations make monitoring for fish especially difficult, and a large number of samples are necessary. Selecting the appropriate equipment for sampling is difficult, since each gear type is designed to operate at a specific depth in the water column. Furthermore, sick fish are likely to be over-represented in catches since healthy ones ordinarily avoid many types of gear. Depending upon the depth and species to be sampled, samples can be taken by otter trawls, gill and trammel nets, various seines, traps, submersibles, divers, and hook-and-line methods. Histopathological examinations of fish that exhibit tissue abnormalities can provide supplemental information on contaminant burdens, population structure and health. Interpretation of fish monitoring results is difficult and requires consideration of natural variability, sometimes far removed from the outfall area, quantities and age cohorts of fish landed, and the level of effort per unit of catch

Phytoplankton populations are transient, successional, and responsive to phosphorus and nitrogen in wastewaters, and to carbon in receiving waters. Programs to monitor increases in plankton production need to consider the time needed for mineralization of organic phosphorus and nitrogen in wastewaters are they move and disperse into larger marine ecosystems (77,80). Mineralization rates are highest in sewage oxidation ponds and shallow coastal lagoons, beyond which they decrease as depth increases. Rates in Santa Monica Bay for surface sediments near the 10-meter deep Hyperion outfall were found to exceed reported rates for natural ocean sediments by one to two orders of magnitude (55,64). Where circulation is restricted, phytoplankton populations can be predictors of anoxia in bottom waters. Field studies have revealed no adverse effect on zooplankton resulting from sewage discharges. For research purposes, chlorophyll may be measured as an indirect indicator of the standing stock of phytoplankton.

Bivalves characteristically concentrate marine pollutants, and are frequently used as "sentinel organisms" to warn of potentially hazardous levels of contaminants. The bivalves are collected from local habitats or placed in cages in the discharge area and at uncontaminated control stations. They are analyzed for contaminant levels and effects and for comparison with data from regional or international activities as Mussel Watch (27) and the Mediterranean Action Plan (39).

10.6 The Infaunal Trophic Index

Descriptive models of the pollution tolerances or opportunistic feeding by marine organisms include species lists, population dynamics, and spatial distribution. A number of pollution indices have been proposed with increasing statistical elegance in multivariate analyses such as ordination scores that are increasingly difficult to explain to policy makers. These are plotted or ranked by distance from mass loadings of suspended solids from a discharge. In the United States, these are generally prepared from data obtained under the long-term compliance monitoring programs mandated under the National Pollutant Elimination Discharge System that, as noted earlier promote financial equity rather than environmental efficiency in wastewater management costs.. For planning and operational purposes, the most useful portion of these data is that obtained during periods of change and recovery of damaged ecosystems.

To meet the need for an operational diagnostic and predictive model, the Infaunal Trophic Index (ITI) was defined, applied, and disseminated in 1978 for the Southern California Coastal Water Research Project by Word (87), Word and Mearns (88), and Bascom, et al (45), respectively. It is a powerful diagnostic and potentially predictive tool (81,83) although it is perfunctorily criticized (74) by some marine biologists because its original publication was in an annual report rather than a peer-reviewed journal   Used in conjunction with quantitative physical and chemical data, it is a tool for predicting effects of other wastewaters (42) in selected areas. The index is based on filter feeding strategies of benthic species in one of four groups:

Group I   - suspension feeding organisms,
Group II - organisms feeding on suspended material or detritus on the sediment surface,
Group III - organisms feeding on surface detritus, and
Group IV - organisms feeding on detritus below the surface.

Numbers of individuals in each group are put into a simple formula   for a value from 0 to 100 that indicates the response of the benthic community to pollution:

$$\text{ITI} = 100 - \{33 - 1/3\,(0n_1 + 1n_2 + 2n_3 + 3n_4)\,/\,(n_1 + n_2 + n_3 + n_4)\}. \quad ..(10.1)$$

where n = the number of individuals in the group and coefficients in the numerator are an arbitrary, evenly increasing scaling factor.

The ITI values can then be lumped into groups that reflect the type of organism that dominates. Group I organisms dominate in a community with an ITI value of 78-100, Group II's dominate ITIs of 58-77; Group III from 25-57; and Group IV from 0-24. Group I organisms predominate in least-affected areas where organic matter does not accumulate. Conversely, Group IV organisms dominate in areas that are most impacted by organic material in the sediments or that have a high concentration of hydrogen sulfide. The dominance of a particular indicator group indicates the degree to which the normal sediments and their associated biota have been altered. These groupings can then be plotted on a bathymetric chart to give a graphic representation of the probable extent of contamination (see Section 11.5).

Figures 10.1 was prepared from benthic monitoring data from the Southern California Bight. It demonstrates the diagnostic properties of the ITI, and its inherent utility in summarizing a large body of scientific and technological data for official and public communications, and for cross-sectoral planning.

Ordination scores from multivariate analysis of more recent Santa Monica Bay data (50) are included for comparison with the ITI values. The slopes of the curves, considering the variance due to locating isopleths on charts (see Section 11.4 for examples), are the same. The ordinal displacements of the two data sets reveal the subjective nature of identifying background conditions.

Figure 10.2 shows the time rates of change due to reductions of mass loadings of solids from the County Sanitation Districts of Orange County. It provides a measure of the time, in this case five or six years, required for recovery of a damaged ecosystem at a depth of about 60 meters. The numerical values relating the arcas of degradation to solids loadings are consistent with those on Figure 10.1, and although their slopes are somewhat greater (points 3a to 3b) the diagnostic and predictive capabilities of the ITI is demonstrated.

10.7   Recovery of Damaged Marine Ecosystems

Recoveries of damaged ecosystems reflect the amount of energy in the system. Recovery rates from large oil spills are highest for rocky headlands and pebble beaches, decreasing for sandy beaches, upper then lower estuaries, and tidal flats (57). Similar gradients are found when waste discharges are removed from shallow waters near the surf zone to, say, 100 m in coastal waters (84).

We argue that the most useful monitoring information is obtained during recovery of damaged because it can be changed, focused, and environmentally bounded as succession proceeds as shown by Figures 10.1 and 10.2 Mandated compliance monitoring does not have this flexibility.

10.8   Post-Audits

Post-audits include the descriptions, purposes, and effects of projects and programs, both intended and unintended. They explain why things are as they are and become an integral of planning for future works. Gilbert White (86) has emphasized that successful ones require first institutional self-confidence and then full cooperation between the auditing and audited entities. They are highly

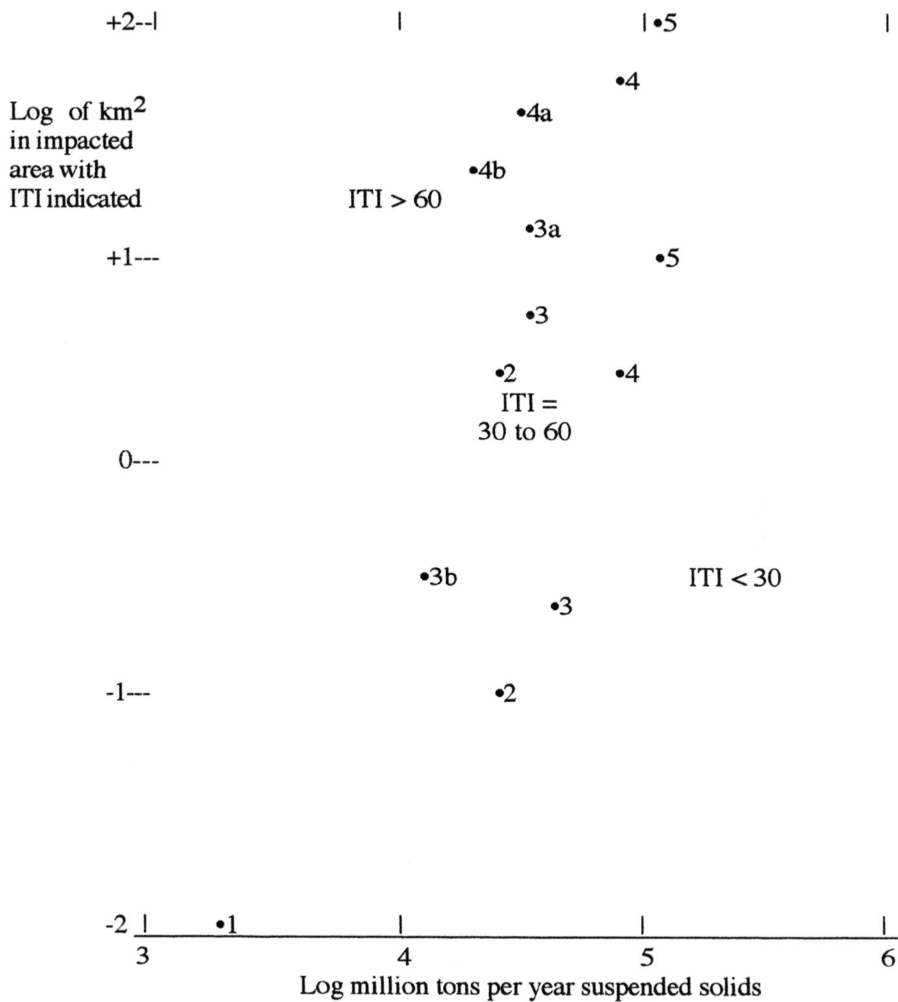

Figure 10.1. Infaunal trophic indices of effects of municipal wastewater discharges in southern California coastal waters. Legend: 1 - Oxnard, 2 - San Diego (Point Loma), 3 - County Sanitation Districts of Orange County (3a and 3b show changes over five years following oxtending outfall to to deep water), 4 - City of Los Angeles Hyperion discharge to Santa Monica Bay (4a and 4b indicazte multivarriate ordination scores in 1948 (4a) before ending sludge discharge and again in 1992 (50) 5 - Los Angeles County Sanitation Districts. Sources: Mearns and Word (65), Word and Mearns (85), Terry Fleming (personal communication,1995).

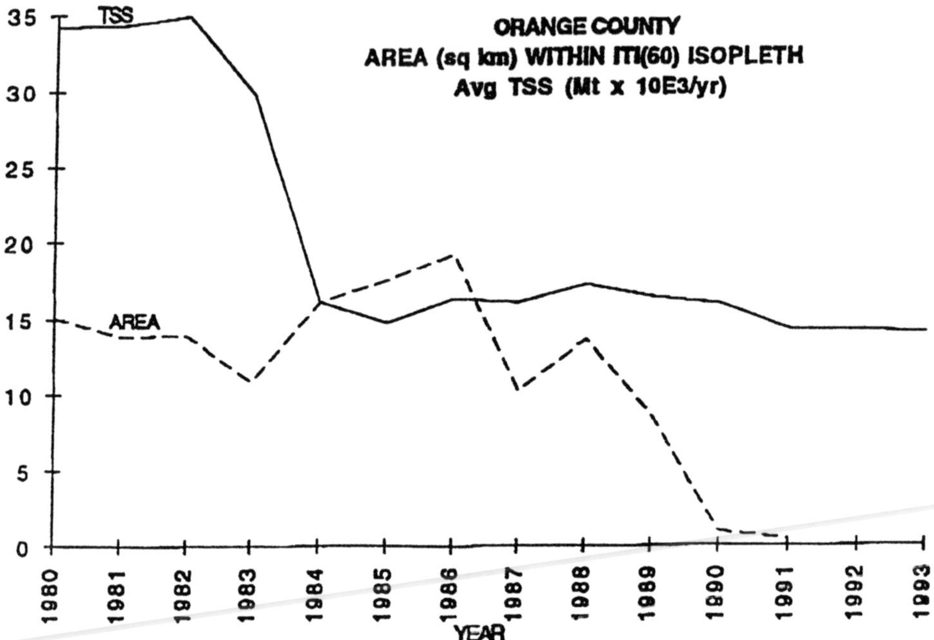

Figure 10.2. The infaunal trophic index as a measure of recovery of a damaged ecosystem .Source: personal communication from Terry Fleming, USEPA Region IX, citing County Sanitation Districts of Orange County data. 1995.

regarded by practicing engineers and policy makers who recognize professional values in seriously assessing their works. In contrast, many public and private practitioners see pose-augits as at best intrusive, and at worst as threatening professional and political pain. There are exceptions (60,73). Using inputs from both internal and external sources, post-audits are routinely and successfully fed back into current operations by the World Bank in setting lending priorities and by the U.S. Army Corps of Engineers in recommending spending priorities.

Post-audits of preventive, predictive, and remedial measures are particularly important to people who live in flood plains, seismic areas, or meteorological basins, and who by geophysical certainty will be subjected to future flooding, earthquakes, or automotive air pollution. They are by definition broader in scope and scale than the original projects. In all cases, cost and demand projections can lead to unexpected consequences partly because they tend to be optimistic and partly because benefits are more difficult to estimate and measure. Special cases include sequential programs and projects where long-range program goals are clear but how short-term project objectives approach those goals are not.

10.9   Regulatory and Zero-Discharge Models

There are economic, environmental, and engineering needs for capacity building in developing countries and market incentives in industrial countries to provide them. In the water and wastes sector, regulations tend to precede their financial feasibility. The disappointing results of transferring inappropriate technologies, regulations, and institutions to developing countries are well documented (46, 60, 63). Repeated references to the interactions between technological, environmental, and regulatory issues are found throughout this book. The cultural concept of zero pollutant discharge in current U.S. litigation and legislation is an example.

From zero risk to zero discharge.   The historical, cultural, and economic roots of U.S. zero discharge doctrines derive from the Protestant Reformation, the colonial period with iits labor shortage and technological innovation, and the financial rewards of conquering one third of a continent. These principles are embedded in Calvinist predestination, manifest destiny, trust in law, mistrust of a government that administers that law, human rights, protection of the individual at the expense of the community, and a preoccupation with competitiveness over consensus. They are not likely to be changed or successfully transferred to other nations, particularly less affluent ones. (53,59,85).

A common wish for zero risk in life became law during the post-WWII period of economic growth with the 1958 Delaney Amendment to the Food, Drug, and Cosmetic Act that mandated "...no ill effects from additives to foods and cosmetics." This was repeated when the 1972 Federal Water Pollution Control Act, Public Law 92-500 introduced the doctrine of zero discharge. Although there were some concerns over the thermodynamic implications of the zero-discharge objective, advocates from scientific, engineering, manufacturing, public interest, and governmental communities successfully competed in the market for funds to approach this receding goal. Initial achievements such as eliminating fires on the Cuyahoga River in Cleveland and reducing stream pollution in the Ohio River Basin using off-the-shelf technologies provided early momentum.

The zero-discharge goal is explicit in the title of USEPA's National Pollutant Discharge Elimination System. The intent of the Act was to ensure financial and legal equity even at the cost of overall environmental impact efficiency. Congress intended that a State governor could tell a threatening industrialist, "Go ahead, move your polluting industry to another State. The Feds are waiting for you there too." (76). This intent of the law to promote uniform costs of doing business has since been confirmed judicially (75).

Zero discharge characteristics. The characteristics of a system requiring 100% removals (or, for that matter, 95 or 90%) are shown on Figure 10.3. Imagine a global community of dischargers represented by triangular buckets. The first-order municipal, agricultural, or industrial discharger is at the bottom of the scale. Postulate that, given sufficient material and energy resources, a zero discharge can be achieved. The sources of the energy and materials required to do this are second-order and third-order dischargers that are operationally identical to the first. And so on. Conservation of mass means that any system for uniformly applied, arbitrary removals or discharge concentrations would do the same. The

220

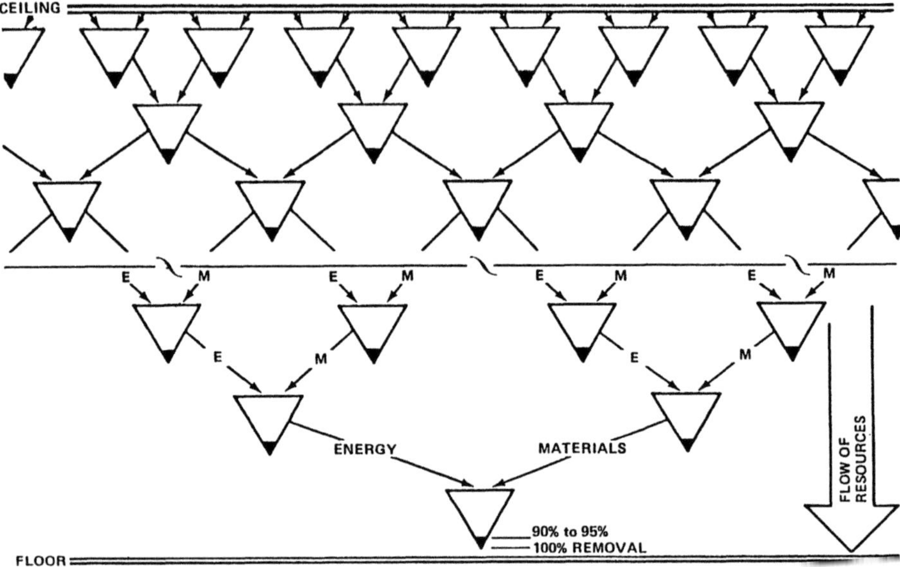

Figure 10.3. The (almost) zero pollutant discharge model with unidirectional resource flows that make it the technological analog of a pyramid scheme.

unidirectional flow of exogenous resources creates the technological equivalent of a financial pyramid scheme. Short-circuiting is inevitable at the lower levels of the scheme. Chain letters and progressive leveraged buyouts both fail early-on because all have similar circles of acquaintances, suppliers, and customers, and because some of the participants cheat. Zero discharge remains an elusive goal. The buckets are leaky, pyramid schemes collapse, and steady-state zero discharge is impossible. A similar argument can be made for zero risk.

Operations under zero-discharge goals. Under the Clean Water Act, receiving water quality standards were first applied. These were replaced by technology-based standards mandating secondary treatment. Next came provisions for waivers The zero-discharge goal remains although, as in most industrial countries, national economies have slowed and marginal benefits are more expensive. These trends have led to redefining zero-discharge in the contemporary term, "virtual elimination " (61). Currently, the trend is towards risk-based approaches based on cost-benefit analysis (Interview, EPA Assistant Administrator for R&D reported in Environmental Science and Technology. March 1995).

10.10   Appendix.   Power Spectrum Analysis.

Analysis of sequential data is greatly facilitated by the use of power spectrum analysis proposed by J.W. Tukey in 1949 ("The sampling theory of power spectrum analysis," Symposium on Application of Autocorrelation Analysis to Physical Problems, ONR, U.S. Navy, Washington, D.C., 47-67).   Originally developed for analyzing noise in communications systems, spectral analysis has since been applied in meteorology, oceanography, and engineering. The benefit of this application, paraphrasing W.H. Munk ("Long Ocean Waves," in The Sea. Interscience, 1962, 647-663), has been the condensation of miles of wiggly curves in the time domain to a few simple traces in the frequency domain.

The number of lags) is basic.. Lags are defined explicitly. Their importance lies in the fact that total computation effort time is a linear function of the record length (number of data points) multiplied by the number of lags. Larger numbers of lags improve the resolution of the calculation. This means that the frequencies (or periods) of cyclic events in the record can be more precisely defined and separated. For example, it becomes possible to measure and compare the effects of the semidiurnal tide with the solar day on dissolved oxygen in estuaries.

The dependence of computation effort on the record length and the number of lags needed to unscramble the record demonstrates that when more data are collected, more work is needed to study them. In this respect, power spectrum analysis is similar to simple averaging. It takes more effort to determine the average of many numbers than for a few numbers.

A time series is a record of repeated observations made at a particular location. Each observation is a momentary summation of the effects of everything happening to the particular parameter. Those effects may be caused by monotonic, cyclic or random phenomena. A trend throughout a given record length may be real or apparent; for example, a short segment of a sine wave will appear as a trend.

Power spectrum analysis identifies the frequencies at which different factors cause the record to vary. The analysis also provides estimates of the variance that derive from each of these factors. Figure 10.4 shows some simple spectra, or

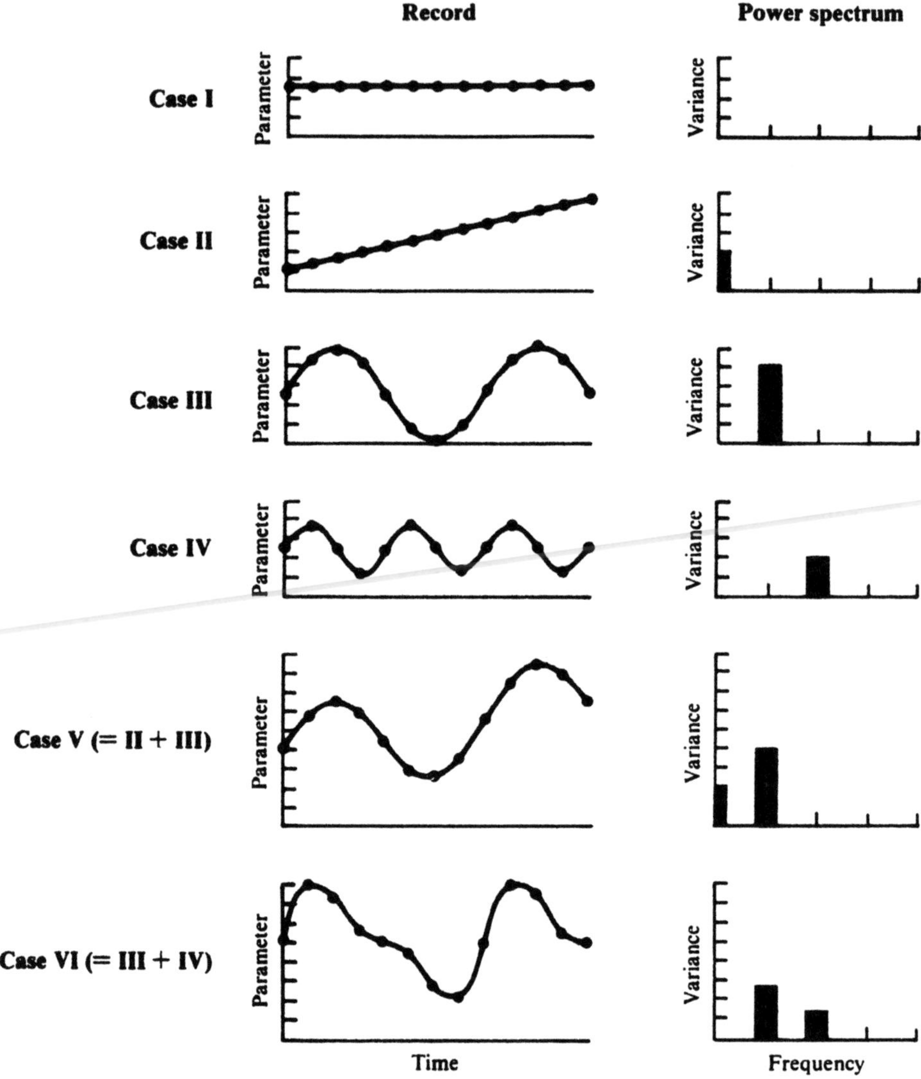

Figure 10.4. Spectral arithmetic. Source: Gunnerson (13).

more precisely, estimates of spectral density) from several types of curves. Where the original curve is made up of more than one cinusoidal component, as in Cases V and VI, the power spectrum clearly reveals their nature. Four steps in the computation of individual power spectra are as follows:

1. The mean and the square of the mean of the record are determined.
2. The autocorrelation function of the record is formed. This operation is basic to spectral analysis and is described in the worked example below.
3. The Fourier cosine transform for each autocorrelation is computed. This defines the difference between spectral analysis and standard Fourier analysis because in the latter, the Fourier transform is applied to the raw record rather than to the autocorrelation function. In spectral analysis, the Fourier transform smoothes out some of the fluctuations present in the autocorrelation function.
4. A second weighting operation provides the estimate of spectral density.

The Autocorrelation Function. The autocorrelation function is obtained by first multiplying each number in the record by another number in the record. From the mean of the sum of these products is subtracted the square of the arithmetic mean of the entire series. The autocorrelation at lag 0 is the record multiplied by itself or variance. The autocorrelations at lag 0, lag 1, lag 2, and lag 3 are computed as shown in Table 10.1.. The entire operation is expressed

$$C_r = \frac{1}{n-r} \sum_1^{n-r} x_t x_{t+r} - \left(\frac{1}{n} \sum_1^n x_t \right)^2 \tag{10.2}$$

where C_r = autocorrelation at lag r ; x_t = record value at t; t = 0,1,2,..n; n = sequential index of values; r = 0,1,2,..m; and m = lag number .

For a pure sine wave, the computation is analogous to looking at a white picket fence through one of a series of vertical gratings. Both the fence and the gratings are constructed so that the bar width equals the slot width. The pickets are a square approximation of the sine wave. The resolution with which the fence may be seen is a function of the spacing between grates. Similarly, the effective resolution (R) of the sine wave is a function of the number of lags (m) used for computing the autocorrelation function and of the sampling interval (T), where R = 1/2mT.

If the spacing of the grates is too large, some of the pickets will not be seen. The total amount of light seen by the viewer will be less than the amount reflected by the fence. When the grating with the optimal spacing is selected, the fence can be described precisely. This also happens when the optimal number of lags is applied to the sine wave record. As the grating spacing is decreased, all of the pickets are still seen, but some light is again lost to the viewer. Soon, only half the reflected light is available, and, eventually, the visual resolution of the fence is destroyed entirely because the grating dimensions approach those of light waves. Therefore, interference and diffraction patterns are set up. Something comparable, although not strictly analogous, happens when the autocorrelation is computed to too many lags for the particular record length. The computation becomes unstable. A common rule of thumb is that the number of lags should not exceed 10 percent of the number of data points. A smaller number of lags will often suffice.

Table 10.1. Sample computation of autocorrelation function for 145 tide observations taken at 4-hour intervals

---

C = Autocorrelation at Lag 0	C = Autocorrelation at Lag 1	C = Autocorrelation at Lag 2
(mean of 145 observed values = 2.43; square of mean = 5.90)		

$1.30 \times 1.30 = 1.69$	$1.30 \times 2.57 = 3.34$	$1.30 \times 3.79 = 4.93$
$2.57 \times 2.57 = 6.60$	$2.57 \times 3.79 = 9.74$	$2.57 \times 1.49 = 3.83$
$3.79 \times 3.79 = 14.36$	$3.79 \times 1.49 = 5.65$	$3.79 \times 2.30 = 8.72$
$1.49 \times 1.49 = 2.22$	$1.49 \times 2.30 = 3.43$	$1.49 \times 4.73 = 7.05$
$2.30 \times 2.30 = 5.29$	$2.30 \times 4.73 = 10.88$	$2.30 \times \ldots \quad \ldots$
$4.73 \times 4.73 = 22.37$	$4,73 \times \ldots \quad \ldots$	$\ldots \quad \ldots \quad \ldots$
$\ldots \quad \ldots \quad \ldots$	$\ldots \quad \ldots \quad \ldots$	$\ldots \quad \ldots \quad \ldots$
$\ldots \quad \ldots \quad \ldots$	$\ldots \quad \ldots \quad \ldots$	$\ldots \quad \ldots \quad \ldots$
$\ldots \quad \ldots \quad \ldots$	$\ldots \quad \ldots \quad \ldots$	$\ldots \quad \ldots \quad \ldots$
$3.10 \times 3.10 - 9.61$ 9.80	$3.10 \times 1.46 = 4.53$	$3.10 \times 3.16 =$
$1.46 \times 1.46 = 2.13$	$1.46 \times 3.16 = 4.61$	$1.46 \times 3.30 = 4.82$
$3.16 \times 3.16 = 9.99$	$3.16 \times 3.30 = 10.43$	$3.16 \times 1.41 = 4.46$
$3.30 \times 3.30 = 10.89$	$3.30 \times 1.41 = 4.65$	$3.30 \times 2.35 = 7.76$
$1.41 \times 1.41 = 1.99$	$1.41 \times 2.35 = 3.31$	$1.41 \quad ;;; \quad \ldots$
$\underline{2.34 \times 2.34 = 5.82}$	$\underline{2.35 \quad \ldots \quad \ldots}$	$\underline{2.35 \quad \ldots \quad \ldots}$
Sum = 1046.9	Sum = 884.67	Sum = 764.62
$\dfrac{1046.9}{145} = 7.22$	$\dfrac{804.67}{144} = 5.588$	$\dfrac{764.62}{143} = 5.347$
$\underline{- 5.90}$	$\underline{- 5.90}$	$\underline{- 5.90}$
$C_0 = 1.32$	$C_1 = -0.312$	$C_2 = -0.553$

---

where U_0 . U_1 , and Um are the power estimates according to respective lags. Each value for U represents a part of the total variance that occurs with a given period (T) corresponding to lag r is:

$$T = \frac{2m \, \Delta r}{r} \tag{10.7}$$

Procedures for calculating the degrees of freedom and confidence levels developed by Blackman R.B., and Tukey, J.W.. in The Measurement of Power Spectra. 1958. Dover, New York, are found in standards texts.

Figure 10.5 shows autocorrelation functions of two pure sine waves and of their sum. Although the example is simpler than those found in nature, it shows how the number of lags (in this case, eight) relates to analysis of the record. It can be seen intuitively that variance due to a secular trend in the record will be measured at lag 0. This is because a secular variation has an infinite period (zero frequency). This variance will be added to that due to the harmonic component.

Figure 10.5 shows autocorrelation functions of two pure sine waves and of their sum. Although the example is simpler than those found in nature, it shows how the number of lags (in this case, eight) relates to analysis of the record. It can be seen intuitively that variance due to a secular trend in the record will be measured at lag 0. This is because a secular variation has an infinite period (zero frequency). This variance will be added to that due to the harmonic component.

Fourier Transform and the Smoothed Spectrum. The Fourier cosine transform is next calculated for each autocorrelation

$$V_t = \frac{k}{m}\left[C_o + C_m \cos r\pi + 2\sum_{q=1}^{m-1} C_q \cos\frac{qr\pi}{m} \right] \tag{10.3}$$

in which V_r = Fourier cosine transform of the autocorrelation at lag r, q = lag number having values between 1 and m-1; k = 1 for r =1,2,...m-1, and k = 1/2 for r = O; and r = m; and the other letters have the definitions previously given.

Computing the smoothed estimate of spectral density. A final weighting function is another weighting operation to moderate the effect of a small sample. This is expressed mathematically as

$$U_0 = 0.54\ (V_0 + V_1 0) \tag{10.4}$$
$$U_r = 0.23 V_{r-1} = 0.54 V_r = 0.23 V_{r+1}\ ,\ \text{for } r = 1,2,3,...m-1,\ \text{and} \tag{10.5}$$
$$U_m = 0.54 V_{m-1} = 0.54\ V_m \tag{10.6}$$

Aliasing. The designs of both the sampling interval and the subsequent statistical or spectral analysis require consideration of aliasing. Aliasing, defined graphically in Figure 10.6, results from the high-frequency events that add variance to the record but that are not "seen" by the particular sampling interval. Figure 10.6 illustrates how the variance from this type of event is folded into the record and reappears at a lower frequency. Where the period of the high-frequency event is known, the period of the aliased record can readily be determined analytically. Figure 10.6 reveals that any cyclic event that occurs at a period less than twice the sampling interval will result in aliasing. Where the period equals twice the sampling interval, the event will never be seen. The same is true for any event whose period (P) is related to the sampling interval (I) by P = 2T/n, in which n is a positive integer. In the design of sampling programs, those cases in which n > 1 are not normally considered. Only when the period exceeds twice the sampling internal can the event be measured. As the period of the cyclic event approaches a value of twice the sampling interval, the record length necessary to describe the event increases. The corresponding frequency, f_N which limits the events seen by the sampling frequency, f_s, is the Nyquist frequency, $f_N = 1/2\ f_s$. For an infinite series, this is also the time-constant for remedial measures.

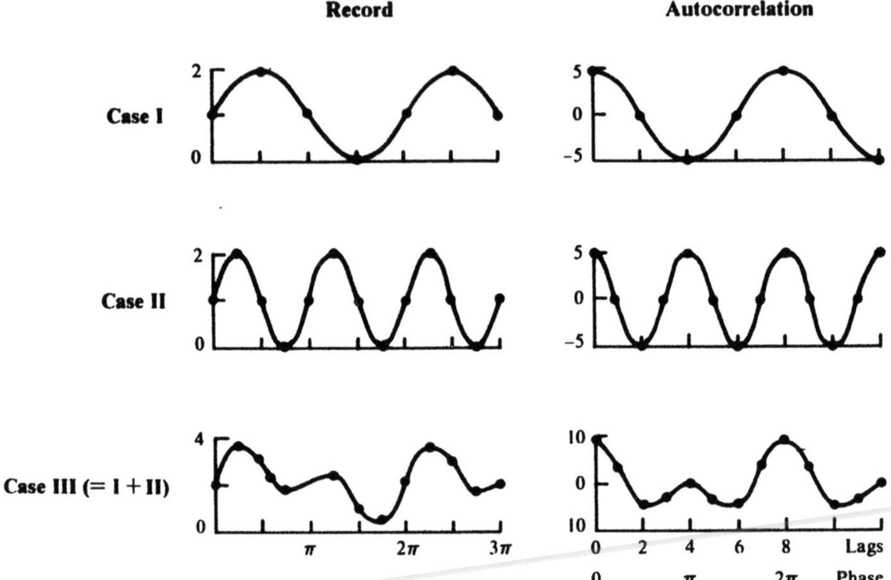

Figure 10.5. Autocorrelation functions of a simple sine waves

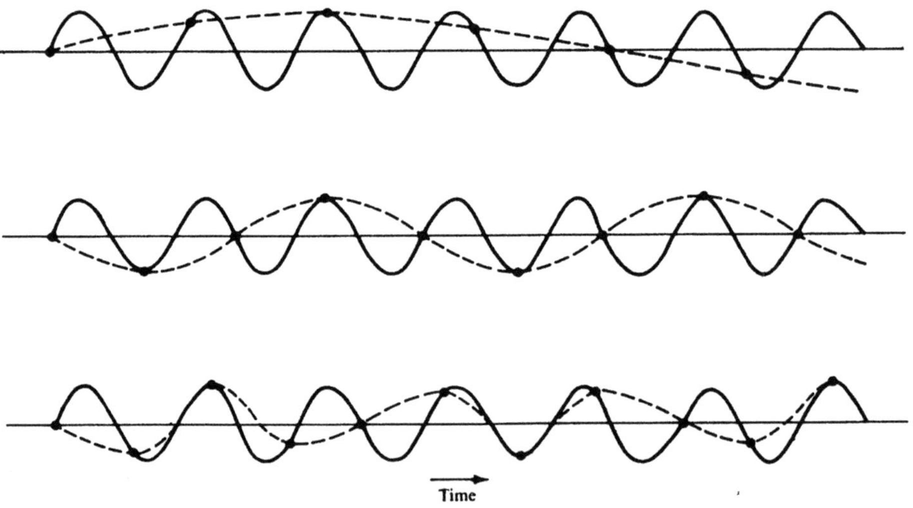

Figure 10.6. Aliasing due to sampling harmonic motion at more than one half its period. Solid line is true record. Dashed line is aliased record. ● is sampling time. Source: Gunnerson: (13).

Cross Spectra. Cross spectra are computed from two simultaneous time series to determine relationships between the two records. The analysis provides a power spectrum for each series, and the cospectrum and quadrature spectrum from which the cross spectra are derived. For each frequency band, the four spectra yield the coherence (analogous to the square of the correlation coefficient), the phase lag (angular time in radians between maxima and minima of the two records) and response function (amount of variation in one record associated with a similar variation in the other. The arithmetic is essentially the same as for individual power spectra plus calculating .their sums and differences. The frequency distributions should be statistically similar so that both are Gaussian, or both skewed in the same direction, or both of similar peakedness (kurtosis) for the response function to be predictive. Details of the computations may be found in any standard work on time-series analysis.

10.11   References

1. American Public Administration. 1980. Standard Methods for the Examination of Water and Wastewater, ed. Washington, D.C.
2. Bascom, W. 1979. Life in the Bottom, in Southern California Coastal Water Research Project Annual Report . Long Beach, California, 57-83.
3. Bascom, W. 1981. Effects on the ecosystem of sewage sludge disposal from a pipeline. in Water Science Technology, Pergamon Press, London, 48.
4. Becker, M., and Cowden, J.W. 1981. Report of Great Lakes regional workshop on ocean pollution monitoring, February 11-13, 1981. U.S. National Oceanic and Atmospheric Administration, Boulder, Colo., 1981.
5. Cabelli, V.J. 1980. Health Effects Criteria for Marine Recreational Waters. Pub.EPA-600/1-80-031. U.S. Environmental Protection Agency, Cincinnati.
6. Cabelli, V.J. 1981. A health effects data base for the derivation of microbial guidelines for Municipal Sewage Effluents, in Coastal Discharges, Thomas Telford, Ltd., London, 51-54.
7. Defant, A. 1961. Physical Oceanography, 2 vols. Pergamon, Oxford.
8. Department of Scientific and Industrial Research. 1964. Effects of Pollution Discharges on the Thames Estuary. Water Poll. Res. Tech. Pap. No. 11, HMSO, London.
9. Ellis, D.V. 1981. Environmental consequences of breaks and interrupted construction at marine outfalls in British Columbia, in Coastal Discharges, Thomas Telford, Ltd., London. 187-190.
10. Flaxman, E.W. 1981. Synopsis of UK experience of modern outfall maintenance, in Coastal Discharges, Thomas Telford, Ltd., London, 181-186.
11. Garber, W.F. 1983. Personal communication, Bureau of Sanitation, City of Los Angeles, California.
12. Gunnerson, C.G., and McCullough, C.A. 1965. Limitations of Rhodamine B and Pontacyl Brilliant Pink B as tracers in estuarine waters. Symposium of Diffusion in Ocean and Fresh Waters, Lamont-Doherty Geological Observatory, Palisades, NY., 53.

228

13. Gunnerson, C.G. 1975. Utilization of data from continuous monitoring networks. in Water Quality Parameters, ASTM Pub. 573, American Society for Testing and Materials, Philadelphia, 456-486.

14. Gunnerson, C.G. 1975. Discharge of sewage from sea outfalls, in A.L.H. Gameson, ed., Discharge of Sewage from Sea Outfalls, Pergamon, Oxford, 415-425.

15. Gunnerson, C.G. 1981. Report of Northeast Regional Workshop on ocean pollution monitoring, Sept. 10-12, 1980. U.S. National Oceanic and Atmospheric Administration, Boulder, CO.

16. Gunnerson, C.G. 1981. The New York Bight Ecosystem, Ch. 14 , R. A. Geyer, ed., Marine Environmental Pollution, Elsevier, Amsterdam.

17. Hakim, K.E. 1978. Study of microbial indicators of health effects at Alexandria bathing beaches. Report to Health Effects Research Laboratory, U.S, Environmental Protection Agency, Cincinnati.

18. Hollings, C.S., ed. 1978. Adaptive Environmental Assessment and Management. IIASA Series on Applied Systems Analysis No. 3, Wiley, New York.

19. Hooper, N.J. 1981. Report of Western Gulf Regional Workshop on ocean pollution monitoring, December 16-17, 1980. U.S. National Oceanic and Atmospheric Administration, Boulder, Colo.

20. Hooper, N.J. 1981. Report of Southeastern Regional Workshop on ocean pollution monitoring, January 27-28, 1981.U.S. National Oceanic and Atmospheric Administration, Boulder, Colo.

21. Hume, N.B., Bargman, R.D,, Gunnerson, C.G., and Imel, C.E. 1961. Operation of a 7-mile digested sludge outfall, Transactions, American Society of Civil Engineers, Vol. 126, 306-331.

22. Iijima, N., ed. 1979. Pollution Japan: a Historical Chronology. Asahi Evening News, Pergamon Press, Oxford.

23 Kamlet, K.S. 1981. Letter from National Wildlife Federation to National Oceanic and Atmospheric Administration, Appendix II in Gunnerson, ref. 15.

24. Kinsman, B. 1965. Wind Waves. Prentice-Hall, Englewood Cliffs, N.J.

25. Ktsanes, V.K., Anderson, A.C., and Diem, J.E. 1979. Health effects of swimming in Lake Pontchartrain at New Orleans. U.S. Environmental Protection Agency, Cincinnati.

26. Moore, B. 1975. The case against microbial standards for bathing waters, in A.L.H. Gameson, ed., Discharge of Sewage From Sea Outfalls, Pergamon, Oxford, 103-109.

27. National Research Council. 1980. The International Mussel Watch. National Academy of Sciences, Washington, D.C.

28. Oakley, H.R. 1981. Site investigation and selection, in Coastal Discharges, Thomas Telford, Ltd., London, 67-73, 97-101.

29. Officer, C.B., and Ryther, J.H. 1981. Swordfish and mercury: a case history."Oceanus, Vol. 24, No. 1, 34-41.

30. Pearson, T.H., and Rosenberg, R. 1976. Macrobenthic succession in relation to organic enrichment and pollution of the marine environment. Oceanography and Marine Biology Annual Review 16, 229-311.

31. Public Health Laboratory Service. 1959. Sewage Contamination of Bathing Beaches in England and Wales, Memo No. 37, HMS0, London.

32. Rittenberg, S.C., Mitwer,, T, and Ivler, D.. 1958. Coliform bacteria in sediments around three marine sewage outfalls. Limnology and Oceanography, Vol. 3, 101-108.
33. Segar, D.A. 1981. An Assessment of Great Lakes and Ocean Pollution Monitoring in the United States. U.S. Dept. of Commerce, National Oceanic and Atmospheric Administration, Boulder, Colo.
34. Serra, R.E., ed. 1981. Report of Southwest Regional Workshop on Ocean Pollution Monitoring, November 18-20, 1980. U.S. National Oceanic and Atmospheric Administration, Boulder, Colo.
35. Serra, R.E. ed. 1981. Report of Northwest Regional Workshop on Ocean Pollution Monitoring, January 6 8, 1981. U.S. National Oceanic and Atmospheric Administration, Boulder, Colo.
36. Shuval, H.I., Tilden, R.L., Perry, B.H., and Grosse, R.N. 1981. Effects of investments in water supply and sanitation on health status, a saturation theory. Bull. World Health Organization, Vol. 59, No. 2. 243-248.
37. Swanson, R.L., Stanford, H.M., O'Connor, J.S., Chanesman, S., Parker, C.A., Eisen, P.A., and Mayer, G.F. 1978. June 1976 Pollution of Long Island ocean beaches. Jour. Environmental Engineering Division, Proc. American Society of Civil Engineers, 104, EE6, 1067-1083.
38. Tetra Tech, Inc. 1982. Design of 301(h) monitoring programs for monitoring wastewater discharges to marine waters. Pub. 430/9-82/0101, U.S. Environmental Protection Agency, Washington, D.C.
39. United Nations Environmental Programme. 1978. Mediterranean Action Plan and the Final Act of the Coastal States of the Mediterranean Region for the Protection of the Mediterranean Sea. Nairobi.
40. Vink, J.K. 1981. Experience with long outfalls -- the Hague, in Coastal Discharges, Thomas Telford, Ltd., London, 191-192.
41. Willis, D.A. 1981. Site investigation and selection--engineering aspects, in Coastal Discharges, Thomas Telford, Ltd., London, 75-80.
42. Word, J.Q. 1978. Infaunal trophic index. in Southern California Coastal Water Research Proiect Annual Report 1978, Long Beach, California, 19-40.
43. Word, J.Q., Striplin, P.L., and Tsukada, D. 1981. Effects of screen size and replication on the infaunal trophic index. In Southern California Coastal Water Research Project Annual Report 1979-1980, Long Beach, California, 123-130.
44. World Health Organization and United Nations Environmental Programme. 1979. Principles and Guidelines for the Discharge of Wastes into the Marine Environment, Pergamon, Oxford.

45. Bascom, W. 1979. Life in the bottom. Annual Report, 1979, Southern California Coastal Water Research Project, Long Beach, CA. 57-83.
46 Baum, W.C/, Tolbert, S.M. 1985. Investing in Development: Lessons of World Bank Experience, World Bank, Washington.
47. Belton, T.J., Ruppel, B.E., Lockwood, K., and Boriek, M. 1983. PCB's in selected finfish caught within New Jersey waters, 1981-82. New Jersey Department of Environmental Protection, Trenton.

48. Carpenter, S.R., et all. 1995. Ecosystem experiments. Science, v. 269, 15 July 1995,324-327.

49. Chen, T.T. 1995. Industrial Pollution Prevention. Springer Verlag, Heidelberg.

50. Dorsey, J.H., Phillips, C.A., Dalkey, A., Roney, J.D., and Deets, G.B. 1995. Changes in assemblages of infaunal organisms around wastewater outfalls in Santa Monica Bay, California. Bull. Southern California Acad. Sci. 94 (1) 46-64.

51. Environment Department. 1991. Environmental Assessment Sourcebook. Volume 1. Policies, Procedures, and Cross-Sectoral Issues. Technical Paper 139. World Bank, Washington.

52. Environment Department. 1991. Environmental Assesssment Sourcebook. Vol. 2. Sectoral Guidelines. Technical Paper 140. World Bank, Washington.

53. Faure, G.O., and Rubin, J.A. 1993, Culture and Negotiation. Sage, Newbury Park, CA.

54. Gillespie, C.C. 1942. Report on a Pollution Survey of Santa Monica Bay Beaches in 1942. Bureau of Sanitary Engineering, California State Board of Public Health, Sacramento.

55. Gunnerson, C.G. 1963. Mineralization of organic matter in Santa Monica Bay, California. In Oppenheimer, C.H., Editor. Symposium on Marine Microbiology. C.C. Thomas, Springfield, IL, 641-653.

56. Gunnerson, C.G., editor. 1989. Post-Audits of Environmental Programs and Projects. Amer. Soc. of Civil Engrs. 106p. New York. 1989.

57. Hazardous Materials Response and Assessment Division. 1992. Oil Spill Case Histories, 1967-1991. National Oceanic and Atmospheric Adm, Seattle.

58. Holling, C.S. 1995. Sustainability: the cross-scale dimension. Munasinghe M, and Shearer, W., editors. Defining and Measuring Sustainability: the Biogeophysical Foundations. World Bank, Washington. pp 65-75.

59. Horsman, R. 1981. Race and Manifest Destiny. Harvard, Cambridge.

60. Interagency Floodplain Management Review Committee, Sharing the Challenge: Floodplain Management into the 21st Century. U.S. Army Corps of Engineers, Washington. 1994.

61. IJC 1989. Great Lakes Water Quality Agreement of 1978, as amended, and 1992. Sixth Biennial Report on Great Lakes Water Quality. United States-Canada International Joint Commission Washington and Ottawa.

62. Jacobsen, J.L. and S.W., Schwartz, P.M., Flin, G.G., and Dowler, J.K. 1984. Prenatal Exposure to an Environmental Toxin: a Test of the Multiple Effects Model." Developmental Psychology. 20. 4, pp. 523-532.

63. Kalbermatten, J.M, Julius, D.S., and Gunnerson, C.G. 1982. Appropriate Sanitation Alternatives. A Technical and Economic Appraisal. Johns Hopkins, Baltimore.

64. Lalli, C.M., and Parsons, T.R.. 1994. Biological Oceanography: an Introduction. Pergamon/Elsevier, New York.

65. Mearns, A..S. and Word, J.Q. 1982. Forecasting effects of sewage solids on marine benthic communities. In Mayer, G.F., editor. Ecological Stress in the New York Bight, Estuarine Research Foundation, Columbia, SC. 495-512.

66 Moore, B. 1975. The case against microbial standards in bathing waters. In Gameson, A.L.H., editor. Discharge of Sewage from Sea Outfalls, Pergamon, Oxford.

67. National Research Council 1990.. Monitoring Southern California's Coastal Waters. National Academy Press, Washington.

68. National Research Council 1993. Managing Wastewater in Coastal Urban A Areas. National Academy Press, Washington.

69. National Research Council. 1995. Understanding Marine Biodiversity: a Research Agenda for   the Nation. National Academy Press, Washington.

70. Open University . 1989. The Ocean Basins: their Structure and Evolution. Pergamon Oceanography Series, Volume 1. Oxford, Pergamon Press.

71. Open University. 1989. Seawater: its Composition, Properties, and Behavior. Pergamon Oceanography Series, Volume 2 Oxford, Pergamon

72. Open University. 1989a. Ocean Circulation. Pergamon Oceanography Series, Volume 3. Oxford, Pergamon Press.

73. Operations Evaluation Department. 1992 Water Supply and Sanitation Projects: the Bank's Experience, 1967-1989. Processed. World Bank. Washington.

74. Peterson, C.H. 1993. Improvement of environmental impact analysis by application of principles derived from manipulative ecology: lessons from coastal marine case studies. Australian Journal. of Ecology 18 (1) 21-52.

75. Pregerson, Harry. (1987). U.S. District Court in Los Angeles, Amended Consent Decree, Docket No. CV 77-3047-HP decision on Hyperion sludge discharges to Santa Monica Bay asserting that the legislation placed financial equity superior to environmental efficiency. Affirmed by the 9th U.S. Circuit Court of Appeals and the United States Supreme Court

76. Quigley, James. 1974. Personal communication. from his position as former Congressman sponsoring 1972 Water Pollution Control Act, later the first administrator of the Federal Water Pollution Control Administration now the U.S. Environmental Protection Agency.

77. Rapport, D.J., Gaudet, C.L., and Calow, P. 1995. Evaluating and Monitoring the Health of Large-Scale Ecosystems..Springer-Verlag. Berlin Heidelberg.

78. Sherman, K., Alexander, and Gold, B.D. 1995. Large Marine Ecosystems. AAAS Press. Washington.

70. Sherman, K. 1995. Large marine ecosystems and fisheries. In Munansinghe, M, and Shearer, W., editors. Defining and Measuring Sustainability, World Bank, Washington. 207-233

80. Smith, B.J. 1984. PCB Levels in human fluids. Sea Giant Institute, University of Wisconsin, Madison.

81. Smith, R.W., 1995. Numerical tools for assessing benthic monitoring data. Report to USEPA Region IX, San Francisco. Ecoanalysis, Inc., and Southern California coastal Water Research Project, Long Beach, CA.

82. Stommel, H. 1958. The Gulf Stream. Univ. of Calif. Press, Berkeley

83. Tetratech 1994. Amended Section 301(h) Technical Support Document. U.S. Environmental Protection Agency, Washington.

84. Thompson, B. 1991. Recovery of Santa Monica Bay from Sludge Discharge. Technical Report #349, Southern California Coastal Water Research Project, Long Beach, CA.

85. Weber, Max. 1904, 1905. The Protestant ethic and the spirit of capitalism. Archiv für Sozialwissenschaft und Soziopolitik. Reprinted 1948, Allen and Unwin, London, Tranm. 1970 J.E.T. Edlredge, Michael Johnson, London.

86 White, G.F. 1988, When may a post-audit teach lessons? In Rosen, H., and Reuss, M., Editors. The Flood Control Challenge: Past, Present, and Future. Public Works Historical Society, Am. Public Works Assn, Chicago.

87. Word, J.Q. 1978. The infaunal trophic index.Annual Report, 1978, Southern California Coastal Water Research Project, Long Beach, CA. 19-40.

88. Word, J.Q., and Mearns, A.S. 1978. The 60-meter control survey. Annual Report, 1978, Southern California Coastal Water Research Project, Long Beach, CA. 41-56.

89. World Resources Institute. 1990 World Resources, 1990-91 Washington.

90. Cosper, E.M., Brice, V.M., and Carpenter, E.J. 1989. Novel Phytoplankton Blooms. Coastal and Estuarine Studies 35, Springer-Verlag, Heidelberg.

91. World Bank (1995) Monitoring Environmental Progress. Environmentally Sustainable Development Publication. World Bank, Washington.

11   Case Studies

There is no shortage of solutions to sanitation problems. Ancient records include the admonition in Deuteronomy (21) to bury the stuff, Homer's eighth-century B.C. Herakles' diverting a river to flush wastes from the Augeian stables (19), and Socrates' fifth-century B.C. zoning on night-soil dumping (87). Flushing continues now to the sea. Case studies summarized below are about different responses to dissimilar ecological, hydrographic, historical, and wastewater characteristics. They include estuarine waters of the Yangtze and Thames rivers and of Boston Harbor, the open coastal waters of the Southern California Bight, and the Bosporus and Sea of Marimara portions of the Turkish Straits.

11.1   Scope of Case Studies

Shanghai is undertaking a very large scale Yangtze River wastewater manageement progream with cooperative leadership shared by the Municipality of Shanghai and the World Bank and bilateral assistance from official development agencies and consultants from Australia, Canada, Denmark, France, Norway, and the U.K. Section 11.2 presents the technological, institutional and financial integration of a major international development project. Conceptual designs, initial calibrations and extensive computations for state-of-the-art numerical modeling of estuarine hydrography, sedimentation, and water quality are briefly summarized.   .

The Thames estuary story is one of continuing ecysystem recovery in an area long damaged by wastewater dsscharges. The basic 2-dimensional mathematical model described in 1965 was intended only to assure enough dissolved oxygen for fish. The damaged ecosystem below London Bridge has been largely recovered.

Upper layer ucrrents in the Bosporus and Sea of Marmara are from the Black Sea. The lower layer is from the Mediterranean and Aegean Seas. Mixing is marginally affected by bydraulic jumps at the interface. Sewage discharged to the lower layer thus goes mostly into the lower layer of the Black Sea. Institutional factors in environmental, engineering, and economic decisions provide secondary benefits.

Boston Harbor's outfall program demonstrates how engineering ingenuity and state-of-the-art technology requires even more social than financial innovation as projects increase in size and elegance..

The oceanographic edge of southern California's is changed by importing more people and water into what has been called the "Cadillac Desert," Sec 11.6). where an infaunal feeding model can describe recovery of damaged marine ecosystems.

11.2 The Yangtze River Estuary: The Second Shanghai Sewerage Project (SSPII)[1]

11.2.1 Introduction

Shanghai is one of East Asia's most important economic centres and has a population of about 14 million. The highly-varying facilities for collecting and treating domestic and industrial waste water, solid waste and night soil are due to historic reasons. This has caused widespread pollution of water courses in the city environs, and the continuing discharge of untreated industrial and domestic waste water into the Huangpu River and its tributaries has created an ever-increasing pollution problem. This problem is exacerbated by a growing population, expanding industrial development, urbanisation, and improved living standards.

The present development policies pursued by the Shanghai Municipal Government (SMG) recognize that protection of the environment is a prerequisite for long-term economic growth and consider appropriate investments in environmental protection to be of the highest priority. As part of these policies the SMG is undertaking a phased investment program to improve waste water interception, conveyance, treatment, and disposal. The first phase (SSPI) dealt with the waste water flows (about 1,400,000 m^3/d, ADWF: Average Dry Weather Flow) along the Suzhou Creek, which are discharged into the Zhu yuan area of the Yangtze River Estuary. This phase, which was financed by SMG and supported by the World Bank, has cost about 350 million US$. The second phase (SSPII) will build upon the experience of the first phase and focus efforts on the waste water situation of the rapidly developing Pudong and Puxi areas of Shanghai. This phase will comprise a conveyor and treatment system discharging effluent into the downstream of the Yangtze River Estuary in the vicinity of Bailong Gang (Figure11.1). Details of the master plan are shown in Section 11.2.3.

The SSPII is one of the largest projects for sewage discharge in the world, concerning a capacity of 1,700,000 m^3/d (ADWF) for Stage 1 and 5,000,000 m^3/d (ADWF) of waste water for the Ultimate Stage. Bailong Gang is the proposed location of the outfall for the SSPII. In order to show that the environmental requirements of national standards for both discharge and receiving water are met, it is necessary to conduct studies on ourfall design and environmental Future studies are to be carried out outfall layout design and on the environmental impacts of the sewage discharge into the Yangtze River Estuary.

---

[1] This is a 1995 status report prepared by Qian-ming Lu and Jacob Steen Møller, Danish Hydraulic Institute, Jørgen Færch Knudsen, Søren D. Eskesen, COWIconsult, Denmark, and Zhou Yucheng, Shanghai Sewerage Project Construction Company with the guidance and encouragement of Geoffrey Read, World Bank, Washington.

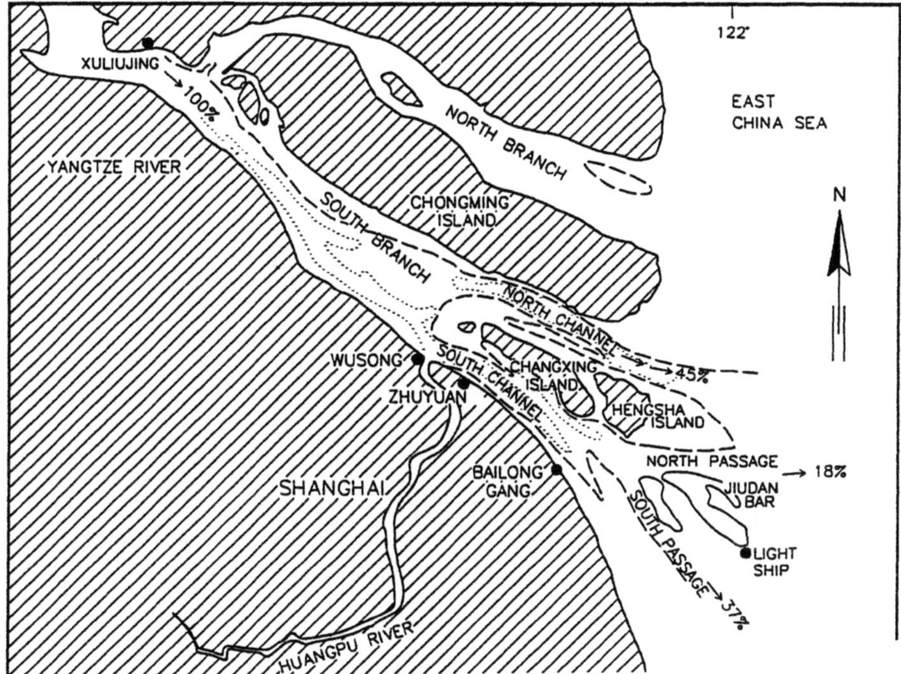

Figure. 11.1. Locations of outfall of Shanghai Sewerage Project, Phases 1 & 2.

11.2.2 Organization of Feasibility Studies

The project is carried out by the Shanghai Sewerage Project Construction Company (SSPCC).

As this is such a large project in the Yangtze River Estuary, a comprehensive feasibility study for SSPII is necessary and is organized by the SMG with support from Shanghai and several bilateral and multilateral agencies. The feasibility studies consists of master planning, outfall studies, treatment plants, and management.

Four consultant teams have participated in the studies and the details of the organizations are shown in Figure 11.2. A large group of companies cooperated very openly to carry out the comprehensive and arduous studies. During the project, meetings were held to review project reports and to exchange comments and opinions. A very helpful international technical workshop on the Yangtze River Estuary south bank outfall and associated environmental impact studies was also organized in Shanghai to enable a wide group of experts to further review the results from the feasibility studies. About 100 people from the Chinese NEPA, SMG, consultant companies, universities, World Bank and construction companies participated in the workshop. Based on the efficient organization work, all the feasibility studies were completed as scheduled and with valuable results.

11.2.3 Master Plan

For various reasons, the construction of wastewater facilities lags behind the development of the city, and the major rivers in the urban area are seriously polluted. The Suzhou Creek and other tributaries of the Huangpu River have turned black and smell all year round. The water quality of the Huangpu River is declining. Therefore, the Shanghai Municipal Government (SMG) has decided to improve the situation by carrying out wastewater management on a large scale.

In 1983, SMG, the World Bank and the Australian Development Assistance Bureau (AIDAB) agreed to carry out a joint study on wastewater management in Shanghai. With financial assistance from the Australian government, consultants from Binnie and Partners of Australia, together with the local experts, submitted a 'Liquid Waste Management Strategy Report' for the management of wastewater from the urban area after more than one year's joint study. The option proposed was to construct interception facilities for collecting the wastewater, delivering it to and discharging it into the Yangtze River so as to dilute and diffuse it after preliminary treatment. The project is to be carried out in phases. The first phase of the project is to improve the water quality of the Suzhou Creek.

The SSPCC was founded to construct the sewerage project. After five years' planning and preparation, SSPI commenced construction in 1988. SSPI's service area is 70.57 km^2 with a service population of 2.55 million. SSPI's task is to collect the wastewater from the 44 catchments along the Suzhou Creek and the main conveyor and to deliver and discharge it into the Yangtze River Estuary at Zhuyuan after preliminary treatment. The planned ADWF (planned until the year 2000) is 1.4 million m^3/d. SSPI's construction costs amount to 1.6 billion RMB. The project was supported by the World Bank. The main body of the project was completed to deliver the first flow at the end of 1993. Today, SSPI collects and discharges 1-1.2 million m^3 wastewater per day.

In 1994 SSPCC started the advanced work for SSPII. SSPCC has engaged the Shanghai Urban Construction Design Institute (SUCDI), Shanghai Municipal Engineering Design Institute (SMEDI) and Shanghai Tunnel Engineering Design Institute (STEDI) assisted by Interconsult (Norway) in collaboration with Mott MacDonald (UK) and SOGREAH (France) for the planning and design. A Strategic Planning Report for all Shanghai's wastewater has been prepared. The report was reviewed and accommodated as appropriate the findings of the previous Liquid Waste Management Strategy Report for urban area wastewater and the Shanghai Sewerage Professional Plan prepared by the SUCDI in 1994. SSPCC was again assisted by the World Bank in obtaining financial support from Norway, Denmark, France and Canada. Shanghai Municipality is divided into five areas (Figure11.3) with regard to strategic planning /10/:

    Area A1 includes the urban area, Baoshan and Jiading;

    Area A2 includes Quingpu County and Songjiang County;

    Area B1 includes part of Nanhui County, Fengxian County and Jinshan County;

    Are B2 includes the other part of Nanhui County, Fengxian County and Jinshan County, and

    Area C includes the Pudong New Area and Minhang Area.

It is assumed that the wastewater from Area A1 will be discharged into the Yangtze River Estuary, the wastewater from Areas A2 and B2 will be discharged into the Huangpu River, B1 will be discharged into the Hangzhou Bay and Area C will be discharged into either the Hangzhou Bay, the Huangpu River or the Yangtze River. To decide the optimum receiving water for area C, eight scenarios were provided to calculate the impact on the receiving water using models. The cost of each scenario was also roughly estimated (Table 11.1).

The comparisons below show that Scenarios 6,7 and 8 are acceptable for the receiving water. In terms of further reducing the total phosphorus load so as to reduce the nutrient level in the vast water body, only Scenarios 6 and 8 remain to be considered. However, Scenario 6 requires the construction of a large treatment works in the urban area, which will be difficult. Therefore, priority should be given to consideration of Scenario 8.

Based on some initial studies, the proposed strategy is, therefore, that the wastewater from Areas A2, B1 and B2 need secondary treatment and that wastewater from Area A1 should be given preliminary treatment. If, in the future, nutrients need to be reduced, the level of treatment should be enhanced. The treatment facilities for the wastewater from Area C can be implemented step by step.

The SSPII project is part of the Master Plan. When the SSPI project is completed, problems with wastewater coming from the Xuhui District, the Luwan District, the Huangpu District, the Hanshi District, the Yangpu District and the Honkou District (all six districts are part of Shanghai's urban area) still need to be resolved. With the development of the city, there is also an urgent need to solve the problem with wastewater from the Pudong New Area, the Wujing area and the Minhang area. Therefore, the total service area of SSPII is 332 km^2 and the service population is 5,408 million. The planned ADWF of SSPII (until the year 2020) is 4.934 million m^3/d. In addition to the existing southern interceptor, it is proposed that three conveyors be constructed to intercept the wastewater from the areas (Figure11.4), and deliver and discharge it into the Yangtze River Estuary after proper treatment. Due to the large scale of the project, an overall planning and implementation-by-stages methodology was adopted. At present, one conveyer, one outfall and preliminary treatment works for the first stage C with 1.7 mil m^3/d will be constructed, including the sewers and interception facilities in the Xuhui and Luwan Districts of Puxi.

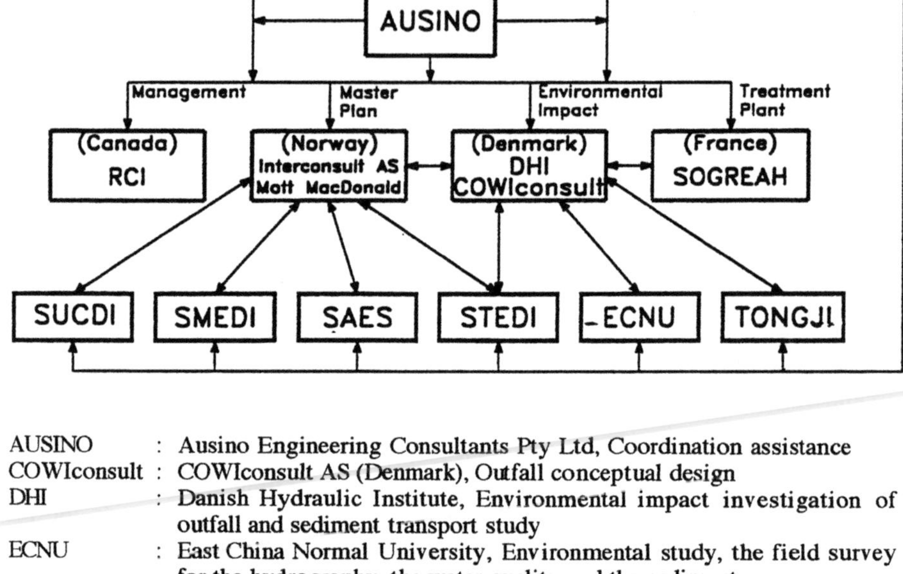

AUSINO	:	Ausino Engineering Consultants Pty Ltd, Coordination assistance
COWIconsult	:	COWIconsult AS (Denmark), Outfall conceptual design
DHI	:	Danish Hydraulic Institute, Environmental impact investigation of outfall and sediment transport study
ECNU	:	East China Normal University, Environmental study, the field survey for the hydrography, the water quality and the sediment
IC & MMD	:	Interconsult AS (Norway) and Mott MacDonald (UK), Design review
RCI	:	Reid Crowther International Ltd (Canada), Management assistance
SAES	:	Shanghai Academy of Environmental Sciences, Environmental assessment
SMEDI	:	Shanghai Municipal Engineering Design Institute, Part of the design of the pipeline and the pumping station in the treatment plant and the review of structural design
SOGREAH	:	Treatment plant feasibility (France)
SSPCC	:	Shanghai Sewerage Project Construction Company
STEDI	:	Shanghai Tunnel Engineering Design Institute, Outfall design
SUCDI	:	Shanghai Urban Construction Design Institute, Design of pipeline and pumping station in the treatment plant
TONGJI	:	Tongji University, Physical model test of the outfall
WB	:	World Bank

Figure 11.2. Organization Plan for SSPII's feasibility studies.

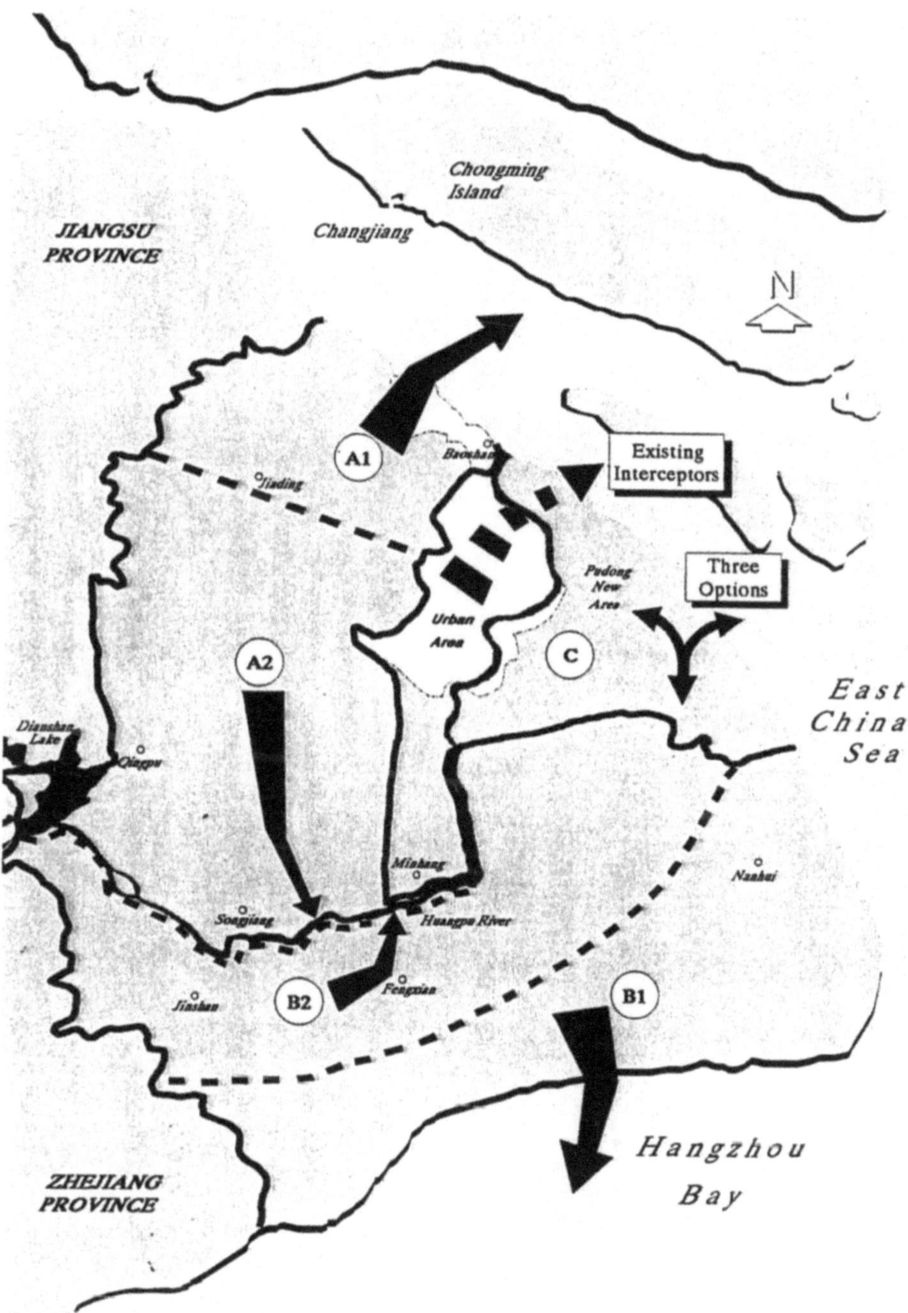

Figure 11.3. Options for the waste water drainage, see also Figure 11.1. Figure from Ref. /10/.

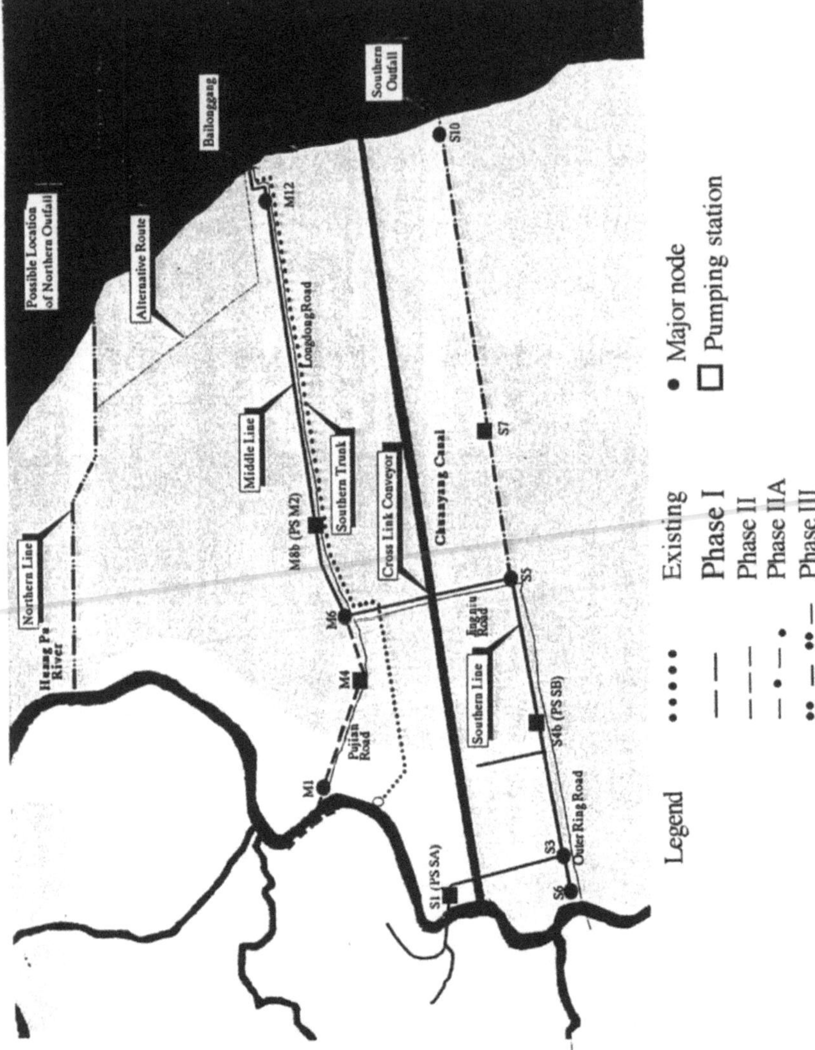

Figure 11.4. Propposed wastewater convenance system for SPII (10).

Table 11.1. Comparison of water quality effects and Net Present Value (NPV) for the 8 scenarios. A1 - to the Yangtze River, B1 - to the Hongzhou Bay, A2/B2 to the Huangpu River, C - as shown in Figure11.3.

Scen		Treatment				Effect on Receiving Water			NPV Rank
		A1	B1	A2/B2	C	Hangzhou Bay	Huangpu River	Yangtze River	
C to HZB	1	P	P	P	P	3	3	1	6
	2	P	P	P	S	3	3	1	7
	3	P	S	P	S	2	3	1	8
C to Huang River	4	P	S	P	P	1	3	1	1
	5	P	S	P	S	1	3	1	3
	6	P	S	S	S	1	2	1	4
C to Yangtze River	7	P	S	S	P	1	2	1	2
	8	P	S	S	E	1	2	1	5

P   Preliminary treatment
E   Enhanced primary treatment
S   Secondary treatment with nitrification and denitrification
1   Likely to provide an environmentally acceptable solution, providing detailed studies confirm strategic plan conclusions
2   Possibly causing unacceptable pollution - further study required if selected as recommended solution
3   Likely to cause unacceptable pollution in receiving water

The preparatory work for the engineering part of the project mainly involves: 1) the master plan and the feasibility study of the engineering design options; 2) the feasibility study of wastewater discharge into the Yangtze River Estuary at Bailong Gang. The latter especially is the key issue to the feasibility of the projet implementation.What will the impact be on the near and far fields due to the wastewater discharge into the Yangtze River. Should the treatment levels be based on the wastewater quality and the environmental impact? How should the outfall facilities be designed and arranged? The study has to answer all these issues.

11.2.4 Layout of Outfall

Outfall Siting. The proposed site for Phase 1 of the SSPII outfall is at Bailong Gang on the south bank of the Yangtze River Mouth. Phase 1 considers discharging the sewage from this area and from part of Shanghai centre. The far river bank is shallow and the water depth at the proposed outfall is 8m

Bathymetry. The south bank of the Yangtze River is shallow with depth contours parallel to the shore (Figure 11.5). The Yangtze River Estuary bed is, in general, unstable. Major dredging and reclamation work influences the bathymetry of the estuary. The effects of these topographical changes are discussed in section 11.2.6.

Geology. The Yangtze River delta is a tide and fluvial delta consisting of delta sand, silt and clay layers to a depth of >50 m of Holocene deposits. To establish a basis for the design of the outfall structure site investigations comprising seismic traverses and geotechnical boreholes along the proposed alignment were performed. Based on the results of the site investigations the stratification (Figure11.6) has been established:

- S1. Fill and mud, grey, 0-1.5 m thick
- S2. Sandy silt, grey saturated, loose, inhomogeneous thin clay bands, 1-8 m thick
- S3. Very soft silty clay, grey - slush, saturated, 1.5 - 5 m thick
- S4. Very soft clay, grey with thin layers of silt 10 - 14 m thick
- S5. Clay, grey - with layers of silty clay and sandy silt, thickness above 20 m
- S7. Sandy silt, silty sand, grey - thickness unknown.

All strata are recent sedimentary soft deposits. For the strata S3 and S4 the natural water content is 40-50 %, which is above the liquid limit indicating that the strata are liquified. For the design of the outfall structure the following parametres may be utilised:

Natural density	γ =	18 kN/m^3
Cohesion	c =	10 kPa for S3, 11 kPa for S4, 12 kPa for S5
Internal angle of friction	φ =	12° for S3, 7° for S4, 12° for S5
Modulus of elasticity	E =	3 MPa for S3, 2 MPa for S4, 4 MPa for S5
Poisons ratio	ν =	0,3
Permeability	k =	10^{-1} m/s

Currents. The receiving water at Bailong Gang is influenced by hydrographic mechanisms in the estuary and the large flow of water in the Yangtze River. The tidal-influenced current velocities up and down the Yangtze River near Bailong Gang are approximately the same magnitude and parallel to the shoreline.

The surface currents are larger than the bottom currents that average 1.0 and 0.8 m/s with maximum velocities of 2.3 and 2.0 m/s, respectively. Average and design maximum wave heights at the hydrometric station at Wai Gaoqiao 15 km upstream from Bailong Gang (Figure 11.1) are 0.2 m and 1.5 m, respectively. An extreme maximum wave height of 3.2 m was recorded during a typhoon.

Densities. The relatively shallow water in the Yangtze River Estuary, combined with the tide and the large flow in the river, contributes to the fact that the estuary normally is mixed from surface to bottom. The densities reported by East China Normal Universityvary between 995 - 1000 kg/m^3 .

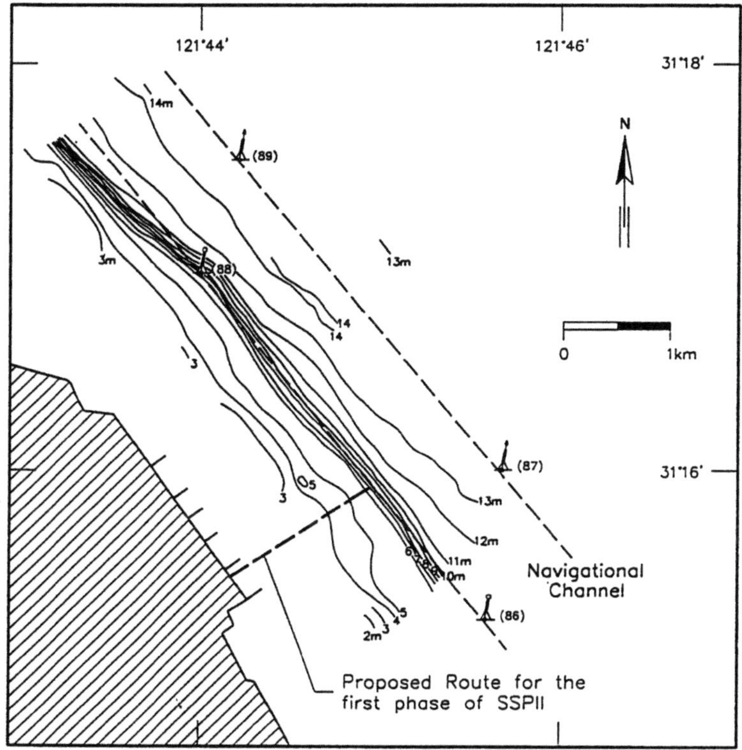

Figure   11.5. Bathymetry in Bailong Gang area. Depths are relative to Wusong Datum.

Sediment Transport. Transport of suspended sediments in the Yangtze River varies between 340 million tons/year and 580 million tons/year as recorded at Datong. The grain size of bed sediments and suspended sediments is very fine and characterized by a median diameter (d_{50}) of 0.004 - 0.009 mm and a maximum of up to 0.063 mm.

Water Levels. The water levels in the Yangtze River Estuary are influenced by flood and ebb in the East China Sea and the seasonal flow variation in the river. The tidal variations above Wusong datum recorded at Waigaoqiao are on average 2.4 m and maximum 4.7 m. The Mean Water Level is 2.04 m above datum. The recorded maximum MWL with a recurrence period of once every 100 years is about 3.74 m above the MSL and the recorded maximum MWL with a recurrence period of once every five years is about 3.18 m and the recorded minimum is 2.5 m below MWL. The outfall will be designed to meet the environmental requirements at the average discharge flow when the water level is MWL, whereas the overall design of the outfall structure, onshore head tank etc. will meet the requirements for the high water level of 3.18 m above MWL.

Sewage Flows. The sewage generated in the new development area of Pudong contributes with 45% of the sewage, which in Phase 1 will be discharged through the outfall at Bailong Gang. The remaining 55% is generated in the Pu Xi area. The average dry weather flow (ADWF) is assessed at 20 m^3/s, whereas the design wet weather flow (DWWF) is assessed at 30 m^3/s due to a peak factor of 1.3 together with a contribution from storm water entering the combined sewerage system in the Pu Xi area.

As all the sewage and storm water conveyed from Shanghai to the outfall enters at a level below the water level in the river it will have to be pumped through the outfall.

Navigational Constraints. Due to the very busy Navigational Channel parallel to the shoreline only 1600 m from the shore, the Naval Authority has stated that no part of the outfall structure can be located closer to the Channel than 100 m and that no riser is allowed to protrude more than 2.5 m above the river bed. The length of the outfall is therefore limited to 1500 m off shore.

Design Criteria. Given the physical and other constraints, the outfall should be designed to achieve the highest possible dilution in the near-field as well as in the far-field. The head loss required to discharge the sewage through the outfall should also be minimised.

Furthermore, the outfall should be designed to discharge all particles with a diameter smaller than 0.4 - 0.5 mm. Larger particles will normally settle in onshore basins before being discharged into the tunnel.

Alternative Diffuser Layouts for Phase 1. In this study two different riser layouts were investigated together with four different diffuser configurations. The objective was to see the influence that different spacing of risers would have on the dilution of the discharged sewage. The risers were either provided with four ports pointing in diagonally different directions or they were provided with a circumferential slot.

The diffuser configurations consisted of 1) four risers with 40 m spacing (concentrated), 2) four risers with 160 m spacing (separated), 3) eight risers spaced 40 m and twin outfalls with four risers spaced 40 m and offset by 40 m. All configurations complied with the overall criteria that the total discharge area is 0.5 - 0.7 times the outfall main area. The layouts are illustrated in Figures 11.7.

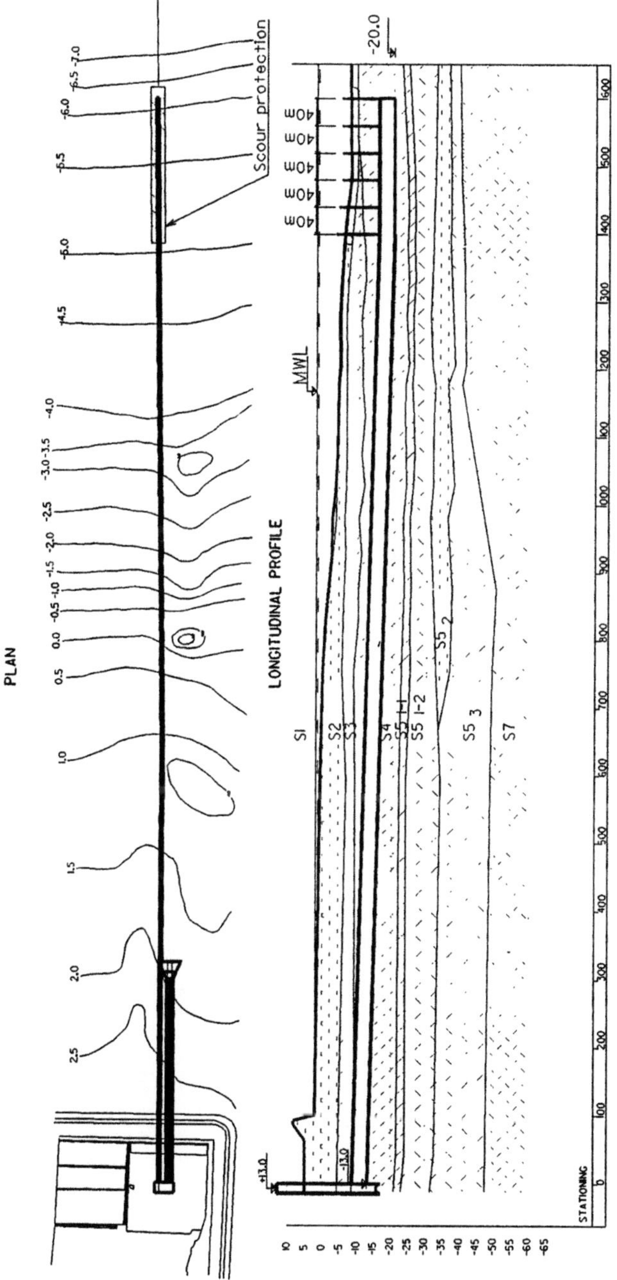

NOTES: Contour lines are based on levels measured by ECNU for levels deeper than -4.0m and by STEDI for levels higher than -4.0m. Levels in metres refer to Wusong Datum. MWL is 2.0m above Wusong Datum.

Figure 11.6. Layout of outfall tunnel structures.

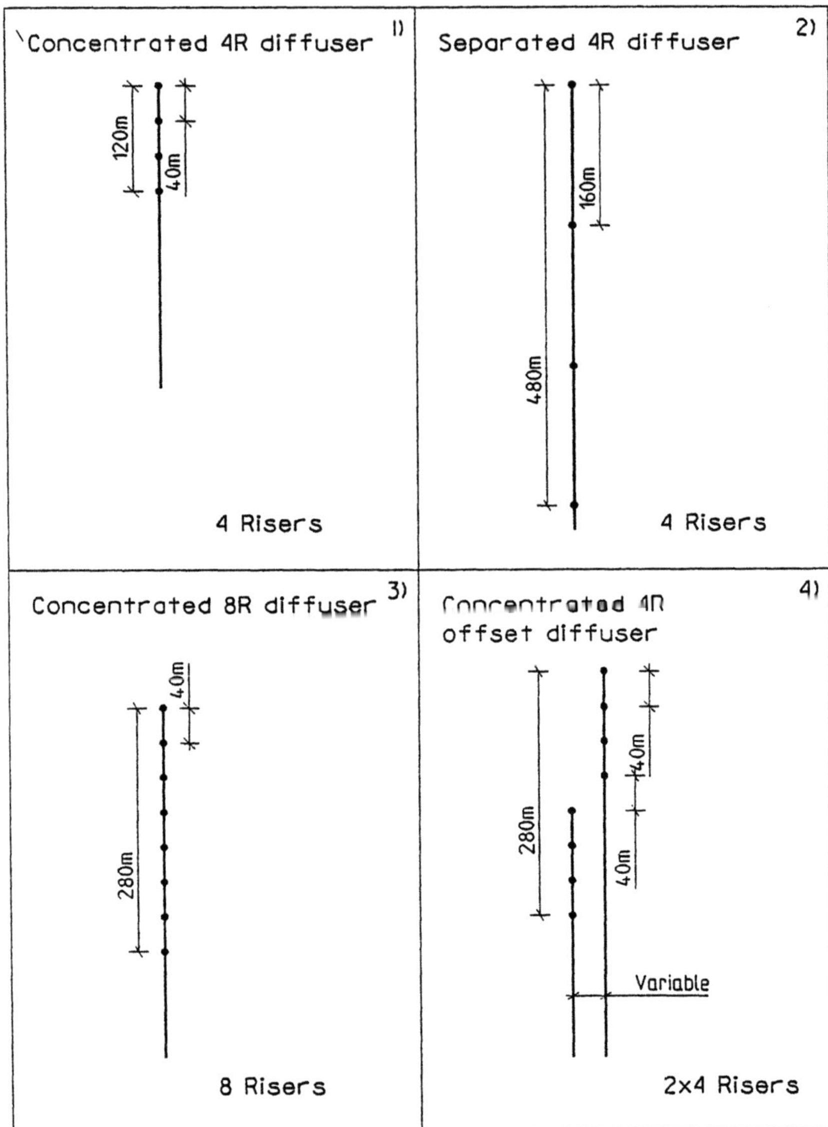

Figure   11.7. Diffuser configuration.

Alternative Structures for Phase 1. Three alternative structures have been investigated for the Phase 1 outfall: bored tunnel, immersed tunnel and pipe supported on a dam and piles (Figure 11.8). Furthermore the feasibility of a single outfall structure or twin outfall structures was assessed.

The bored tunnel option would consist of either one tunnel of diameter 4 m or two tunnels each of diameter 2.85 m. The lining consists of a precast concrete segmentally-bolted lining sealed by means of neoprene gaskets. The tunnel lining should be backgrouted to ensure full ground support. Soil investigations indicate that methane pockets exist in the vicinity of the proposed route for the outfall. The Tunnel Boring Machine (TBM) should, therefore, have facilities for probing ahead. The risers will be jacked from inside the tunnel.

The immersed tunnels could in both cases be constructed by 15 numbers of tube tunnel elements internally sized 4.0 m x 3.2 m (single) and 4.7 m x 3.0 m with a partition wall (twin).

The immersed tunnel is fabricated in 100 m long elements, either in a dry dock or on an elevating platform. The elements are provided with temporary bulkheads and are floated out by barge and placed in a pre-dredged trench. After sinking the elements the trench is backfilled and protected against erosion.

For the solution of the pipe supported on a dam and piles, the tube is either a dia 4 m pipe (single) or a box culvert 4.7 m x 3.0 m with a partition wall (twin). The pipe/culvert near the shore is placed in a backfilled trench on a reclaimed dam. The dam extends 700 m from the shore. This solution was suggested due to plans, at a later stage, to reclaim land up to 700 m from the shore. The outer section of 800 m was placed on pile-supported bearings above a specific high water level. The outfall structures could be constructed as precast elements each 50 m long or in-situ cast structures with expansion joints.

Alternative Structures for the Ultimate Phase. In a later stage the required discharge capacity will increase to a ADWF of 57 m^3/s and 5 million m^3/day. During the study three schemes of outfall locations were evaluated for the increased flow. The schemes include discharge locations 2 km and 10 km upstream and downstream from Bailong Gang.

Cost Estimates. To compare the various outfall structures in Phase 1 the construction costs and the operating costs of pumping amd maintenance were assessed. The latter were based on the average flow, ADWF.

The present values for the alternatives are given in Table 11.2. The comparison was based on comparative interest rates of 5% and 10%. It was also assumed that the outfall would be constructed during the first year and the time horizon was defined to be 30 years.

The present value, including construction and operating costs for the average single outfalls, is 10% lower than the twin immersed tunnels and culverts on dam/piles and 30% lower than twin bored tunnels. However, the single bored tunnel was slightly cheaper.

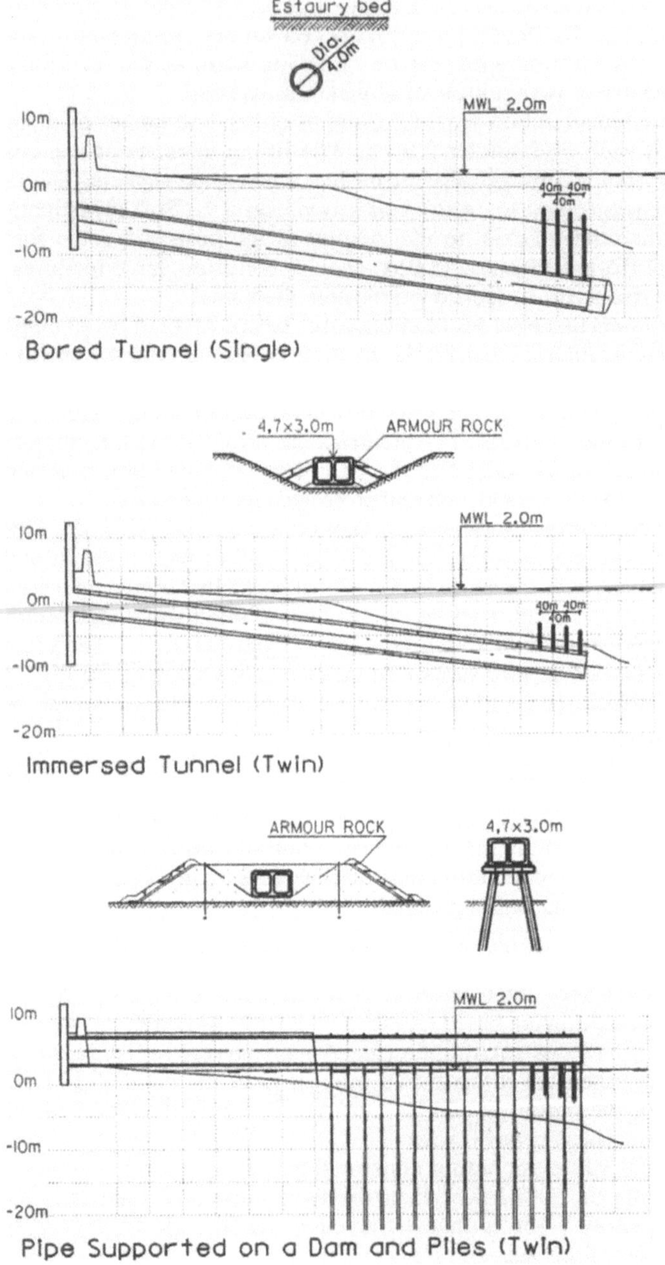

Figure **11.8.** Alternative structural layouts.

Table 11.2 Comparative Present Values.

| | | Present Value in Million RMB | | | | |
| | | PV of running costs | | PV of construction costs | Total PV | |
		5%	10%		5% pa	10% pa
Single Outfall	Bored tunnel	112	68	150	262	218
	Immersed tunnel	112	68	159	271	227
	Pipe on dam/piles	148	91	140	288	231
Twin Outfall	Bored tunnel	122	75	240	362	315
	Immersed tunnel	122	75	170	292	245
	Culvert on dam/piles	159	97	150	309	247

1 US$ = 8.7 RMB. For further explanation see Section "Cost Estimates".

Conceptual Design of Phase 1. Outfall Tunnel. A tunnel with a diameter of 4.2 m was chosen for the conceptual design taking into consideration the cost, operation, maintenance and local experience of design and construction. The length of the outfall is 1600 m (100 m onshore and 1500 m off shore). A longitudinal profile of the outfall is shown in Figure11.6. Due to the poor soil conditions it was decided to recommend tunnel construction in a layer of soft clay and to avoid passing through the silty and sandy layers overlaying the soft clay. The invert level of the tunnel at the head tank should therefore be -13.0 m. From here the tunnel slopes downwards at 0.5% to the diffuser section with an invert level of -20.0 m approximately 10 to 12 m below river bed.

Diffuser. The diffuser was furnished with six risers each with an internal diameter of 1.4 m. Due to structural performance reasons the diffuser section was the same diameter as the main tunnel. The spacing between the risers was to be 40 m. During detailed design studies should be initiated to improve the basis for a possible reduction of the spacing to 25 - 30 m.

The diffuser ports were recommended to be of the slot type as the total head loss was found to be 20% smaller than the traditional type with 4 circular openings. Given the riser diameter of 1.4 m the slot height would be 0.42 m (Figure11.9). The geometric ratio between the port area and the tunnel area is 0.8, which is higher than normally recommended. However, the size of the openings reflects that currents in the river to some extent will prevent the discharge of sewage either upstream or downstream depending on the current direction. So the effective area of the discharge ports will comply with the general rule. During detailed design it is recommended to perform physical model tests to evaluate and decide upon this important and complex issue.

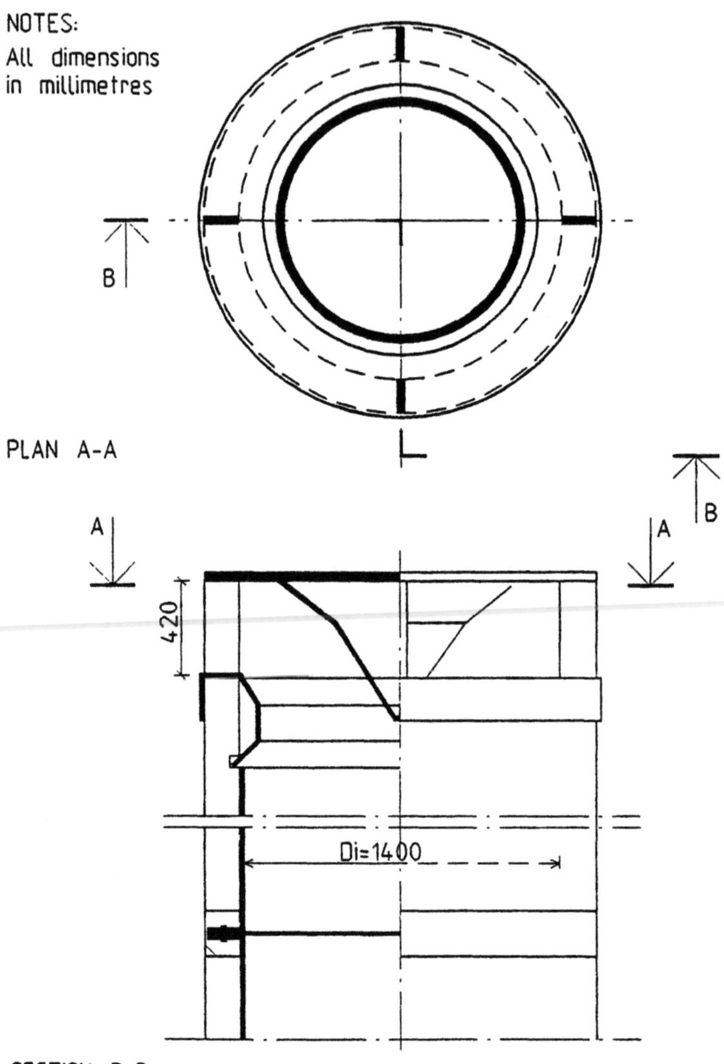

Figure 11.9. Riser layout, slot concept.

Emergency Outfall. An emergency outfall is probided to discharge sewage in case of failure in the tunnel or the diffuser section or when design conditions are exceeded. The emergency outfall was proposed to consist of twin 3 m pipes jacked partly through and under the dyke, which must be protected due to the risk of flooding of low lieing areas behind the dyke. The emergency outfall will extend 900 m from the river bank in order to protect the reclamation scheme and the fish migration route along the river bank.

Head Tank. The head tank is constructed in the shaft from where the tunnel is driven. The head tank protects the outfall tunnel from possible surges which will be introduced in the tunnel in case of start and sudden stop of flow from the pumps. The tank also connects the main outfall to the emergency outfall. The tank consists of a shaft with the bottom level at -13.0 m and it protrudes 8.0 m above ground level having the roof of the tank at level 13.0 m.

The shaft is 13 m x 10 m. The overflow chamber has a bottom level of 0.00 m and an area of 15 m x 13 m. The tank is furnished with large motor-operated gates to direct the water from the chamber to the shaft or vice versa. The water variation in the tank was found to vary between level -6.4 m and 10.5 m. However, the normal variation will only be 8 m between level -3 m and 5 m.

Cost Estimate. A cost estimate has been estalbished based on the afore-mentioned structures and the fact that local labour, materials and machinery will also be available for international tenderers. The prime costs consisting of construction costs, supply of equipment together with tax and duties are estimated at 151 million RMB (= 17.4 million US$). The capital costs, including prime costs, management costs and contingencies of 50% at the present stage, amount to 318 million RMB (= 36.5 million US$).

11.2.5 Environmental Impact

It is essential to have detailed knowledge of the current field in the Yangtze River Estuary in order to assess the environmental impact in the area. The current field in this area is very complicated, being mainly dominated by tidal range, discharge from the Yangtze River, monsoon winds and the Japan Current (Kuroshio). The nearshore current is variable in time due to the tide, the wind and the discharge from the Yangtze River, and also in space due to varying water depths.

On the rising tide, the water levels in the Hongzhou Bay and the Yangtze River Estuary are set up. On the falling tide the currents generated by the high water levels combine the discharges from the rivers running towards to the ocean. These tidal currents are the main components of the hydrodynamics in the area.

Fifth largest in the world, the Yangtze River discharge is the second largest component of the hydrodynamics in the area. The peak discharge (45,000 m^3/s) usually occurs from late June to August and the minimum discharge (10,000 m^3/s) occurs in January and February resulting in annual mean discharge of 30,000 m^3/s, /1/.

The Yangtze River empties into the East China Sea through the Yangtze River Estuary. According to the previous study /2/, the Yangtze River discharges mostly fresh water through three channels; approximately 45% of the water through the North Channel, and the rest through the North and South Passages (Fig.11.1).

Currents driven by monsoon winds are additional components. Generally, the north-easterly winter monsoon begins in late October and lasts four months, The south-westerly summer monsoon from late May ends early September. Average winter and summer monsoon speeds are 8 m/s 5 m/s, respectively.

There are also several generally recognized currents just outside the Yangtze River Estuary: the Yellow Sea Coastal Current (YSCC), the East Sea Coastal Current (ESCC), and the Taiwan Warm Current (TWC) caused by the Japan Current (JC) /3/. All these currents, combined with the monsoon wind actions, form the typical monsoon currents off the coast of China.

As mentioned above, the hydrodynamic situation in the Yangtze River Estuary is very complicated. Detailed reliable current fields in the area can only be obtained through mathematic model simulation. It is very important to have correct boundary conditions for the simulation in order to obtained reliable current fields. Thus, to assess the environmental impact in the Yangtze River Estuary due to the disposal of a large amount of waste water from Shanghai, a suite of comprehensive mathematical models has been used including 1D, 2D and 3D model descriptions (Figure11.10). The 1D river model was set up from Datong (600 km from the mouth of the Yangtze River and with non-tidal effect) to the end of the Yangtze River Estuary, in order to supply correct upstream dynamic boundary conditions to the running 2D model. Furthermore, according to our experience from current simulations in the Far East /4/, it is also necessary to first set up a relatively large model to supply the correct boundary conditions for other fine grid models.

Three models (2D and 3D) have been set up for this study: a regional model (M1), an estuary model (M2) inlaid within the M1, and a near-field model (M3) inlaid within the M2. A summary of the models' specifications is given in Table 11.3. The model origin is to be found in the left-hand corner of the southern boundaries. The time steps are determined based on the bathymetry, the typical velocity in the area and the grid size. The coefficients for different models were determined from the model calibrations.

Two typical periods, summer and winter, have been considered in this study. Based on the available measurements, the simulated periods were determined as: 02/03/91 -16/03/91 as the winter period, and 08/07/92 - 22/07/92 as the summer period. It should be mentioned that the 1D model supplies the boundary conditions for the M1 simulations. All the boundary conditions for the M2 simulations are obtained from the M1 simulations and the boundary conditions for the M3 simulations are obtained from the M2 simulations.

A series of calibrations for different models was carried out first. The hydrodynamic calibrations were divided into two steps: a pure tide calibration and a calibration with the combined actions of tide, wind and net flow. The first step is to obtain a correct tidal flow pattern and correct surface elevations generated by the tide. The second step is to obtain a reasonably up-to-date current field in the area. The calibration was carried out using three comparisons: the general tidal pattern, the water surface elevation and the current velocity.

Based on the hydrodynamic models, the advection-dispersion models were calibrated by salinity comparisons between the simulated results and the measured data Ref. /5/. Through the calibrations, the dispersion coefficients were determined for the present study as follows:

for 2D model: $D_x = 80$ m^2/s and $D_y = 60$ m^2/s;
for 3D model: $D_x = 19.2$ m^2/s, $D_y = 19.2$ m^2/s and $D_z = 0.0025$ m^2/s.

Table 11.3. Summary of the models' specifications and the grid relations.

	M1	M2	M3
Model Area	280 km x 340 km	106 km x 61 km	20.8 km x 5 km
Grid Size	1000 m	250 m	40m (horizontal) 2m (vertical)
Latitude of Origin	29°20'40"	31°13'50"	31° 19' 05"
Longitude of Origin	120°20'59"	121°9'34"	121° 39' 41"
Origin in the M1 (j,k)	(0,0)	(78,209)	(130,217)
Origin in the M2 (j,k)		(0,0)	(145,130)
Turning Angle (from true north)	0°	30°	55°
Time Step	120 sec	60 sec	20 sec

Following the calibration of the advection-dispersion models, the 2D water quality and heavy metal models in the far-field were also calibrated. Six components considered in the water quality calibration are: Coliform, BOD, DO, Ammonia, Nitrate and Phosphate; and three components considered in the heavy metal calibration are: dissolved metal in water, suspended metal in water and suspended matter. All the rivers and outfalls as the loading points in the M2 are listed in Table 11.4, while the applied concentrations at the open boundaries are listed in Tables 11.5 & 11.6. For these simulations constant salinity and temperature were selected, i.e. salinity 4.0 PSU and temperature 29°C for the wet season and salinity 6.0 PSU and temperature 9.5°C for the dry season. These values give an oxygen saturation concentration of 7.4 mg/ℓ and 11.0 mg/ℓ respectively. These have been chosen as the initial value for the simulation and as the boundary value at offshore boundaries.

Normalized water quality calibration parameters for biologically significant parametersand stream classification are:

- Degradation constant for dissolved BOD = 0.5 d^{-1} at 20°C with temperature coefficients of 1.02 at 20 to 29°C and 1.07 at 9.5 to 20°C. .
- Ammonia degradation constant = 0.4 d^{-1} with temperature coefficients of 1.1 for winter and 1.2 for summer.
- Nitrate degradation constant = 0.1 d^{-1} with temperature coefficient of 1.16
- Release of ammonia by BOD = 0.065 g/g for both winter and summer

254

Adjustments of the concentration values at the western boundary (upstream of the Yangtze River Estuary) have also been included in the calibration. Both the water quality and heavy metal simulations reach steady state after three days' simulation. The basic water quality coefficients determined at the calibration stage are listed in Table 11.7 below.

The heavy metal module describes the adsorption/desorption of metals to suspended matter, the sedimentation of adsorbed metals to the bed as well as resuspension of settled metal. The details of the calibrations are referred to in Ref. /6/.

A large number of the simulations have been carried out combining all the components listed above. A few points about the study should be explained before discussions of the preliminary study.

1) Since the exact loading concentrations for SSPII were not determined at the beginning of this study, the study had to be based on data supplied by SSPCC during the study. Therefore, we got two groups of loading concentrations, ie. the first estimated group obtained for an alternative layout study and the second estimated group obtained later for a conceptual design study. Details are shown in Table 11.4.

2) Except for the riser layout, Con. 6R, and the second estimated group of loading concentrations, other components listed above are applied in the alternative layout study.

3) In the conceptual design study, only Phase 1 with the riser layout of Con. 6R and the second estimated group & the standard group of the loading concentrations are considered.

4) Because all nine water quality and heavy metal components have a similar tendency of the concentration distributions based on the simulations, the discussions in this publication will mainly be focused on the BOD concentrations. The other components are only discussed generally.

For the study of the alternative layouts, Figure 11.11 shows the BOD concentration distributions from the far-field simulations for Phase 1 with the different loading concentrations. Figure 11.12 shows the BOD concentration distributions from the far-field simulations for the different schemes and Figure 11.13 shows the BOD concentration distributions from the near-field simulations. For the study of the conceptual design, Figure 11.14 shows the BOD concentration distributions from the near-field simulations, and Figure 11.15 shows the concentration distributions of Cu. and Zn. from the near-field simulations. The data shown in the Figures are the temporal arithmetic maximum values for each simulation period. It is not possible to compare the 2D results with the 3D results directly, as the two models have different resolutions and describe different hydrodynamical processes. The 3D model is used to investigate the diffuser disposal and the pollution situation in the near-field and the 2D model is used to investigate the entire scheme and the environmental impacts in the far-field.

Besides the nine components considered in the water quality and heavy metal simulations mentioned above, the preliminary environmental study also includes:

Two Periods:

    Period 1: Summer and

    Period 2: Winter

Two Phases:

    Phase 1: a discharge capacity of 1.7 mill m^3/day and

    Ultimate Phase: a discharge capacity of 5.0 mill m^3/day.

Three Schemes:

    Scheme I : all outfalls are located in the Bailong Gang area,·

    Scheme II : phase 1 outfalls are located in the Bailong Gang area, while the outfalls for the later phase are located 2 km up and down stream of the Bailong Gang, and

    Scheme III: phase 1 outfalls are located in the Bailong Gang area, while the outfalls for the later phase are located 11 km upstream and 10 km downstream of Bailong Gang.

Four Riser Layouts:

    Concentrated 4R: 4 risers with a distance of 40 m between each other,

    Concentrated 6R: 6 risers with a distance of 40 m between each other,

    Concentrated 8R: 8 risers with a distance of 40 m between each other, and

    Separated   4R: 4 risers with a distance of 160 m between each other.

Three Groups of Loading Concentrations:

    I : the first estimated group from SSPCC,

    II : the second estimated group from SSPCC, and

    III: the standard values from the standard /7/.

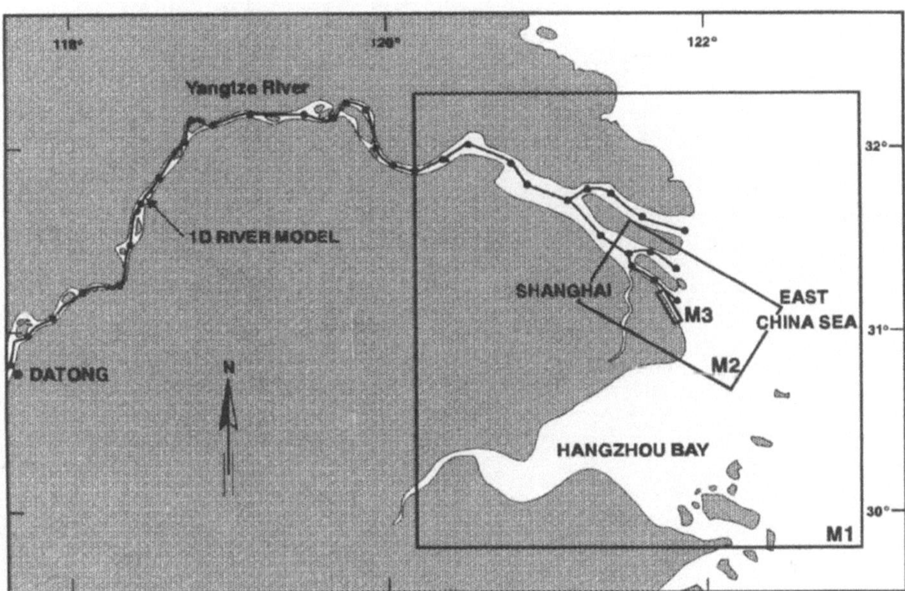

Figure   11.10. The model domains.

256

The water body in the Yangtze River Estuary is neither purely fresh water nor marine water, and there are also some fishing areas in the region. All the standards for the different water bodies give different reference values. There is no standard for the near-field mixing of sewage outfall at the moment and, therefore, the discussion of the environmental impact will only be related to the Chinese standard values of fresh water.

Table 11.4. Concentrations at sources, outfalls, and rivers.

No		Coli / 100 ml	BOD D mg/l	D.O. mg/l	NH_3 mg/l	NO_3 mg/l	$P)_4$ mg/l	Cu mg/l	Zn mg/l	SS mg/l
1	Huang-pu	4.3×10^5	3.0	4.3	1.5	0	0.5	.04	.05	350
2	Westrn outfall	1.5×10^6	377	0	25	5.27	5.5	2.1	2.8	718
3	Southn outfll [2]	4.3×10^5	218	0	33.6	10.7	6.4	.85	1.0	139
4	Zhuyn outfall	7.5×10^4	160	0	22	0.3	1.5	.02	.42	216
5	Yeulng nferric	0	0	0	0	0	0	0	1.7	85
6	Baos-han	0	0	0	0	0	0	0	.53	9
7	SSPII 1st est	1.0×10^6	200	0	24.48	11.7	6.4	.36	1.7	300
8	SSPII 2nd est	1.0×10^6	129	0	24.3	11.7	1.6	.36	1.7	178
9	Stan-daad [1]	1.0×10^6	100	0	25	20	2.0	1.0	5.0	400
10	Limited value [3]		80		40		2.0	1.0	5.0	250

Note: (1) Standard: reference /7/, (2) When SSPII works, the southern outfall .#3, will stop, (3) from N. Standard: Ref. /8/.

Table 11.5. Concentrations at open model boundaries, March (dry season).

Bound-ary	Coli form / 100ml	BOD mg/l	D.O mg/l	NH$_3$ mg/l	NiO$_3$ mg/l	PO$_4$ mg/l	Cu mg/l	Zn mg/m	SS mg/l
Western	9180	2.6	11	1.3	1.7	0	0.03	0.095	330
Northrn	0	0,5	11	0	0	0	0.0005	0.001	10
Eastern	0	0,5	11	0	0	0	0.0005	0.001	10
Southrn	0	0,5	11	0	0	0	0.0005	0.001	10

Table 11.6. Concentrations at open model boundaries, July (Wet season).

Bound-ary	Coli form / 100ml	BOD mg/l	D.O mg/l	NH$_3$ mg/l	NiO$_3$ mg/l	PO$_4$ mg/l	Cu mg/l	Zn mg/m	SS mg/l
Western	9180	2.7	7.4	0.2	9	0	0.03	0.10	330
Northern	0	0.5	7.4	0	0	0	.00048	.001	10
Eastern	0	0.5	7.4	0	0	0	.00048	.001	10
Southern	0	0.5	7.4	0	0	0	.00048	.001	10

Table 11.7. List of calibrated coefficients for the water quality model.

Coefficients	DRY SEASON	WET SEASON
Coliform degradation (/day)	0.5	1.8
BOD degradation (/day)	0.25	0.6
Temperature (°C)	9.5	29.0
Salinity (PSU)	6.0	4.0
Ammonia degradation (/day)	0.15	2.1
Yield factor for release of NH$_3$N by BOD (mg NH^3-N / mg BOD)	0.065	0.065
Nitrate degradation (/day)	0.02	0.38
Yield factor for release of phosphate by BOD (mg P/mg BOD)	0.003	0.003

The Different Phases. The Phase 1 sewage discharge (1.7 mil. m³/day) will be at the Bailong Gang area. The outfalls will be about 1500 m off the coast and the diffuser section will be at a depth of 6-9 m. In general, all the simulated results show that the Phase 1 discharge with the first estimated loading concentrations from SSPCC will cause a polluted area in the near-field. The plume lengths are quite limited and the edge of the plume (a BOD concentration is higher than 3 mg/ℓ, the second grade of Chinese fresh water standard) never comes closer to the river bank than 700 m. This means that the plumes do not reach the bank and there is a passageway for fish migration.

The Ultimate Phase is with a discharge of 5 mil. m³/day. If this amount of sewage is to be totally discharged in the Bailong Gang area (like Scheme I), it will generate heavy pollution in the near-field and a significant environmental impact in the far-field. The plumes will reach the river bank.

The Different Riser Layouts. Three different types of diffusers have been tested for the alternative layout studies, concentrated 4R, concentrated 8R and separated 4R. The 2D results show that the different types of diffusers will not influence the environmental impacts in the far-field. It is clear from the 3D results that the most important factor for the near-field dilution in the Bailong Gang area is the position of the diffuser. In general, a separate disposal of diffusers (i.e. extending the length of the diffusor section) should enhance dispersion and result in lower concentrations. However, this does not happen in the Bailong Gang area. The main reasons are that the water in this area is shallow and there is a wide tidal flat. Thus, any sewage discharged into the shallow area will be difficult to dilute in a small surrounding water body, added to the fact that the advection is weak due to the low currents in the shallow area. The 3D results also show that the type with 8 risers will not give a better dilution than the type with 4 risers. On the contrary, due to the short distance of deep waters, some risers with the type of concentrated 8R may be located in a relatively shallow position, and as a result the 8R type causes higher pollution than the 4R type at the coast.

The Different Schemes. As mentioned above, the main difference between the schemes is the distance between the outfalls. If the 5 mil. m³/day is discharged totally into the Bailong Gang area, heavy pollution could be caused as the assimilative capacity of the surrounding water body is limited. Based on the present simulated results, the environmental impacts from the different schemes can be summarized as follows:
1) Scheme I will cause high concentrations of pollutants in the Bailong Gang area.
2) Scheme II can significantly reduce the high concentrations in the Bailong Gang area and also form a narrow passageway between the edge of the plumes and the river bank for fish migration.
3) Because the distances between the outfalls of Scheme III are sufficiently long for an effective dilution, the effect between the outfalls is quite small. Thus, the environmental impacts of Scheme III can be considered similar to the situation of Phase 1 at three different areas. Therefore, Scheme III gives the smallest environmental impact when compared to the other two schemes.

However, further study is recommended of the schemes for the Ultimate Phase both on the environmental impacts and on the economic comparisons.

The Different Loadings. The two groups of loadings have been used in the studies of the alternative layouts. The first group estimated loading concentrations from SSPCC and the standard values. The BOD concentration from the first estimated loadings is higher than the value from the standard, while the concentration values of Cu and Zn from the first estimated loadings are lower than the standard values.

The simulated results show that the water quality (for BOD and DO) will be improved if the BOD loading concentration can be kept below the standard value. As the present concentration values of Cu and Zn are much lower than the standard values, the present estimated pollution situation could become worse in view of future industrial development.

The Different Seasons. The main differences between the two seasons are 1) the current fields are different due to the different discharges from the Yangtze River; and 2) the salinity and temperature vary with the season. These differences affect the plume length, degradations of the different components, and the contents of Dissolved Oxygen.

Simulated results reveal differences between the summer and winter periods plume sizes and the concentration distributions. If directly comparing the absolute distribution of pollution areas from the seasonal calculations, the degree of pollution during the winter period is larger than that during the summer period. Since the background and the boundary values are also different for the two seasons, there are no clear concluding remarks for the seasonal comparison.

The Conceptual Design. The present conceptual design considers only Phase 1 of SSPII. There are two main differences compared with the alternative study of Phase 1, i.e. 1) the conceptual design is with six risers, and 2) the second group of the estimated loading concentrations for SSPII has been used for the environmental evaluation. Since the second group of the loading concentrations is smaller than the first group, the environmental impact due to the sewage discharge from the outfall could be improved. For example, the areas with a BOD concentration larger than the 4 mg/ℓ that is a standard value for Grade III fresh water in China (11) are

listed as follows:

Simulation with the first group loading (200 mg/ℓ):	1.7 km^2
Simulation with the second group loading (129 mg/ℓ):	0.6 km^2
Simulation with the value (100 mg/ℓ):	0.36 km^2

Chinese environmental quality standards for surface waters provide for five classifications as follows:

I   very clean waters for natural protection of spawning and nursery areas
II   (untresated) drinking water sources
III treated drinking water supplies
IV recreation, boating, fishing, and industrial water supplies
V   irrigation and cooling water
with representative maximum concentrations in mg/ℓ unless otherwise noted

Constituent	Class I	Class II	Class III	Class IV	Class V
pH	6.5 - 8.5	6.5 - 8.5	6.5 - 8.5	6.5 - 8.5	6.0 - 9.0
D.O. saturation	>90%	>6	>5	>3	>2
BOD	<3	3	4	6	1
COD	<15	<15	15	20	26
Kjehdahl nitrogen	0.5	0.5	1	2	2
Lead	0.01	0.05	0.05	0.05	0.1
Cyanide	0.005	0.05	0.2	0.2	0.2
Phosphate	0.02	0.025	0.05	0.2	0.2

Short Concluding Remarks on the Conceptual Design. The present conceptual design only considers the sewage discharge from the Phase 1. The simulated results show that there is no strong environmental impact from the Phase I of SSPII. During both the summer and winter periods, the BOD distributions in a large Bailong Gang area can meet the third grade of the Chinese fresh water standard (4 mg/ℓ), and only in a very small area (less than 1 km^2) exceed the third grade.

The Coliform distributions generally exceeded the Chinese fresh water standard (1,000/100ml). The background values of Coliform in the Bailong Gang area are already higher than the standard value.

The Phosphate distributions in the entire Bailong Gang area can meet the second grade of the Chinese fresh water standard (0.1 mg/ℓ) during both summer and winter.

The DO contributions in the entire Bailong Gang area can meet the second grade of the Chinese fresh water standard (6 mg/ℓ) during both summer and winter.

Both the Cu and Zn distributions in the entire Bailong Gang area meet the second grade of the Chinese fresh water standard (1.0 mg/ℓ) during boththe summer and winter periods.

11.2.6 Other Studies

The following three topics have been focused on in this stage: the strategy of the treatment plant, the environmental effect from a general plan of the Yangtze River Estuary and sediment transport problems.

Treatment Plant Requirement. The degree of treatment for the SSPII waste water depends on two factors: the environmental impact of the discharge in the Yangtze River Estuary and the loading concentrations of the discharge. Of course, financial consideration is another very important factor.

According to the results from the present studies, the effect of the layout with a discharge of 1.7 mil. m^3/day from the conceptual design will only cause a limited environmental impact both for the near-field and far-field.

In order to protect the main outfall from sediment problems, it is required that any grit larger than 0.5 mm has to be removed before the sewage reaches the out-fall.

Preliminary Treatment including screening and grit removal is suggested based on the technical of view of the discharge of 1.7 million m^3/day. In the year 2005,Advanced Primary Treatment (with chemical dose) will be used for 50% of discharged sewage. will be used for 50% of discharged sewage. This will reduce BOD by by 57%, SS by 83% and Phosphate by 56%.

Effect of Dredging and Reclamation. The present topography of the Yangtze River Estuary (called existing topographic situation) is to be changed by a general realignment of the Yangtze River Estuary and by local reclamation around the location of the proposed outfall (the changed topography of the estuary due to the general plan and the local reclamation is called the future topographic situation, Ref. /9/).

Such changes would affect the current fields in the estuary and result in a different pollution distribution. Therefore, the proposed disposal of the SSPII is checked, taking these two factors into consideration. The environmental results obtained from the studies with the existing topographic situation should be checked for the future topographic situation, which includes:

- the local/near field environmental impacts of the proposed discharge of waste water with the future topographic situation; and
- the far field environmental impacts of the proposed discharge of waste water with the future topographic situation.

Furthermore, the effect of the three gorges project on the Yangtze River Estuary is also estimated.

Morphological Analysis. The morphological evolution of the Yangtze Estuary is governed by the hydrographic conditions and the huge amounts of fine sediments discharged from the Yangtze River. The sediments from the river either deposit in the estuary area and in the mouth of the estuary or spread into the sea. Due to the ongoing discharge of material, the morphology is developing by natural means. Local bed level changes (erosion) of the order of magnitude 1 - 2 m are registered over a period of 3 years at the location of the planned outfall. This natural variation might be enhanced locally by the construction of the reclamation. The total erosion could influence the stability of the outfall.

In connection with the planned reclamation at the location of the outfall, increased current speeds on top of the pipes can be expected. This increase might also cause additional erosion. The initial morphological analysis is separated into two tasks:

1) Sediment transport study in the near field of the outfall; called local morphological analysis. The local morphological analysis consists of evaluation of the natural morphological evolution around the location of the outfall, the local effects of the proposed reclamation on the local bathymetry and evaluation of the need for bed protection and its design.

All existing knowledge on the ongoing natural evolution is first reviewed and the additional erosion due to enhanced currents is estimated. The estimate of additional erosion is based on modelling of the local currents and waves. The sediment transport pattern is evaluated by a model for mud transport. At the moment mud transport models are based on empirical relations for erosion rates, flocculation, consolidation, etc.

Due to the sparse amount of available data, it is suggested to use either the estimated transport fields or the calculated fields of bed shear stresses to evaluate the risks of enhanced erosion. Both normal and severe hydrodynamic events with respect to mud transport shall be considered.

It may be concluded, that the pipe should be protected against erosion. The design of the protection should be carried out based on results from the hydrodynamic models. Available information of the extreme storm events will be used as input to the wave model. Extreme wave and current conditions shall be simulated and the size of stones for protection shall be given.

2) Initial morphological evolution of the entire estuary. From the information available on the evolution over the last decades it is clear that the entire estuary is far from being stable but large natural changes take place over short periods of time.

This overall morphological evolution is extremely complex and influenced by several parameters: hydrodynamics, river outflow and sediment discharge. The behaviour of the fine cohesive sediments, e.g. flocculation, influences the settling velocity and thereby the deposition pattern. Flocculation is a function of concentration of suspended matter, temperature and salinity. Especially in the outer part of the estuary wave action influences the transport, but also inside the estuary locally generated wind waves might be sufficient to stir up the fine material.

This natural evolution should be studied together with the possible changes due to large reclamations, extensive dredging of navigation channels and possible establishment of training walls.

Based on the flow fields, possible changes in sediment transport patterns, and thereby morphological evolution, could be qualitatively discussed and recommendations for future topographic investigations could be given.

It should also be noted, that in connection with planning of navigation channels, even if these do not influence the overall morphological evolution, the backfilling rates and thereby the maintenance dredging costs should be considered.

a)   Without SSPII outfall, with southern outfall
b)   1.7 mil. m^3/d waste water with a BOD concentration: 100 mg/l
c)   1.7 mil. m^3/d waste water with a BOD concentration: 200 mg/l

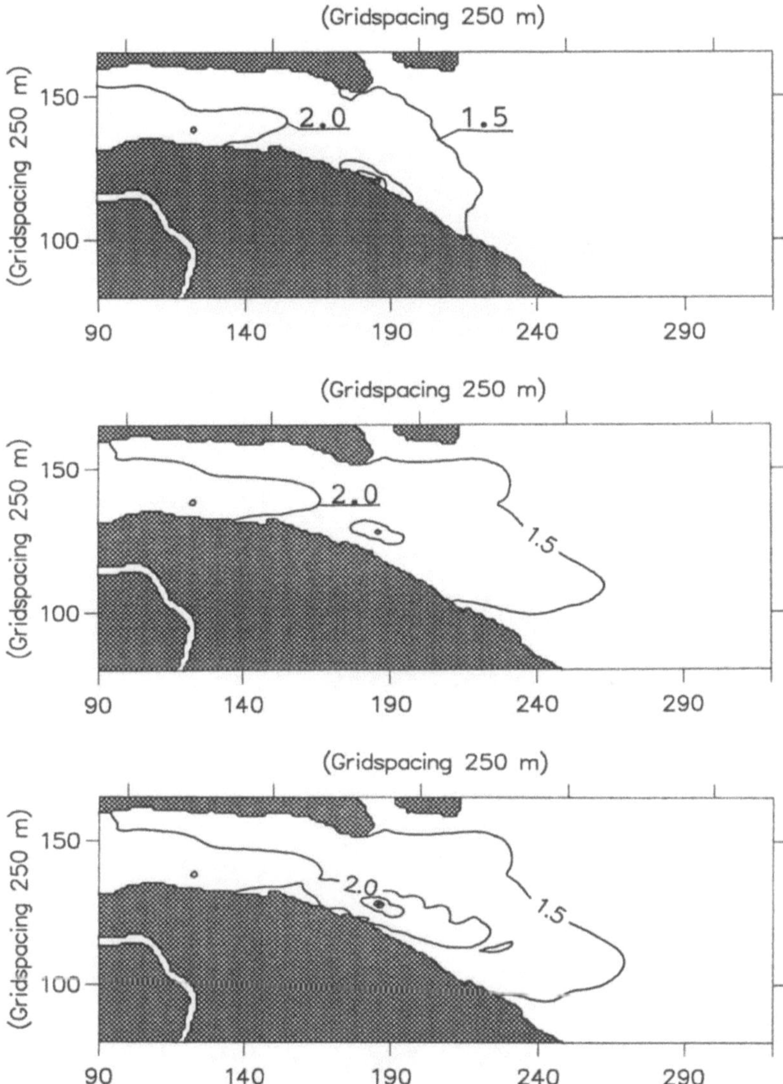

Figure 11.11. The BOD concentration distributions simulated for Phase 1 with the different loading concentrations in the far-field during the summer period. The unit in this figure is mg/l. The isolines are with 1.5, 2.0, 2.5, 3.0, 3.5 and 4.0.

264

a) 5.0 mil. m^3/d waste water with a BOD concentration: 200 mg/l, Scheme I
b) 5.0 mil. m^3/d waste water with a BOD concentration: 200 mg/l, Scheme II
c) 5.0 mil. m^3/d waste water with a BOD concentration: 200 mg/l, Scheme III

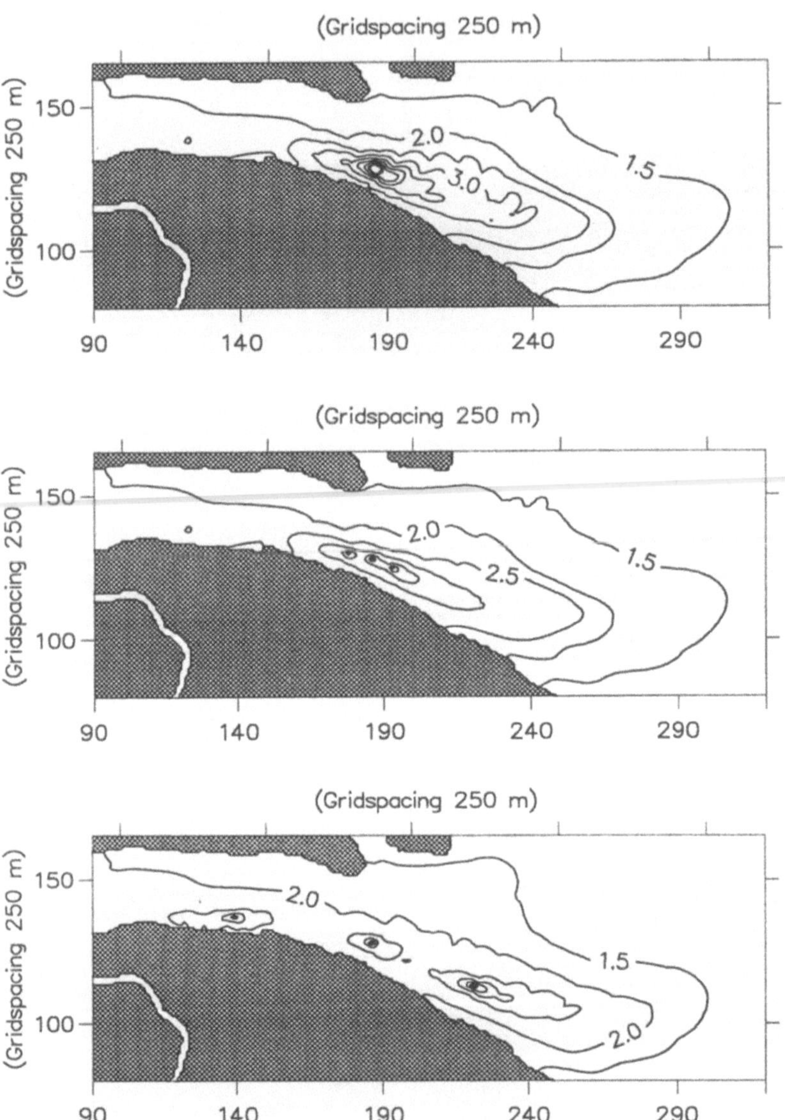

Figure 11.12. The BOD concentration distributions simulated for the Ultimate Phase with the different schemes of the layouts in the far-field during the summer period. The unit in this figure is mg/l. The isolines are with 1.5, 2.0, 2.5, 3.0, 3.5, 4.0, 4.5 and 5.0.

a) 1.7 mil. m^3/d waste water with a BOD loading concentration: 200 mg/l, Concentrated 4R

b) 1.7 mil. m^3/d waste water with a BOD loading concentration: 200 mg/l, Separated 4R

c) 1.7 mil. m^3/d waste water with a BOD loading concentration: 200 mg/l, Concentrated 8R

d) 5.0 mil. m^3/d waste water with a BOD loading concentration: 200 mg/l, Concentrated 4R

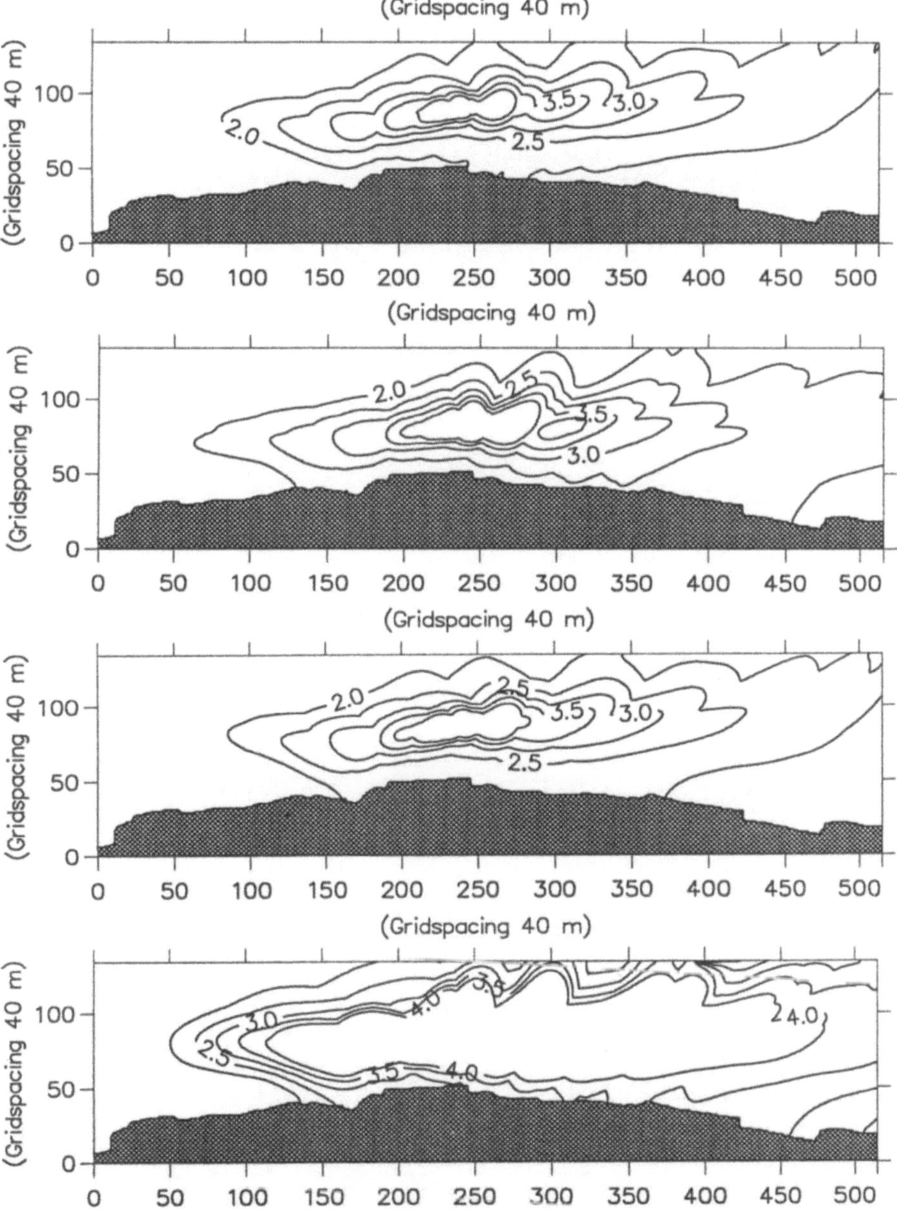

Figure 11.13. The BOD concentration distributions simulated for the alternative study in the near-field during the summer period. The unit in this figure is mg/l. The isolines are with 2.0, 2.5, 3.0, 3.5 and 4.0.

a) Summer period simulation with a BOD loading concentration: 129 mg/l, 6 Risers
b) Summer period simulation with a BOD loading concentration: 100 mg/l, 6 Risers
c) Winter period simulation with a BOD loading concentration: 129 mg/l, 6 Risers
d) Winter period simulation with a BOD loading concentration: 100 mg/l, 6 Risers

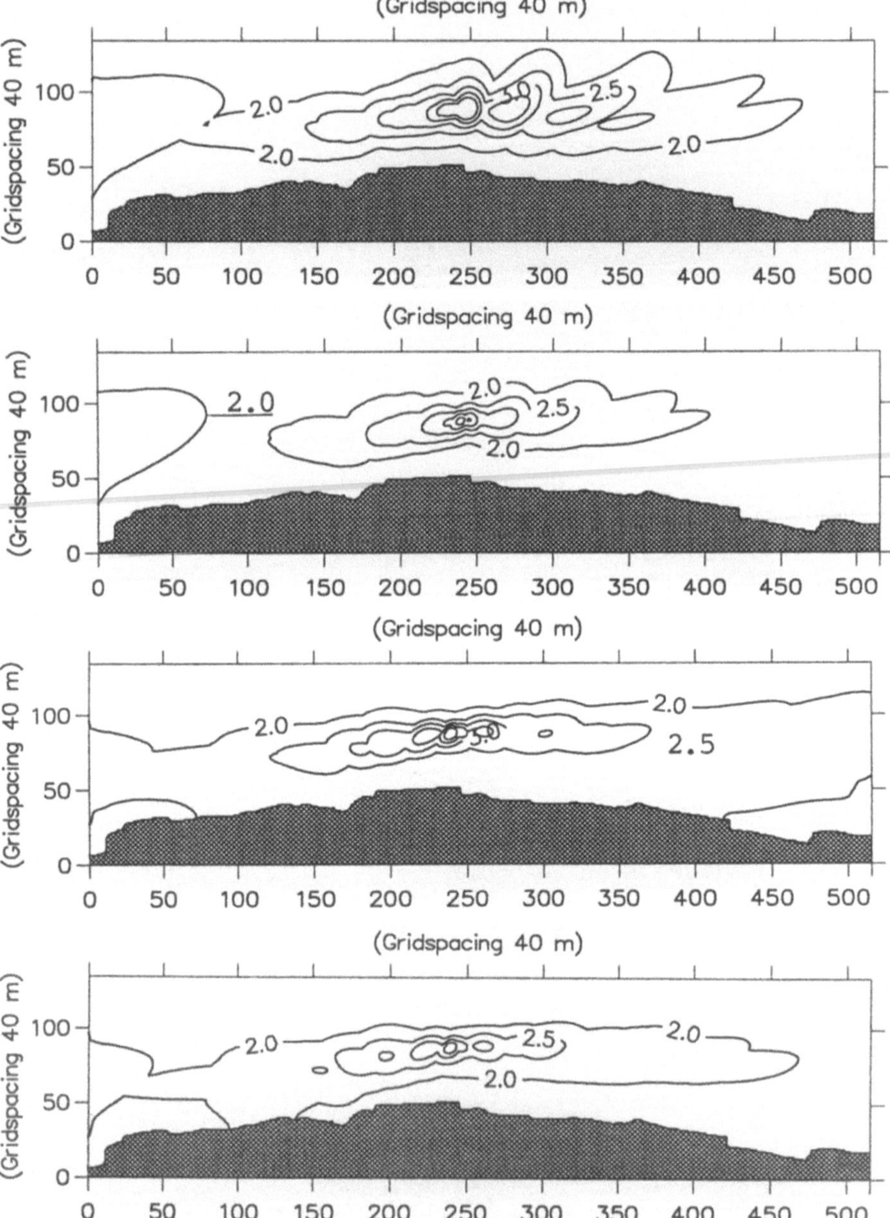

Figure 11.14. The BOD concentration distributions simulated for the study of the conceptual design with the sewage discharge of 1.7 mil. m³/day in the near-field. The unit in this figure is mg/l. The isolines are with 2.0, 2.5, 3.0, 3.5 and 4.0.

a)   Summer period simulation with a Cu loading concentration: 0.364 mg/l, 6 risers
b)   Summer period simulation with a Cu loading concentration: 1.0 mg/l, 6 risers
c)   Winter period simulation with a Zn loading concentration: 1.735 mg/l, 6 risers
d)   Winter period simulation with a Zn loading concentration: 5.0 mg/l, 6 risers

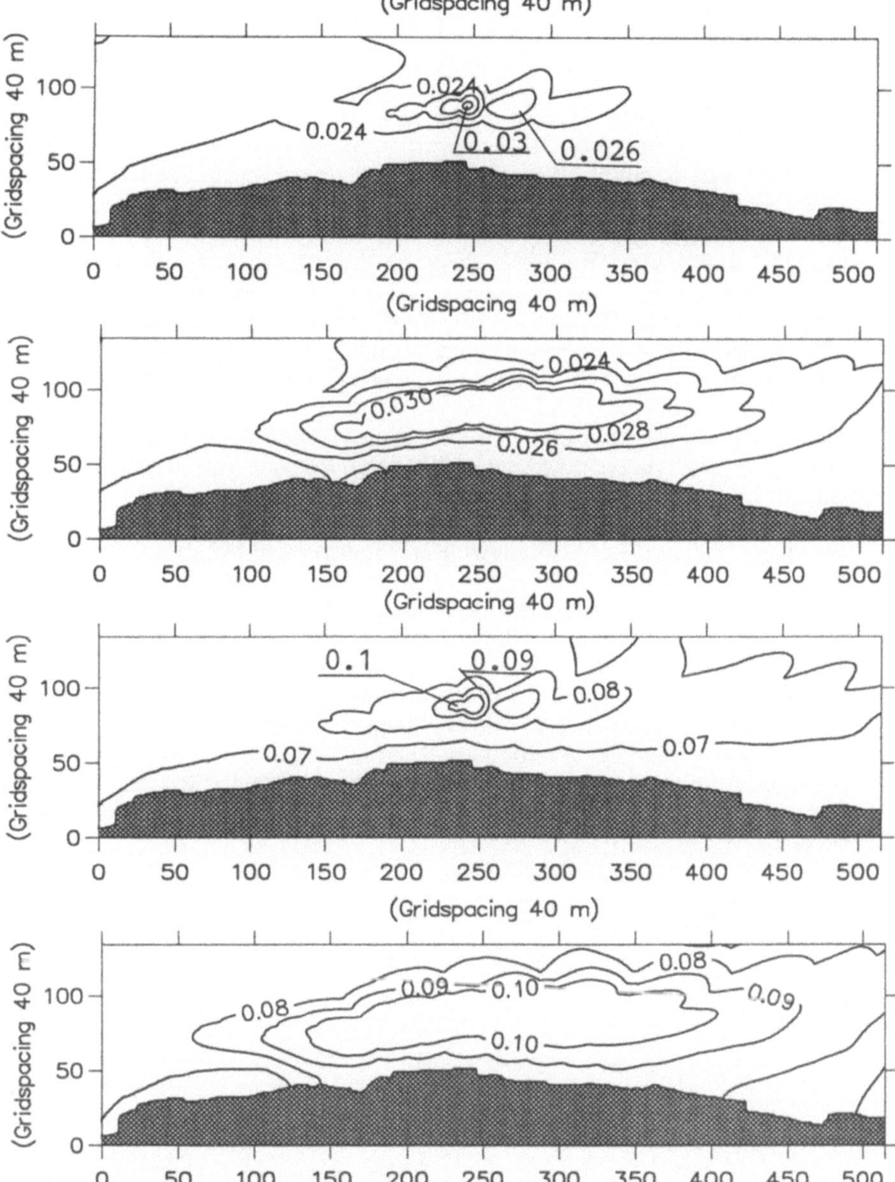

Figure 11.15. The concentration distributions of Cu. and Zn. simulated for the study of the conceptual design with the sewage discharge of 1.7 mil. m³/day in the near-field. The unit in this figure is mg/l. The isolines are with 0.024, 0.026, 0.028 and 0.03 for Cu and with 0.07, 0.08, 0.09 and 0.1 for Zn.

11.2.7 References

1. Beardsley, R. C., R. Limeburner, H. Yu and G. A. Cannon (1985). "Discharge of the Yangtze River into the East China Sea". Continental Shelf Research Vol. 4, Nos. 1/2, pp. 57-76.
2. Wang, K., J. Su, and L. Dong (1983). "Hydrographic Features of the Changjiang Estuary". In Proceedings of International Symposium on Sedimentation on the Continental Shelf with Special Reference to the East China Sea, Hangzhou, China, pp. 125-133.
3. Chen, C., R. C. Beardsley, R. Limerurner and K. Kim (1994). "Comparison of Winter and Summer Hydrographic Observations in the Yellow and East China Seas and Adjacent Kuroshio during 1986". Continental Shelf Research, Vol. 14, No. 7/8, pp. 909-929..
4. Lu, Q.M. and Warren, R. (1992). "Current Simulations in the Taiwan Strait". XIV Conference on Ocean Engineering, Taiwan, pp. 128-147.
5. Ning X., D. Vanlot, Z. Liu, Z. Liu (1988). "Standing Stock and Production of Phytoplankton in the Estuary of the Changjiang (Yangtze River) and the Adjacent East China Sea". Marine Ecology - Progress Series, Vol. 49: pp 141-150.
6. Danish Hydraulic Institute (1995). "Supporting Modelling Studies for Feasibility of Proposed Disposal of Shanghai Waste Water to Sea". Interim Report.
7. Shanghai Research Institute of Environmental Protection (1992). "Sewerage Discharge Standard for the Yangtze River Estuary and Hangzhou Bay". Draft version.
8. Chinese National Environmental Protection Bureau (1989). "National Standard of the Peoples Republic of China, Standard for Discharging of Combined Sewerage". GB8978-88.
9. Danish Hydraulic Institute (1995). "Supporting Modelling Studies for Feasibility of Proposed Disposal of Shanghai Waste Water to the Yangtze River Estuary, Phase II". Preliminary Report.
10. Interconsult in association with Mott MacDonald (1995) "Strategic Planning Report". For Shanghai Planning Report". For Shanghai Second Sewerage Project.

11.3   The Thames Estuary

The Thames River estuary is an ongoing case study of the restoration of a damaged ecosystem. From about 1920 to 1964, most of the estuary was devoid of fish life. Now, some 120 species of fish can be found there. This recovery is attributed to (1) identification, quantification, and modelling of the factors controlling concentrations of dissolved oxygen and other determinants of survival of the biota, (2) determination of the pollutant inputs of individual discharges to the Thames, and (3) effective control through waste regulation and treatment based on minimum dissolved oxygen requirements for segments of the river.

11.3.1   Recent History of a Maturing Remedial System

Noxious odors from the Thames have been documented since Elizabethan times (1, 2). Even so, there was a large fishery based on whitebait, shad, smelt, salmon, and sea trout. However, nineteenth century increases in population, importation of water, and the convenience of the flush toilet caused accelerating degradation of the estuary and its environs. The fishery was failing and drinking water from the river was a source of cholera. Remedial actions responded to aesthetic rather than health reasons. The sulphurous odors coming from the river in the summer, caused people to complain of headaches and nausea, and in London the situation was severe enough to disrupt the workings of government. Gameson and Wheeler (2) relate how by the mid-nineteenth century sheets soaked in disinfectant were being hung in the Houses of Parliament in an attempt to counteract the stench.

In 1856, the Metropolitan Board of Works was established by Parliament and was charged with preventing any sewage from flowing into the river within the Metropolitan District. A comprehensive system of drainage was constructed in 1865 to diverte sewage downstream to be discharged during the ebb tide from outfalls at Beckton on the north bank and Crossness on the south bank (see Figure 11.16 and Table 11.8). This improved conditions within London proper where the old outfalls had been located, and moved the waste downstream to the vicinity of the new outfalls. (Unfortunately, discharging twice the quantities of wastes into estuaries during half the time into ebb tides that become flood tides during the other half increases hydraulic requirements and disposal costs without corresponding benefits. The sea is another source within which previously discharged pollutants are returned upstream and average concentrations are the same as for a continuous discharge.) In 1882, a further attempt was made to rectify the situation with the creation of two sedimentation channels at the outfall sites, and the sewage was treated with coagulants so that the solids would settle and not disperse. The treated sewage sludge was then periodically dredged, transported to sea, and dumped there. Although the water quality of the Thames improved slightly, there were still numerous complaints of offensive odors, especially during the dry summer period. Some fish life reestablished itself, with whitebait reappearing at Gravesend in 1892 and Greenwich in 1895; in the latter year, flounders were caught in the upper reaches for the first time in twelve years (2).

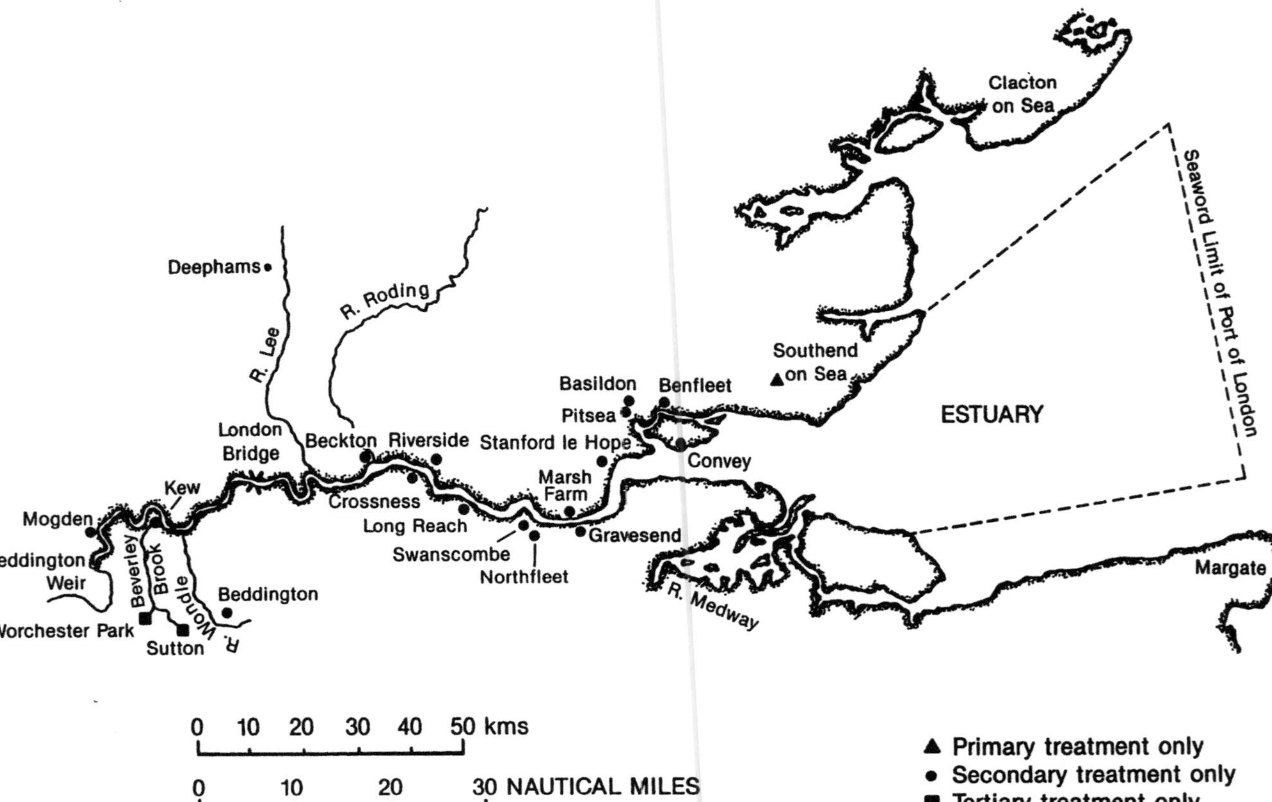

Figure 11.16. The Thames Estuary and locations of major sewage discharges.

Table 11.8. Principal mmicipal sswage treeatment tlants and discharges to the Thames Estuary

Treatment Works	Year commi-sioned	Loca-tion [1]	Flow, m^3/d thousands		EOD [2] tons/day		Treatment 1994
			1983	1994	1983	1994	
Mogden	1936	6	446	470	18.8	8.4	Secondary
Worcester Pk	Early 20th Century	---	16	24	0.8	0.7	Tertiary
Sutton			8	-----	0.4	----	Tertiary
Beddington	1970	—	86	85	3.8	3.0	Secondary
Deephams	1957	—	185	214	5.2	2.2	Secondary
Beckton	1935	52	818	1181	23.5	26.7	Secondary
Crossness	1964	56	500	642	32.8	25.1	Secondary
Kew	1963	14	34	35	0.8	0.6	Secondary
Riverside	1935	58	87	104	14.8	10.9	Secondary
Long Reach	1920	65	100	197	96.8	15.3	Secondary
Basildon			18	28	2.1	2.4	Secondary
Benfleet			6	6	0.5	0.5	Secondary
Canvey			4	13	1.1	0.7	Secondary
Marsh Farm/Tilbury			32	32	16.0	1.1	Secondary
Pitsea			6	6	0.3	0.5	Secondary
Southend			33	43	15.0	12.8	Primary
Stanford-le-Hope			5	5[3]	0.6	0.4[3]	Secondary
Gravesend			7	7[3]	3.1	1.1[3]	Secondary
Northfleet			7	7[3]	1.1	1.1[3]	Secondary
Swanscombe			1	1[3]	0.2	0.2[3]	Secondary
TOTAL			2485	3100	237.7	113.7	

Sources: Hoffman, (3), Jones (8), Norton (5), Woodcosk (3), personal communications, 11995

Note to Table 11.8.   (1) Distance from Teddington Weir, km. (2) Effective oxygen demand = 1.5 (B + 3N),   where B = 5-day BOD and   N = oxidizable nitrogen. (3) estimated.

Freshwater fish also began to move downstream, but several attempts to reestablish the commercially important salmon fishery were unsuccessful.

Interceptor sewers were constructed between 1900 and 1910, and the emphasis of pollution control shifted to improving the quality of treatment. The first secondary treatment plant at Beckton, a paddle-aeratior activated-sludge plant was built in the 1930s to treat about one-quarter of the total flow. In 1936, a new sewage works was built upstream at Mogden to replace twenty-eight smaller treatment works. Despite these improvements, the period from 1930 to 1950 was one of progressive deterioration in water quality, particularly dissolved oxygen levels. This was caused by (1) the population was still increasing, and thus inputs of sewage also increased; (2) during World War II bombs damaged the sewage treatment works; and (3) non-degradable detergents were introduced that increased the nutrient loading and reduced the capacity of the river for self-purification.

After the war, improvements in sewerage facilities were postponed for several years while the city concentrated on other reconstruction efforts. Then, in 1955, a new, efficient primary settlement plant was constructed at Beckton, followed by additional activated-sludge plants, which brought the portion of total flow receiving biological treatment to about 50 percent by 1960 (2). With further improvements in treatment since that time, the water quality in the Thames Estuary has become much better.

11.3.2   The DSIR Dissolved Oxygen Model

In 1949, the Water Pollution Research Laboratory of the Department of Scientific and Industrial Research (DSIR) began a fifteen-year study of the factors affecting the water quality of the Thames Estuary (1)

This effort was stimulated by complaints of offensive sulphurous odors and reports that the fumes were causing brass to tarnish rapidly and lead-based paints to discolor. Because sulfide was present in the water only when anaerobic conditions prevailed, the investigation was aimed primarily at studying the factors affecting the distribution of dissolved oxygen (78). The study had three parts: (1) river water was sampled and analyzed throughout the lengthof the estuary; (2) water quality records were examined to reveal changes in the condition of the estuary during the previous fifty years; and (3) models were developed to predict the movements of effluents discharged into the estuary.

As these models were developed, they were validated by comparisons between predicted and observed values of dissolved oxygen. Later, attempts were made to predict the condition of the estuary if certain changes were made in a number of the variables (e.g., amounts of pollutants discharged, source locations, different freshwater flows). The model resulting from this study has withstood the test of time--the agency responsible for water quality (presently, the Thames Water Authority)-- continues to use it in managing the estuary.

The mathematical model that is used assumes the estuary to be a barrier-free tidal river with boundaries at the tidal limit (Teddington Weir) and the sea. This 150 km reach is divided into 3.3-km segments, each considered to be of uniform chemical composition throughout. Composition is determined by (1) the inputs of pollutants from all sources; (2) the movements of these pollutants by diffusion and advection into freshwater sources, including the river and groundwater, and by tidal movements and exchange with seawater; and (3) the rate of decay or removal of the constituents by chemical, physical, or biological means (1, 5).

Although simple in nature and selective in the variables it takes into consideration, the model works. It accurately predicts dissolved oxygen, ammonia, and oxidized nitrogen concentrations, as well as the temperature of the river along its length, all on the basis of established decay characteristics of these water quality parameters. Current interest lies in modern physical and umerical models that have been developed along with advanced computer technology of the Thames estuary for designing sediment transport and other harbor improvements (6).

11.3.3 Hydrography of the Thames Estuary

Water movements and the distribution of dissolved oxygen in the Thames are dominated by freshwater flow from upstream and by the effects of t;des. Low freshwater flows lead to lower oxygen reserves and reduce the estuary's assimilative capacity for sewage effluents.

Over a fi3fty-year period (1925-74), river flows measured daily ranged from 0.9 to 709 m /s. Withdrawals from upstream of Teddington for municipal water supplies have increased from 4.3 m3/s in 1885 to 17.0 m~/s in 1970-74. During periods of low river flow, this removal significantly reduces the freshwater flow into the estuary, and causes oxygen concentrations to drop (1, 2).

The average tidal range at Teddington is 2 m; it increases to 6 m at London Bridge, and then gradually decreases to 4m at Southend. Depending on the freshwater flow, tidal state, and tidal range, pollutant inputs to the estuary are dispersed considerable distances upstream and downstream during successive tidal cycles. Recognizing that the dissolved oxygen contents of two samples taken at the same point in the estuary under different tidal states would be likely to differ considerably, the DSIR model reduces the dissolved oxygen data to a common predicted tidal state by replacing the true sampling position with the location of the water at "half-tite." Half-tide is defined as the instant when the volume upstream (to Tedtington Wier (see Fig. 11.8) is the mean value for the average tidal cycle . The result is that samples taken at low water are, in effect, moved upstream ant samples taken at high water are moved downstream. Statistical analysis of tissolvet oxygen measurements has revealed an essentially linear relationship between dissolved oxygen and freshwater flow. Effects of temperature photosynthesis, and seasonal loading upon dissolved oxygen content have been more difficult to determine.

DSIR studies found that when the oxygen concentration falls below 10 percent saturation, nitrification (oxidation of nitrogen compounds to nitrites and nitrates) ceases, and denitrification (reduction of nitrates to nitrogen) occurs. When reserves of nitrates are exhausted and fully anaerobic conditions are established, sulfates are reduced and give rise to offensive odors. Dissolved oxygen levels have continually improved since 1950 and have essentially never fallen below 10 percent saturation in somce the mid-1970s.. Although sewage effluents still exert the greatest polluting load to the tidal Thames, improved treatment has reduced the average daily oxygen demand from over 800 tons/day in the early 1950s to 300-400 tons/day in the early 1970s and to the present 130 tons /day . This has worked without complaints of offensive odors from the river for many years and with dissolved oxygen levels above those necessary for passage of migratory fish (5).

11.3.4 Fish Populations

Early fish population data are sparse and based largely on incidental observations. There is no evidence of any fish in the late 1950s for some 68 km upstream of Gravesend, except for eels that breathe at the surface. There were no commercial fisheries on the river and sportfishing was limited to only a few isolated areas. Evidence that the river was finally on its way to recovery first appeared in 1964 with the capture of fish on power station cooling water intake screens within this previously fishless zone just above the mouth of the estuary. The number of freshwater and marine species has increased (Table 11.9) to well over 100. Seasonal lows in dissolved oxygen levels decrease the range of distribution of all species during these warmer months. Many migratory euryhaline species capable of survival in a wide range of salinities are commercially important.

Table 11.9. Cumulative fish species numbers caught on the intake screens of the Thurock power plant near the mouth of Thames Estuary, 1963-1991. Source:Thomas (7).

1963	1	1973	71	1983	106
1964	2	1974	76	1984	108
1965	4	1975	87	1985	109
1966	11	1976	91	1986	110
1967	38	1977	96	1987	110
1968	50	1968	97	1988	111
1969	55	1969	98	1989	112
1970	56	1980	99	1990	113
1971	57	1981	99	1991	114
1972	68	1982	103		

The introduction during the early 1980s of young salmon into the upper reaches of the river has proven successful in that a number of marked fish have been taken from the tidal Thames on their return to spawn (5).

11.3.5 Principal Findings and Conclutions

We conclude that improvement in the Thames Estuary did not come about by applying administratively and legally simple uniform treatment standards. Rather, the needed reductions of inputs in specific segments of the estuary, and alternative levels of biological treatment of sewage have been applied to domestic and industrial effluents throughout the estuary. Inputs have been regulated only to the extent that assures adequate oxygen supplies for fish life. Scheduling of treatment plant improvements to achieve the desired water quality was determined by the use of the DSIR model. The model predicted that significantly improved dissolved oxygen levels would be achieved by higher levels of treatment at the Beckton plant rather than at the Long Reach plant that was upgraded several years later. Application of model findings has resulted in desired water quality levels throughout the river basin using different levels at different plants. Still, generally higher levels of treatment to are anticipated to control toxic industrial wastes.

11.3.6   References

1. Gameson, A.L.H., et al. 1964. Effects of Pollution Discharges on the Thames Estuary. Water Pollution Research Technical Paper No. 11. Department of Sciontific and Industrial Research. H.M. Stationeery Office, London.
2. Gameson, A.L.H. and Wheeler, A. 1977. Restoration and recovery of the Thames Estuary. In Cairns, J., Jr. et al, Editors. Recovery and Restoration of Damaged Ecosystems. Charlottesville. University of Virginia Press.
3. Hoffman, M.R.. 1995. Personal communication. Thames Water. London.
4. Jones, L.D. 1995. Personal communication. National Rivers Authority, Thames Region, Reading, U.K.
5. Norton, M.G. 1983. Experiencess in the U.K. on the control of discharges of sewage and sewage sludee to esstruaries and coastal waters. In Myers, E.P. Ocean Disposal of Municipal Wastewater: Impacts on the Coastal Environment. Sea Grant Program, Massachusetts Institute of Technology, Cambridge. 947-1023.
6   Price, W.A., and Thorn, M.F.C. 1994. Physical models of estuaries. In Abbott, M.B., and Price, W.A. , editors.   Coastal, Estuarial and Harbout Engineers' Reference Book. E & FN Spon, London. 275-288.
7. Thomas, M. 1996. Temporal changes iin the movements and abundance of Thames Estuary fish populations. 1996. In Attrill, M.J., and Trett, M.W.. A Rehabilitated Estuarine Ecosystem: The Thames Estuary, Its Environment and Biology. Chapman and Hall, London.
8. Woodcock, P. 1995. Personal communication. Amglian Water Company, Histon, Cambridgeshire, U.K.

11.4   The Bosporus and the Sea of Marmara[1]

Istanbul, the largest city in Turkey, was founded by Megarian Greeks in 657 B.C.
as fabled Byzantium. In A.D. 330 it became Constantinople, capital of the Eastern
Roman (later Byzantine) Empire and in 1453 of the Ottoman Empire. The city is
on the northern shore of the Sea of Marmara and lies on both sides of the
Bosporus (*İstanbul Boğaz*),a 31-km-long strait from the Marmara to the Black
Sea at the northern end Turkish Straits between Europe and Asia.

European Istanbul is divided by the Golden Horn (*Haliç),* aa tidal estuary and
excellent harbor. The Golden Horn watershed includes rolling hills, valleyfloor
villages, and the Kagithane and Alibey riverbasin (26,35). Westward, along the
Marmara, the terrain ascends gently from the coast to a plateau cut by narrow
valleys perpendicular to the shoreline. Northward, along the Bosporus, the
coastline is steep and cut by sharp, narrow valleys.

Asian Istanbul has a more rugged topography. The steep Bosporus coastline
continues southeasterly along the Marmara for about 20 km to flat coastal areas of
varying widths. The five Prince's Islands lie about 7 km offshore.

The city lies in the transition zone from the Mediterranean to the humid sub-
continental climates with average summer and winter air temperatures of 50° C
and 25° C, respectively. Approximately 70 percent of the average rainfall (726
mm/y) occurs from October through March.

Istanbul lies in a second degree seismically active zone and the city has suffered
extensive earthquake damage in the past. Two faults have been located within the
city and a third, showing signs of recent activity, crosses the Bosporus 5 km
south of the Black Sea.

Most of the population growth has occurred since 1920, when it was around
500,000. It reached 1 million in 1940, 2.8 million in 1970, and some 7.5 million
by 1995 over an area of about 300 km^2. Rural in-migration since 1960 has
resulted in the tenured establishment of large squatter settlements (*gecekondu,*)
meaning overnight-built) that meet certain legal criteria and that are slowly
upgraded by the owners ahd the municipality. The growth is accompanied by
industrial expansion from 1,100 ha in 1970 to about 7,000 ha by 1995. The
population is expected to reach 9 million by the year 2000.

11.4.1 Regional Geography and Oceanography

The unique geographical and oceanographic situation of Istanbul establishes the
environmental design of its sewerage system The Turkish Straits extend about
300 km from the Aegean Sea through the Dardanelles (60 km), Sea of Marmara
(210 km), and Bosporus (31 km) to the Black Sea. A two-layer current system
develops. Mediterranean water increases in density from the excess of evaporation

---

[1] We retain the original 8th Century B.C. geographic spelling, "Bosпopus" and its
commonly used European rendering, "Bosporus" adopted by the American Board of
Geographic Names. The French "Bosphorus" is preferred by a number of authors and
agencies cited in the References. Turkish orthography is used for most proper names
and places in the text and throughout Section 11.4.8, References.

over precipitation during its circulation from the Strait of Gibraltar along the African and Levantine coasts to the Aegean Sea (Figure 11.17). From the Aegean Sea, this heavy, highly saline water flows northerly through the Turkish Straits to the anaerobic lower portion of the Black Sea. Less dense, brackish surface waters carry runoff from tributaries to the Black Sea (6). Of these, the most notable are the Danube and Dneiper Rivers, which drain areas of 840,000 km^2 and 502,000 km^2 with average flows of 6,200 m /s and 1,700 m3/s , respectively (18, 20, 22). The outflow from the Black Sea also includes any excess of precipitation over evaporation from the 423,000 km^2 of the sea itself. The salinity of the outflow is due to vertical mixing within the Black Sea of salts from the Bosporus undercurrent inflow.

The large-scale features of this two-layer circulation are shown in Figure 11.18. Bottom waters move north through the Bosporus into the anaerobic lower portion of the Black Sea. This lower portion has been anaerobic for some 7,000 years and at present occupies 88 percent of the 543,000-km^3 volume of the sea (14, 15).

11.4.2   Oceanography of the Bosporus

Focus during the 1966-71 DAMOC study was on the coastal Sea of Marmara, the southerly two-thirds of the Bosporus and the Golden Horn where most of the population and pollution are. Current emphasis includes the north-eastern Sea of Marmara and southwestern Black Sea approaches to the Bosporus.

The Bosporus is a meandering strait (Figure 11.19) about 31 km in length. It varies in widths from 0.7 km to 3.5 km and averages about 1.6 km Average and maximum depths are 36 m and 110 m. A sill between 32 m and 34 m depth about 3 km from the southerly entrance strongly affects the two-layer current system.

Dry northerly to northeasterly winds prevail approximately 50 percent of the time in winter and 80 percent in summer. Southerly to southwesterly winds occur about 2 percent of the time in August and bring warm, humid weather to the area. During winter they occur approximately one-third of the time and, when they are sufficiently strong, bring storm waves and increased sea levels to the Marmara coastline, the southerly entrance to the Bosporus, and the Golden Horn.

Long-term mean air temperatures vary from about 5° C in winter to 25° C in summer. From July 1966 to December 1967, mean air temperatures varied from about 26±5° C down to 3±1° C. Mean surface water temperatures rose from 4° C in winter to 25° C in summer and followed mean air temperatures by about 1 month.. Occasionally, low winter temperatures cause drift ice, most recently in 1954 when the Bosporus was so blocked with ice that one could walk across it (27). Chihatchef (11) reported that ice formation was even more severe in earlier times. The most extensive freezing recorded was in A.D. 732 when the Black Sea froze solid to within 90 km of the Bosporus and drift ice reached the Dardanelles.

Southerly flows of water from the Black Sea are revealed by water surface slopes between the northern and southern ends of the Bosporus. Sea level data from Úsküdar, Çubuklu, and Kavak at 7, 15, and 24 km, respectively, from the southerly entrance to theBosporus showed effects of tides, winds, and seasonal

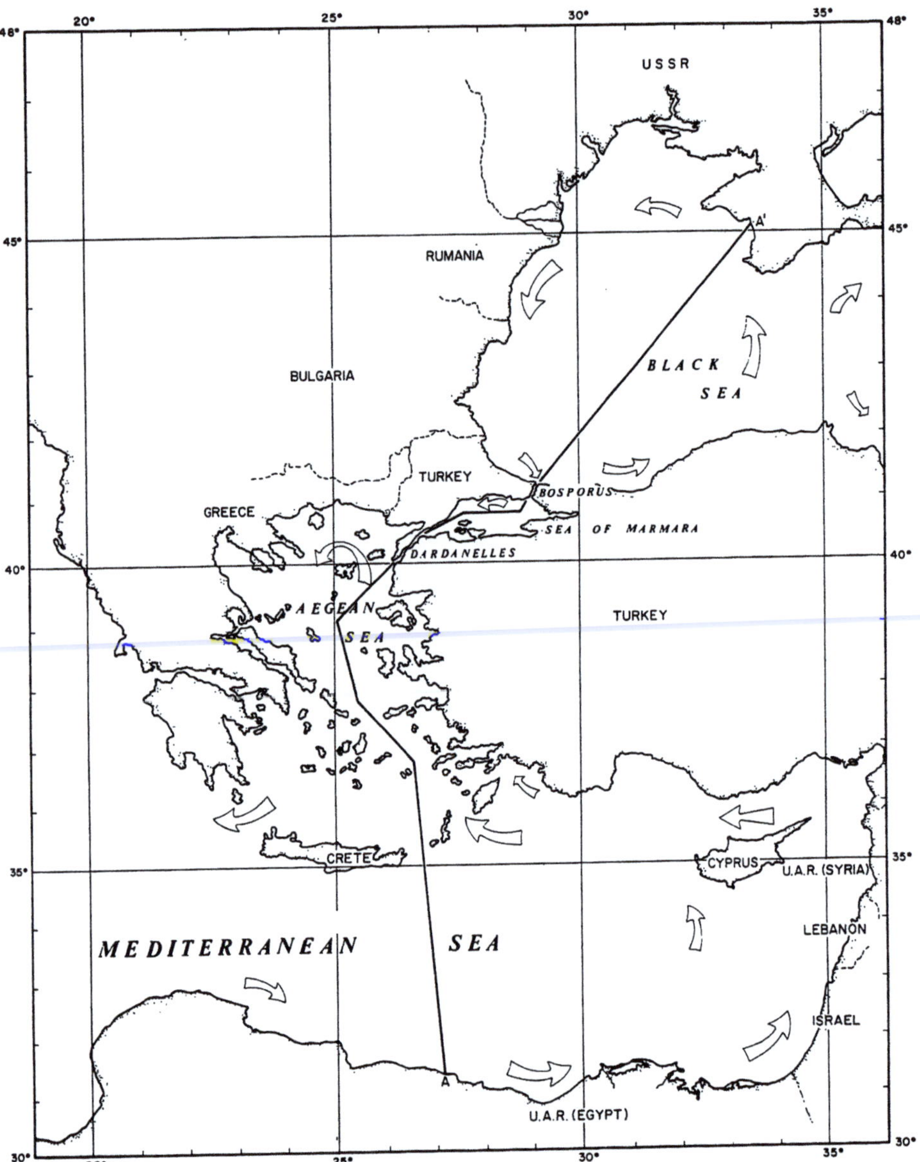

Figure 11.17. Regional marine geography of the Eastern Mediterranean, Agean, Marmara, and Black Seas. Source: Gunnerson (35).

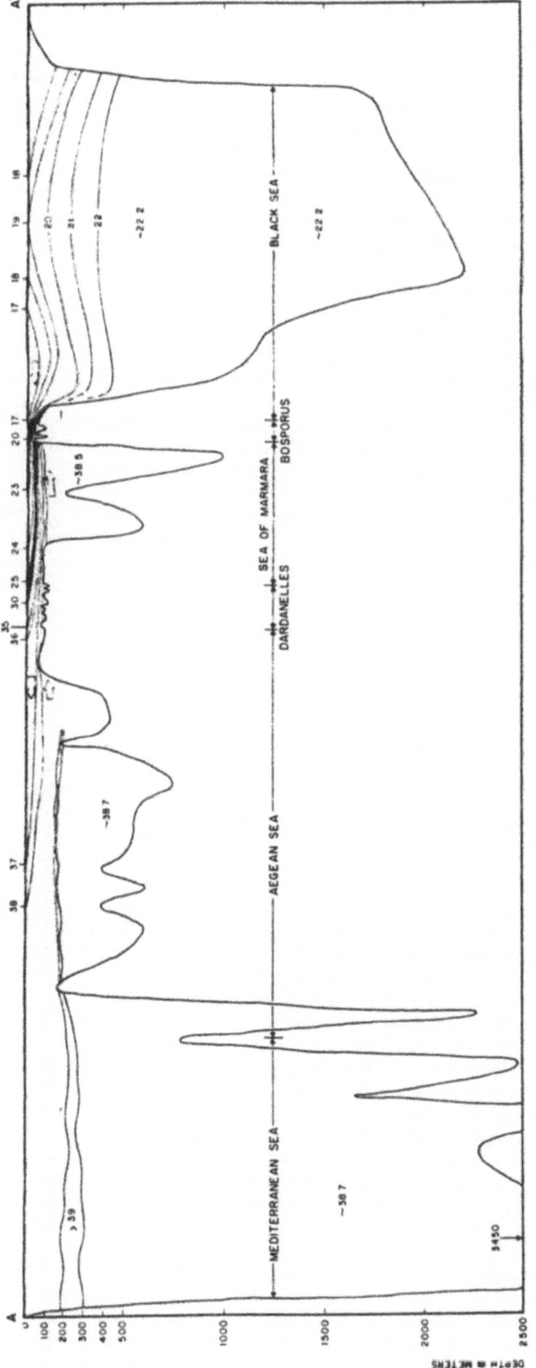

Figure 11.18. Vertical distribution of salinity in parts per thousand in the Eastern Mediterranean, Aegean, Turkish Straits, and Black Sea. Source: Bunnerson (35).

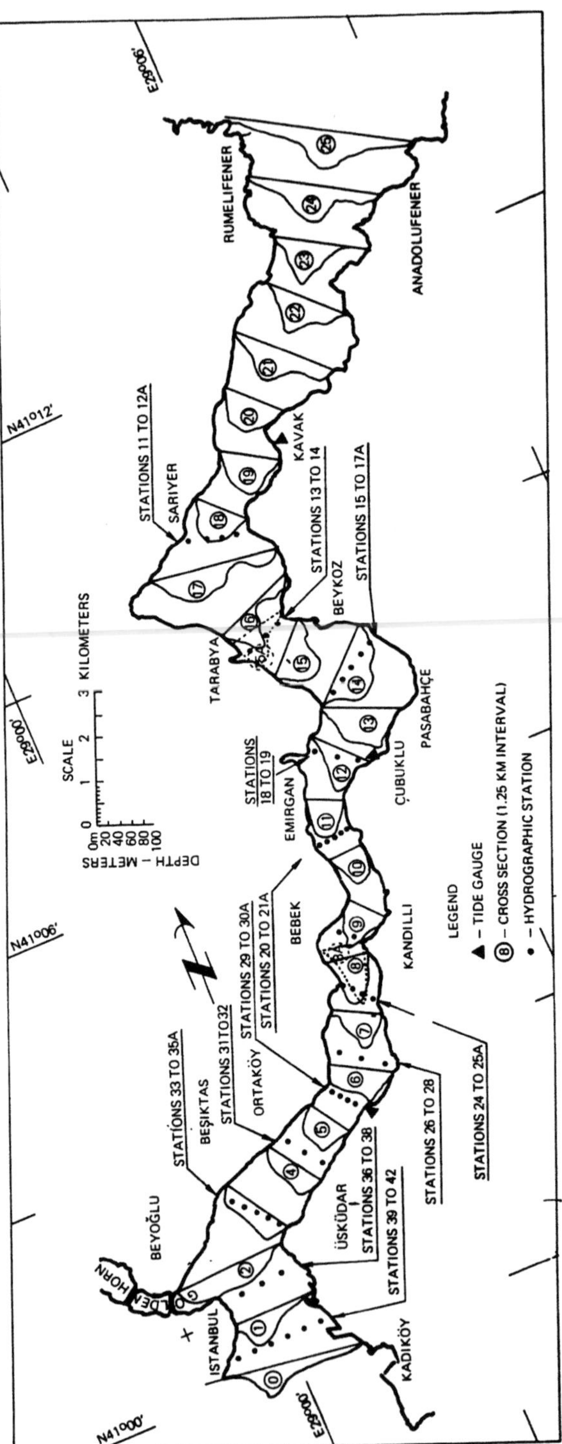

Figure 11.19. Bosporus morphology and sampling stations. Source: Gunnerson and Özturgut (20).

changes in outflow from the Black Sea. In the Bosporus, the semidiurnal tide has a range of about 2.1 cm with a large diurnal inequality. The lunar fortnightly tide has an estimated range of 5 to 20 cm. These tides are oftenbobscured by wind setup and storm tides.

During July 1966 to December 1967, monthly average water surface elevations varied from about 40 cm in late autumn to 55 cm in July 1967 at Kavak, and from 38 cm to 52 cm for the same months at Çubuklu. These elevations and the surface salinities at Çubuklu reflect seasonal stream runoff into the Black Sea. Short-term variations at the three locations are due mostly to winds. Bosporus surface currents increase from close to 100 cm/s near the Black Sea to 250-350 cm/s near Istanbul.

The average salinity of Black Sea waters at Çubuklu from July 1966 to December 1967 was approximately 17.5 parts per thousand (‰). The 16 to 17‰ values between July 19, 1967 and August 23, 1967 reflect peak runoff from Black Sea tributaries. During the same period in 1966, 17.0 to17.5‰ salinities indicated lower seasonal precipitation and runoff. Higher average salinities of 17.5 to 19.0‰were found throughout the rest of the year, with occasional values up to 25‰ during winter months. Surface salinity increases by an average 2‰ in the Bosporus, mostly in the southerly 10 km, where currents and mixing are greatest (18, 20). There are similar increases in the southerly aproach to the Bosporus.

Infrequent, short-lived reversals of surface slopes and currents are marked by high salinities at Çubuklu. They follow strong, persistent southerly winds during winter when Black Sea outflows are low. Surface slope reversals are followed within one day by salinity increases corresponding to wind stresses. Ordinarily, the wind changes after a day or so, and the sea surface slopes and salinity return to normal. Even if the wind continues, salinities at Çubuklu begin to decline by the third day, and thus indicate a new equilibrium slope.

The wind-driven current reversals that bring Sea of Marmara surface water into the Bosporus may locally increase vertical mixing due to wave action. In all cases, some high-salinity Mediterranean water remains in the deepest parts of the Strait.

The Golden Horn is approximately 7 km long. Its maximum depths is 1 m at the upper end and 40 m at the mouth; its hydrography has been summarized by Kor (26). Planned development of local water resources will reduce tidal flushing of the upper reach of the Golden Horn by about half, and will increase the need for removal of wastewaters from the estuary.

11.4.3   Two-Layer Current System in the Turkish Straits

Bosporus currents, salinity, and morphometry are intimately related. Möller's (28) estimates of 6,100 m^3/s in the lower layer and 12,600 m^3/s in the upper layer have been accepted as a working estimate of average flows. Flow variability is revealed by data published by the United States Navy Oceanographic Office (41) of 3,000 m^3/s to 30,000 m^3/s in each layer. es of 1959-60, and the 1961 Turkish Navy study (37).

During winter months, regional and seasonal changes in atmospheric pressure, upper layer water temperatures, and Black Sea fresh water inflows reduce the cross-sectional area of the upper layer is reduced by about half (20) The relation between this and average net velocities in the southerly 24 km of the Bosporus are only beginning to be known (5, 31, 39) .

In the lower layer, cross-sectional areas are reduced from $15\text{-}23\text{x}10^3$ m^2 at Saray Burnu $3\text{-}4\text{x}10^3$ m^2 at Tarabya, 20 km to the north. This implies a five- to sixfold increase in average velocity. While Möller (28) reported currents in the lower layer of 1.0 m/s to 1.5 m/s, Carruthers (9) found bottom velocities of 4 cm/s at the southerly entrance to the Bosporus, which increased to as much as 75 cm/s 1 km or more to the north. Long-term effective velocities are probably best determined from grain-size distributions of bottom sediments (12,25).

Many elements of Möller's analyses of the data from A. Merz' 1918 observations (28) have been empirically verified by e the Turkish Navy (37), Özturgut (31), DAMOC (12), Beşiktepe (5), Akyarlı, et al (2,3,13), and Ünlüata, et al (38). Following contributionds to Abbott and Price (1), these hese have been extended by using numerical model simulations by Bach, Hansen, Orhan and their co-workers (4, 22, 30). The empirical works include details of the two-layer flow separated by the 10 m hydraulic jump in the lower layer first observed by A. Merz in 1918 (28) and explained by Ünlüata, et al in 1990 (38). It is just northerly (downstream) of the 32 m sill located about 3 km from the southerly entrance to the Borporus in a zone of rapid mixing between the two layers. A second 2m internal hydraulic jump downstream from the 50 m ridge at the entrance to the Black Sea is of much interest to proponents of the Riva project to collect all sewage from the Asiatic side of the Borporus for tertiary treatment and discharge into the Black Sea through Either Riva Creek or a deep outfall (24,30).

Effective cross-sectional areas for both upper and lower layers are further reduced, particularly in embayments, by stationary eddies. Here, near-shore countercurrents up to 25 cm/s are followed to advantage by ferry and other small boat traffic.

For steady-state salinities in the lower and upper layers of the Black Sea, the long-term ratios of inflow to outflow quantities of water will be the inverse of inflow to outflow salinities. The ratios from Möller's (28) average flows and the Turkish Navy's average salinities were 12,600 m^3/s $\div$ 6,100 m^3/s = 2.07 for flows and 17.5‰ $\div$ 38.5‰ = 1/2.20 for salinity, amounting to a 6% net error. (17,41). Corresponding ratios have been calculated from from recently reported focused field work (Ünlüata (38) and Beşiktepe, (5), and iterative simulations by the Danish Hydraulic Institute reported by Hansen, et al, (22) Bach, et al (4) and summarized by Orhan (30). Their data yield remarkably precise ratios of 19,100 m^3/s$\div$9,600 m^3/s =1.99 for flows and 17.86 ‰$\div$35.54‰$\div$1.99 for salinities (30)

The stability of the interface and mixing between the two layers has attracted much scientific and environmentalist interest. As noted above, about 10 percent of the dissolved solids mixing takes place in the southerly 10 km of the Bosporus where surface salinities increase by about 2‰. A similar increase takes place in a 10 to 15 km reach of the approach to the Bosporus in an area named the Bosporus-Marmara Junction (BMJ). Here, the slow-moving Marmara bottom currents converge and the Ahirkapı outfall discharges the effluent from the service area located between the Golden Horn and Yenikapı. The dynamics of this mixing are a subject of continuing environmentalist concerns. An essential but insufficient element of these anxieties is the conjecture that the sewage suspended solids with their BOD and bacterial constituents may be entrained and subject to the same forces as the dissolved solids and salinity throughout the lower layer. During travel throughout the Bosporus., an estimated average 10 percent of the salinity in the lower mixes into the upper layer. Applying this to the 100:1 initial diluton of sewage rising in time from the lower layer to the upper layer yields a net dilution of 1000:1 for conservative constituents, a dilution equivalent to a cupful in a bathtub Since decay coefficients for coliforms result in 90 percent reductions during each 1.1 hours of travel time, determining these travel and mixing times is essential for quantifying risks of exposure. Estimateing risks of actual infection requires credible epidemiological studies (see Table 3.1).

11.4.4   Evolving Environmental Engineering Design Criteria

The ancient history of Istanbul municipal water and sanitation is preserved in Roman, Byzantine, and Ottoman cisterns, aqueducts, reservoirs, fountains, baths, latrines, drains and "black channels". Recent history includes the works of the Turkish State Hydraulic Works, the Istanbul Water Works Administration, the Istanbul Municipality Sewerage Section and, most recently, the Istanbul Water and Sewerage Administration General Directorate (ISKI). Since the founding of the Republic, urban growth has focused increasing attention on sanitation . German consultants, Ing. A. Wild from 1925-37 who endorsed combined sewers, was followed by Prof-Dr. D. Kehr, 1959-66 who advanced separaate sewers, activated sludge treatment, and discharge to surface waters.

In 1964, the Government of Turkey requested funds for engineering feasibility studies from UNDP that in 1966 were contracted by WHO to the DAMOC consortium of Daniel, Mann, Johnson & Mendenhall, Los Angeles: Alvord Burdick and Howson, Chicago; Motor-Columbus, Baden,and Chechi & Co., Washington. The DAMOC report has been the basis for subsequent works by Scandia Consult on storm drainage, and other expatriate engineering firms, including CDM International; the Danish Hydraulic Institute,:Motor-Columbus, Watson Hawksley, Taylor, Binnie and Partners, and Nedeco.   Turkish entities included Tekser, Uluslararası Birleşmiş Muşavirler, Temel Muhendislık, Istanbul Technical University, Marmara University, Middle Eastern Technical University. and the Institute of Marine Science and Technology.

During the 1970s and 80s, there were ncreasing rates of urban growth, state-of-the-art advances in construction technology and elegance in design, changes in national and local administrations, and availability of external development funds. These are reflected in information and technology transfer through joint ventures with increasing participation by Turkish engineering firms. Turkish firms have always been dominant in the construction of sanitation works.

Initial design criteria (17,12) provided for orderly investments for matching system capacity to demand and retaining financial and political flexibility for future needs and systems changes, particularly for reclaiming wastewater. This led to operational criteria including (1) staged construction and evaluations prior to extensions of system components, (2) discharging to the lower layer oof the Bosporus or Sea of Marmara so that the near-term destination for some 90 percent of the wastewater would be the lower layer of the Black Sea, (3) individual drainage basins, (4) minimum-length gravity outfalls with open-ended diameters of about 1 m, and (5) effective treatment costs and benefits (21). Coastal interceptors are more expensive than outfalls. Combining two service areas into one would increase costs by from about a third more to twice as much as for the two separate drainage areas (12).

Imaginations were fired by ideas of modern tunneling technology that surfaced in 1968. This led to the inevitable choice of an alignment from the Golden Horn to the south under Fatih in old Stamboul and on to the Marmara coastline at Yenikapı where there was room for the expanded pretreatment facilities.

Planners envirioned a tunnel system with interceptors for service areas along each side of the Golden Horn, initially even including parts of Beyoğlu and the design attraction of a pipeline catenary in the shifting bottom muds of the estuary. There would be a long outfall from Yenikapı southeasterly into the lower layer of the approach to the Bosporus. Thus there would be increased average incremental costs during the period of unused system capacity, and a discounted diseconomy of scale in sanitation costs. Then in due course, 90 percent or more of the treated effluent would, with geophysical aplomb, flow north past the area of its origin.
. With some differences in detail, the Fatih tunnel system was eventually designed and constructed. The higher costs have arguably been offset by an enhancd political status and by the opportunity for international academic and scientific recognition accorded to cutting-edge environmental technologies (23,24). There are also investment benefits in technology transfer to local constructors whose state-of-the-art skills can be used in other local and international pipeline projects.

During the mid-1970s, more changes in design criteria were adopted with Bosporus outfalls extended to the thalweg, diffusers to increase initial dilution, and numerical modeling of BOD concentrations in the lower layer ((8). Further evolution is expected in elegance of hydraulic modeling and design, extensions and consolidation of service areas, treatment, and in monitoring and cost-benefit audits. If funds materialize permit, popular and political goals for elimination of all direct discharges to marine waters may be realized (24). Meanwhile, the first phase of the Istanbul Sewerage Project has been renamed The Golden Horn Project (23).

Coliform Bacteria.   Until such time as tertiary treatment becomes universal throughout Istanbul, survival dynamics for coliform bacteria provide information for design and monitoring of systems. Times for 90 percent reduction (T_{90}s) for coliforms in marine receiving waters have been routinely included in outfall design criteria for treatment and disposal ever since the 1955 *in situ* studies for the Los Angeles Hyperion Treatment Plant showed that coliforms disappeared from surface waters by site-specifric dilution, sedimentation, and mortality factors (16). Dilutions can be estimated from continuity considerations as in the Bosporus, predicted from a variety of numerical models of diffusion and dispersion (12,16) or measured as conductivity to concentrations of as little as say, 1 part in 30,000. Average T_{90}s from eleven studies over as much as 7 hours in 10 to 22° C in the Bosporus and Sea of Marmara averaged 1.1 h, consistent with the those for raw sewage around the world (see Table 3.2). T_{90}s from the Golden Horn were 2 to 3 times this value, presumably because of older sewage in the diluting water.

Discharges to the Upper or the Lower Layer? Nearshore ischarges to the upper layer would would require a higher degree of treatment followed by extended diffuser sections providing maximum feasible initial dilutions.In the homogenous upper layer , effluent would rise to the surface regardless of dilution.

In contrast, sewage ($\rho = 0.998$) discharged to the lower layer ($\rho = 1.028$) with an initial dilution of only 2 parts of bottomwater to 1 of sewage will have a density of 1.0167, and would diffuse slowly within and through the stable boundary layer. Theoretical studies (Chapter 4) and field observations of point discharges through pipe openings with diameters on the order of 1 m into well-mixed California surface waters at depths of 10 m typically have shown average initial dilutions of 20:1 or more (7). This is an order of magnitude greater than the 2:1 mixture indicated above to be sufficient. A 20:1 mixture would have a density of 1.022. For design purposes, discharges into 10 m or more of boundary and lower-layer water without a diffuser section would provide for spreading of the sewage:seawater mixture within or below the boundary layer. .

Sewage discharged into the lower layer of the Bosporus disperses within the lower-layer flow. An estimated 90 percent of the flow with attenuated non-conservative constituents reaches the Black Sea within about 18 hours and the rest diffuse into the upper layer. Its impact upon the Bosporus lower layer has been estimated in terms of BOD loadings (8). Total BOD_5 loading for systems discharging into the Bosporus is expected to be 527.6 ton/d by the year 2020. This corresponds to a BOD loading of 173 ton/d, which, for the average flow of 6,100 m /s, would depress oxygen concentrations in the lower layer by up to 0.33 mg/ℓ.

Mediterranean water in the lower and boundary layers containing a 20:1 bottom water:sewage mixture then flows into the Black Sea. Further sinking and spreading of water from the 8osporus into the Black Sea has been detailed by Bogdanova and his associates (6) , Tolmazin (36), and more recently by Ünlüata, dt al (38).

Some Outfall and Treatment Considerations. Locations and numbers of outfalls discharging to the lower layer in either the Bosporus or Sea of Marmara are determined by topography and the availability of space for outfall and headworks construction. The continuing evolution and examples of alternative outfall locations are shown on Figure 11.20. Flotation and grit removal are the first-phase systems whose costs and benefits can be evaluated prior to upgrading. Relative costs of interceptors, pumping plants, energy, land, rights of way, headworks and pretreatment facilities, and outfalls may favor the construction of some other interceptors .

For a conservatively estimated minimum flow of 3,000 m^3/s in the lower layer, and assuming (1) that essentially all discharges into the Sea of Marmara end up in the Bosporus, (2) complete cross-sectional mixing, and (3) and no dispersion into the upper layer, there would be enough dilution water at 20:1 for 52,000,000 people at 250 liters per capita per day (lcd). For a population of 9,000,000, there would be an approximately 110:1 dilution. Concentrations of coliforms, BOD, and other non-servative constituents would be further reduced according to their decay coefficients and travel times.   Practical control of conscrvative toxic

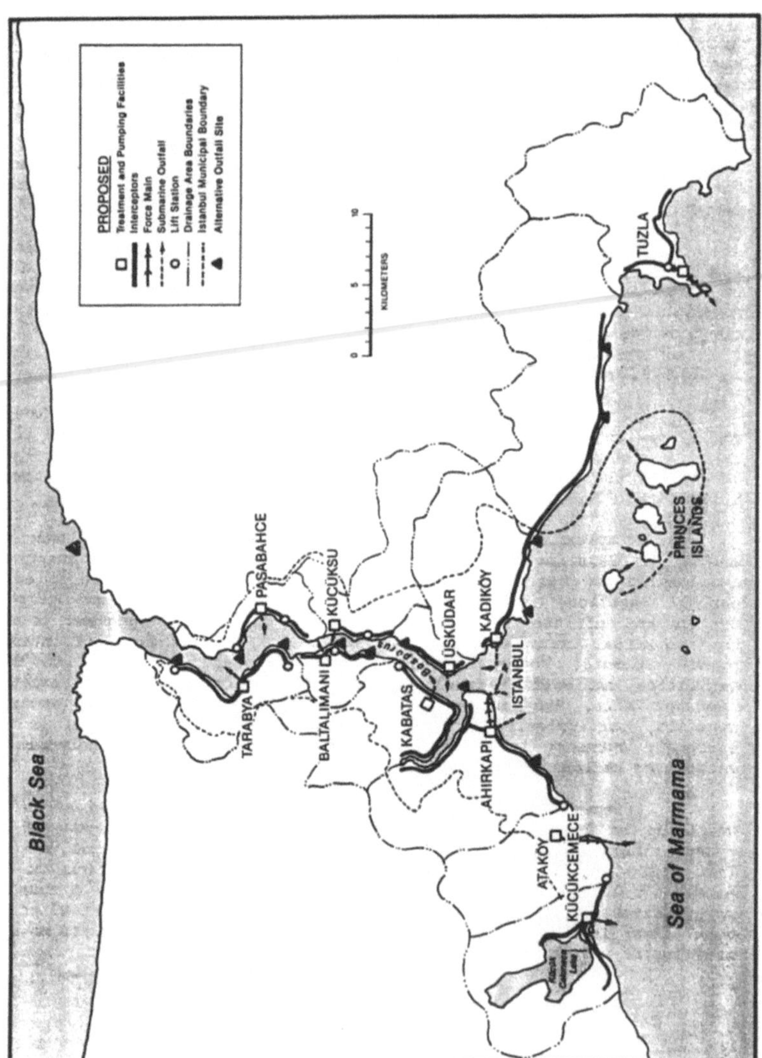

Figure 11.20. Existing and proposed Istanbul outfall locations. Sources: DAMOC (12), ISKI (24), Gunnerson (18), Orhan (30).

constituents is another matter. Industrial wastes can be controlled at their sources. Toxic constituents in consumer wastes require either their removal from the market, or very expensive municipal treatment.

11.4.5   Enviroonmental Impacts of Outfall Alternatives

Effluent Dicposal into the Upper Layer of the Bosporus   The composition and distribution of planktonic and benthic organisms in both the surface and lower layers òf the Sea of Marmara, Bosporus, and Black Sea have been summarized by Caspers (10), Sverdrup et al. (34), and Zenkevitch (43). The gradual lowering of salinity in the surface layer from the Aegean to the Black Sea is accompanied by an impoverishment of both planktonic and benthic organisms. Nevertheless, an important fishery is found entirely within the upper layer of the Black Sea, Bosporus, and Sea of Marmara.

The lower layer of the Black Sea is anaerobic. Large populations and biomasses of anaerobic bacteria occur here in a climate of hydrogen sulphide. There is a stable boundary between the upper and lower layers at depths established by freshwater inflow and by the Coriolis effects on surface circulation patterns. The boundary layer marks the lower limit of plankton and benthos generally at a depth of less than 200 m.

Effluent Dicposal into Lower Layer of the Bosporus.   The lower layer of the Black Sea is anaerobic. Large populations and biomasses of anaerobic bacteria occur here in a climate of hydrogen sulphide. There is a stable boundary between the upper and lower layers at depths established by freshwater inflow and by the Coriolis effects of surface circulation patterns. The boundary layer marks the limit of plant and animal life at about 200 m, below which bacteria thrive.

There is a continuous flux of salinity and of dissolved nutrients into the upper layer. Doubling times for phosphorus concentrations in the lower layer due to Istanbul waste discharges are estimated at 700 yr to 16,000 yr (18) and concentration increases in the upper layer are expected to occur very slowly.

Available data on heavy metals in the Black are also of interest. Studies by Spencer and Brewer (33) indicate that copper and zinc are precipitated from the lower layer as insoluble sulfides. Nickel and cobalt tend to remain in solution, presumably because of their ability to form soluble thio complexes. These studies provide insights into similar fractionation that occurs in anaerobic digestion as shown by relative concentrations of copper, zinc, and nickel in City of Los Angeles other effluents and sludges.

An overall negligible impact of Istanbul sewage discharged into the lower layer of the Bosporus is accordingly indicated. This may support cost-benefit decisions to delay removals of suspended solids from Istanbul sewage, which would then require conventional anaerobic digestion. The solids which remain in the effluent will be stabilized in the largest anaerobic digester in the world--the Black Sea.

Effluent Dicpocal to the Sea of Marmara.   There is increasing public concern over municipal and industrial wastewater discharges to the Sea of Marmara. Southeast of Istanbul, increasing discharges of these wastes into the shallow circulation of Izmit Bay since about 1950 has led to a serious degradation of its waters, which has led to concern that the assimilative capacity of the lower layer of Marmara is similarly limited. This led Camp-Tekser (8) to recommend a 25.5 km long interceptor to eliminate the Kartal outfall and discharge the effluent at a new lower layer location at or near the entrance to the Bosporfus

11.4.6 Proposed and Constructed Outfalls

The continuing evolution of engineering and environmental criteria for Istanbul waster management is revealed in the numbers and locations of facilities since 1968 is shown on Figure 11.20. Final design and scheduling decisions depend upon topography, bathymetry, economic and population growth, and regional internal and external funding exigencies. Meanwhile, three outfalls were constructed by December 1995 with due regard for site-specific conditions. Shore areas near the Baltalimı and Úsküdar outfalls contained historic structures whose preservation required use of flexible joints. Both were constructed by the Turkish Alarko Company, Istanbul. (32).

Ahirkapı. The Ahirkapı outtall was completed in 1989 at a cost of $13.5 million with a capacity of 12.0 m^3/s. The submarine section consists of twin 1.6 m diameter pipes, 1100 m in length,with 600 m diffuser sectiion discharging at depths of f 30 to 37 m depths. The land section leads from the headworks at Yenikapı that provides preliminary treatment for grit and flotables removal. The outfall is buried with a minimum 2 m of cover.

Baltalimı. The Baltalimı outfall was completed in 1994 at a cost of $9.5 million and a capacity of 9.90 m^3/s. It consis of twin 1.7 m diameter pipes, 270 m in length with diffusers discharge at a depth of 75 m. The pipes were pulled by a winch on the opposite shore into in a prepared. trench.

Úsküdar..

The Úsküdar outfall was completed in 1994 at a cost of $5.1 million for a capacity of 1.33 m^3/s. it consists of twin 1.2 m diameter pipes, 300 m in length. The pipes were pulled by an onshore winch against an anchored offshore pulley.

11.4.7   References

1. Abbott, M.B., and Price, W.A., editors. 1994. Coastal, Estuarial, and Harbour Engineers' rRerence Book. E&F Spon (Chapman & Hall), London.
2. Akyarlı, A., and Arisoy, Y. 1993. Oceanographical measurements for the Sea of Marmara crossing of the Hamidabab natural gas pipeline system. Proc third internaaational Offshore and Polar Engineering Conference, Vol 2, Singapore, 702-707.

3. Akyarlı , A., and Arisoy, Y. 1994. Oceanographic measurements for the tube-tunnel crossing of the Bosphorus. Proc, Fourth International Offshore and Polar Engineering conference, OOsaka.

4. Bach, H.K. 1994 . Environmental model studies for the Istanbul master planning. Part II. Water Quality and Eutrophication. Proc. International Specialized Conference on Marine Disposal Systems. İstanbul International Association on Water Quality, London. 167-176.

5. Beşiktepe, ş. T., Özsöy, E., and Latif, M.S. 1994 Sewage outfall plume in the two-layer channel: an example of İstanbul outfall. Proc. International Specialized Conference on Marine Disposal Systems. İstanbul· International Association on Water Quality, London. 95-112.

6. Bogdanova, A. K. 1961. Raspredelerii Sredizemnomozskikh vod n Chonon More. Okeanologiya, 1(6):983-91. Engl Transl. 1963; The distribution of Mediterranean waters in the Black Sea. Deep Sea Research, 10: 665-72.

7. Bureau of Sanitation. 1955. Oceanography of Santa Monica Bay. Final Report. Dept. of Public Works, City of Los Angeles.

8. Camp Tek-Ser. 1975. Istanbul Sewerage Project. Master Plan Revision. Report to Illerbankasi, Government of Turkey. Camp, Dresser, McKee. Boston, Massachusetts.

9. Carruthers, S. N. 1963. The Bosporus Undercurrent. Nature, v. 201, 363-65.

10. Caspers, H. 1957. The Black Sea and Sea of Azov. Treatise on Marine Ecology and Paleccology, Memoir 67, v. 1. Geol. Soc. of America, 801-90.

11. Chihatchef, P. 1855. Consideations historiques sur les Phnomees de Congelation constates dan les Basins de la Mer Noir. Bull des Sciences. Annuaire Meteorologique de France, Paris, vol. 3, pp. 12-37.

12. DAMOC. 1971. Master Plan and Feasibility Report for Water Supply and Sewerage for the Istanbul Region. Daniel, Mann, Johnson, and Mendenhall, Los Angeles, California.

13. De Filippi, G.G., Iovenitti, L., Akyarlı , A. 1986. Current aanlysis iin the Marmara-Bosphorus junction. Proc. 1st AIOMM Congress, Italy. 5-25.

14. Degons, E. T., and Ross, D. A. 1972. Chronology of the Black Sea over the Last 25,000 Years. Publ. WHOI 72-73. Woods Hole Ocenaographic Institution, Woods Hole, Massachusetts.

15. Deuser, W.G. 1973. Evolution of anoxic conditioners in the Black Sea durong the Holocene. In Degens. E.T., and Ross, D.A. Eds. The Black Sea: Geology, Chemistry, and Biology. Memoir 20, American Assoc of Petroleum Geologists, Tulsa Oklahoma.

16. Gunnerson, C. G. 1959. Sewage disposal in Santa Monica Bay. Jour. San. Engr. Div., ASCE, Proc. vol. 84, no. SA1, Paper 1534 (1958), pp. 1-28; Trans. ASCE, vol. 124 (1959), pp. 823-51.

17. Gunnerson, C.G. 1968. Internal Report to Files, DAMOC, Istanbul

18. Gunnerson, C. G. 1974. Environmental design for Istanbul sewage disposal. Jour. Env. Engr. Div., Amer. Soc. Civil Engrs. 100 (EED):101-18.

19. Gunnerson, C. G. 1975. Discharge of sewage from sea outfalls. In A.L.H. Gameson, ed., Proceedings, International Symposium on Discharge of Sewage from Sea Outfalls. Pergamon Press, New York, pp. 415-25.

20. Gunnerson, C. G., and Özturgut, E. 1974. The Bosporus. In D. A. Ross, ed., The Black Sea. Amer. Assoc. Petroleum Geologists, pp. 99-113.

21. Gunnerson, C. G., Sungur, E., Bilal, E., and Özturgut, E. 1972. Sewage disposal in the Turkish Straits. Water Research 6:763-74.

22. Hansen, I.S., Vested, H.J., and Latil, M.A. 1994, Environmental model studies for the Istanbul master planning. Part I, Hydrodynamical design basis and marine disposal of wastewater. Proc. International Specialized Conference on Marine Disposal Systems. İstanbul International Association on Water Quality, London. 167-176.

23. ISKI. n.d. (ca.1991). The Objective is to Save Green and Blue. Istanbul Water and Sewerage General Directorate. Istanbul Municipality.

24. ISKI. n.d. (ca.1991. Marmara Will Survive!. Istanbul Water and Sewerage General Directorate. Istanbul Municipality.

25. Hjülstrom, F. 1939. Transportation of detritus by moving water. In P.D. Trask, ed., Recent Marine Sediments. American Association of Petroleum Geologists, Tulsa, Oklahoma.

26. Kor, N. 1963. Haliçin Kirlinmesi İle İlgili Durunlaren Etüdü (An investigation of the factors which affect the pollution of the Golden Horn Estuary). PhD thesis. Istanbul Technical Univ., Turkey.

27. Merian. 1966. Vol. 15, no. 12. Hoffmann und Campe Verlag, Hamburg.

28. Möller, L. 1928. Alfred Merz Hydrographische Untersuchungen in Bosporus and Dardanellen. Neue Folge A, Heft 18. Veroffenlichungun des Instituts fur Meereskunde an der Universitat Berlin, FRG.

29. Nedeco 1981. Istanbul Sewerage Project Engineering Study. 9 Parts. Istanbul-Amsterdam.

30. Orhan, Derin. 1994. Scientific basis for wastewater treatment and disposal in Istanbul. Proc. International Specialized Conference on Marine Disposal Systems. İstanbul International Association on Water Quality, London. 225-255.

31. Özturgut, E. 1971. İstanbul Boğazinin Fisikel Oşinografik Etüdü (A Physical Oceanography Study of the Bosporus). Dissertation, Inst. Geography, University of İstanbul.

32. Sayınlı, T., and Yiğit, S. 1994. Üskudar and Baltalimanı Sea Outfalls of the İstanbul sewerage project, construction case history, Proc. International Specialized Conference on Marine Disposal Systems. İstanbul International Association on Water Quality, London. 237-244.

33. Spencer, D. W., and Brewer, P. G. 1971. Vertical advection, diffusion, and redox potential as controls on the distribution of manganese and other trace metals dissolved in waters of the Black Sea. Jour. Geophys. Res., 76(24):5877-92.

34. Sverdrup, H. U., Johnson, M. W., and Fleming, R. H. 1942. The Oceans. Prentice-Hall. New York.

35. Tezcan, S. S. Esen, I. I., Curi, K., and Durgunoglu, H. T. 1976. Halic, Sorunlari ve Cozum Yollari Ulusul Senipozyumu Tebliğ leri (Proceedings, Symposium on Pollution of the Golden Horn). Bogazici Universitesi, Bebeh, Istanbul.

36. Tolmazin, D. 1985. Changing coastal cceanography of the Black Sea. Part II, Mediterranean effluent. Progress in Oceanography, 15(4) 277-316.

37. Turkish Navy Hydrographic and Oceanographic Office. Turkish Straits Project. NATO Subcommittee on Oceanographic Research, Technical Report, no. 23, Çubuklu, Istanbul, Turkey.
38. Ünlüata, Ü., Oğuz, T., Latif, M.A., and Özsöy, E. 1990. On the physical oceanography of the Turkish Straits. In Pratt., L.J., editor. The Physical Oceanography of Sea Straits. Kluwer Academic Publishers, Amsterdam..
39. Üslü, O. Orhan, D., Ünlüata, Ù, Filibelli, A. 1991. Fzctors affecting pollution control strategies along the coastal zone of Istanbul. Umweltshutz: eine haaaaaerausforderung an uns alle. Stuttgarter Berichte zur Abfallwirtschaft 43. ISB AN 3-502-02956-7.
40. Üslü, O. 1993. A critical eeevaluation of wastewater treatment and disposal schemes in Turkish municipalities. Proc 1st International Conference on the Mediterranean Coastal Environment, 1, 661-666
41. U.S. Navy Hydrographic Office. 1965. Oceanographic Atlas of the North Atlantic. Publ. 700. Washington, D.C.
47 World Bank data. 1983-1991. Project files plus Reports No. 10114 (PCR) and No. 10852 (PPAR), Istanbul Sewerage Project,World Bank, Washington.
43. Zenkevich, L. A. 1963. Biology of the Seas of the USSR. George Allen and Unwin Std., London, pp. 353-464.

11.5   Boston Harbor

Boston, Massachusetts, was founded in 1630 in an area of a system of rivers and shallow bays, punctuated with many islands and peninsulas of hummocky drumlin glacial deposits, and an excellent harbor (Figure 11.21).   In the 1840s, the opening of important aqueducts significantly improved Boston's water supply. With the introduction of household plumbing and flush toilets came problems of conveying, treating, and suitably discharging as waste the water now so abundantly supplied.

11.5.1   Early History of Boston Sewerage

The first sewers were simple street drains to the nearest shoreline. The harbor waters nearest the city became extremely malodorous. However, by the 1890s, the main drainage system, consisting of very large tunnels, often through rock, began collecting wastewater and discharging it to the harbor at ever-increasing distances from the central city.

By 1895, sewage disposal works had been built at Deer Island, a peninsula at the mouth of Boston Harbor. Treatment consisted chiefly of chambers to trap sand and grit. Discharge was to the vigorous tidal currents at the harbor mouth. A decade later, similar, smaller works at Nut Island trapped sand before releasing the South Shore sewage to the harbor through a pair of outfalls each 1.6 km long, each terminating in an upturned elbow.

11.5.2   Proposals for Long Outfalls

Although well removed from the central city, the discharges were still to the waters of Boston Harbor. In the 1930s, a special commission recommended that the effluents be given partial treatment, and that consideration be given to the possible extension of the outfalls beyond Boston Harbor into Massachusetts Bay. By the 1960s, the plants had been upgraded to primary, but with only minor changes to the outfall system. The digested primary sludge from both plants was dicharged to the vigorous currents at the harbor mouth on the outgoing tide.

In 1971 a master plan recommended upgrading the plants to secondary, but a subsequent environmental review process revealed strong community controversy about costs, land requirements, additional sludge disposal requirements, and the value of secondary treatment for ocean discharge. Issues were not resolved and the plan was not implemented.

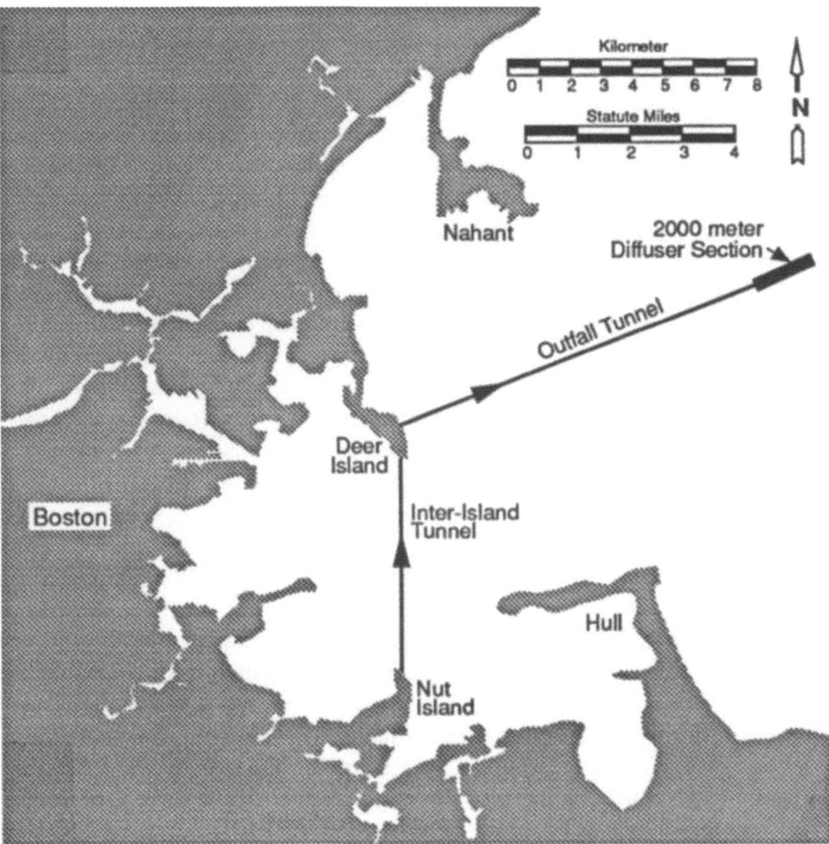

Figure 11. 21. The new cross-harbor sewage tunnel from Nut Island to Deer Island, and the new 8m diameter, 14-km long outfall to Massachusetts Bay. Boston Harbor lies to the southwest of Deer Island and Hull.

In 1977, amendments were made to the 1972 Water Pollution Control Act that had been based on environmental (receiving water) criteria. Renamed the Clean Water Act, its technology=based mandate was for secondary trteatment at all publicly owned sewage treatment works. Subsequent promulgation of the Act's Section 301(h) provided for environmentally-based waivers of the full secondary treatment requirement. In 1979 a proposal with wide public and scientific support to clean up the harbor was made under Sec. 301(h) to combine primary effluents from Nut Island and Deer Island, and discharge them through a 14-km long outfall to Massachusetts Bay. was denied by the USEPA, as was a 1983 re-application. So now (December 1995) litigation continues among Federal, State, and City agencies along with beguiling media discoveries that improvements cost money.

(see, for example, "As Boston Harbor gets cleaned up, rate-payers get cleaned out", Los Angeler Times, March 10, 1992).

By 1984, the existing facilities were clearly inadequate from any point of view: the primary plants failed to remove gross sewage solids from the flow; the effluent was still being released in the harbor or just at its mouth; the waters were brown, unsafe for swimming, and bore debris of obvious sewage origin. The sludge, laced with heavy metals, was not being dispersed well from the harbor despite the outgoing-tide release strategy; many local groundfish and crustacea were diseased. With impetus from a lawsuit filed by the City of Quincy on the harbor's south shore, a Federal District Court ordered that by Year 1999, Metropolitan Boston's sewage be collected, treated, and disposed in a manner consistent with USEPA guidelines.

In this setting, a new water and wastewater agency, the Massachusetts Water Resources Authority, was created. Stronger than its predecessor agency, independent of the State Legislature, and chartered to raise its own revenue, it immediately began a program to comply with the Federal court order.

11.5.3   Selection of a Treatment Plant Site

An immediate task faced by the new Authority was to select a site for the new sewage treatment plant. The only candidates were the existing Deer Island site and Long Island, in Boston Harbor. Each had its own set of problems: Long Island, though centrally located and reachable by bridge, had no existing sewerage tunnels to it, and part of the island is occupied by a hospital. Deer Island had the major advantage that two-thirds of the area's sewage already flows there, and sewage treatment was an established activity on the island. Disadvantages were that the site was limited in size, and that besides the sewage plant it also held a maximum security prison.

Furthermore, the only land route to Deer Island is through the Town of Winthrop. For many years, the residents of Winthrop had complained that they have been asked to bear an unreasonably large share of the unavoidable impacts of a major metropolitan area: low-flying airplanes approaching and leaving Logan Airport; and on nearby Deer Island the prison and the original sewage plant and its poorly-treated sewage, released so close to Winthrop's beaches.

Within a relatively short time, the Authority arrived at its siting decision: a completely new single sewage treatment plant for the entire service area would be built at Deer Island, replacing the existing primary plant there. The selection of Deer Island was agreed to by Winthrop on the following terms:

- During construction of the treatment plant, there would be no traffic to Deer Island through Winthrop, save for a shuttle bus service. All equipment, supplies, and most personnel involved with the construction would be brought to Deer Island by water.

- Power would be brought to Deer Island not via Winthrop but via submarine cable across Boston Harbor.
- The prison would be removed from Deer Island.

The court-ordered timetable was:

- By 1988, facilities plans for secondary treatment and disposal, and for residuals management to be prepared.
- By 1994 (since revised to 1996):

  Rehabilitation of interceptor tunnels in system

  A 7-km tunnel crossing Boston Harbor from Nut Island to Deer Island

  A completely new primary treatment plant, with $4800m^3$/d capacity.

  A new ocean outfall to discharge the effluent
- By 1999, upgrade to secondary treatment.

The Facilities Plan of 1988 provided preliminary layouts of the tunnel from Nut Island to Deer Island; the primary and secondary treatment plant trains on Deer Island; and the ocean outfall.

11.5.4   Siting the Outfall, Hydraulic Design

Five alternative sites for the diffuser section beginning near the existing Deer Island outfall and extending seaward were studied. Extensive current measurements, circulation and transport modelling and nutrient balance considerations indicated that a discharge averaging 2000 m^3/d, even given secondary treatment, would most satisfactorily be discharged to the open waters of Massachusetts Bay, beyond all islands, reefs, and peninsulas of the Boston Harbor system. The result was an outfall 14 km long running ENE from Deer Island to an area beyond the Graves, beyond the Nahant Peninsula, into 31 m of water (Figure 11.22).

At this depth, it is customary to provide about 140 m of diffuser length for each m^3/s of average flow. For Deer Island's roughly 20 m^3/s, this would be a diffuser nearly 3000 m long. However, modeling and analysis of initial dilution of this effluent under these oceanic conditions indicated adequate performance with a diffuser as short as 2000 m, with little improvement at 3000 m of length. Accordingly, the diffuser section is 2000 m long. Special physical model tests showed that adequate initial dilution could be obtained with a 2000-m diffuser consisting of 55 risers from the deep tunnelled diffuser header, each riser terminating in a rose capital diffuser. The spacing between risers is thus 36 m, slightly in excess of the 31-m water depth. Diffuser orientation was dictated by bottom topography in the region targetted for the diffuser (an approximately horizontal 2000-m long alignment was sought), as well as tunnel boring considerations. Current direction was not held as a primary consideration for diffuser orientation, particularly since the current meter studies in the discharge area indicated no strongly predominant direction of tidal currents. As it happened, a suitable horizontal alignment was found that required no change in orientation from that of the main outfall tunnel.

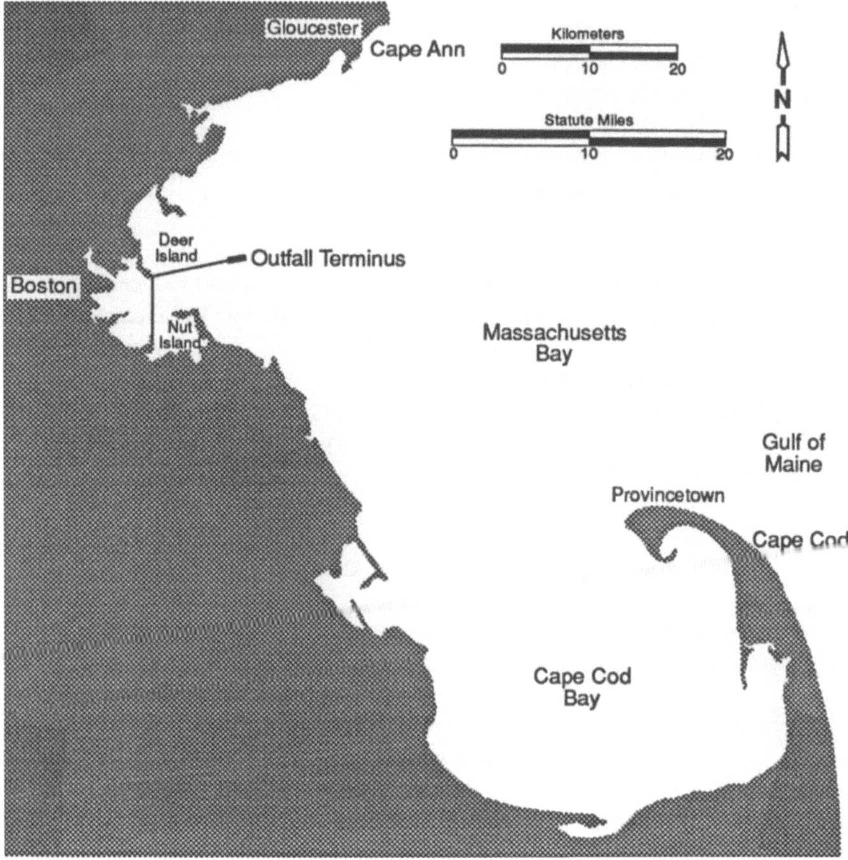

Figure 11.22. The outfall alignment, in relation to Cape Cod, Massachusetts

At Deer Island, the plant hydraulic profile is sufficiently high to permit gravity discharge through the outfall. The resulting finished diameter is 8 m due to the large design flow (55 m^3/s at peak). the great distance, and the 100 year storm surge elevation. There is also the Munro Condition for purging seawater from all risers which requires that the dynamic head in the tunnel exceed the ambient head by at least $H(\Delta\rho/\rho)$ where H is the height of the risers, $\Delta\rho/\rho$ the difference between the effluent and ambient densities and ρ is the effluent density. Here, the dynamic head loss was designed to be sustained partially in the risers, and principally in the exit nozzles. The Munroe Conditionn need not be met at all flows but at a flow reached or exceeded reasonably frequently. For Boston, the condition is met when the flow exceeds about 40 m^3/s, which occurs perhaps once a month. Once all risers are purged of seawater, seawater will not be able to intrude any of the risers unless the outfall discharge falls to less than about 8m^3/s.

In the tunnel just ahead of the diffuser section, a venturi throat is provided to further guard against seawater intruding the main tunnel.

11.5.5 Receiving Water Quality Modelling

The model for predicting receiving water quality demonstrates the dilution and dispersion process by directly simulating the mass loading of many discrete particles representing many different constituents. The simulation is based on first-order disappearance . All constituents were assigned a half life based on available information on chelation, flocculation, adsorption, sedimentation, and other sequestration mechanisms. Only 30-day and 60-day half lives were modeled as limiting values for constituents of concern. Coliforms and other microbiotic parameters were not considered since the effluent will be chlorinated.

Figure 11.23 shows for two outfall locations average concentrations of particles per liter throughout the water column for a discharge of 10,000 particles per second with half-lives of 60 days. This approach passes directly from the mass loading of a constituent to its far-field concentration by-passing entirely such intermediate parameters as effluent concentration and initial dilution. To find the predicted concentration of a 60-day half-life constituent, multiply its effluent mass emission rate in g/sec by the particle concentration indicated on the contour plot, and divide by 10,000.

A companion set of model runs (not shown) was made for suspended solids. Effluent suspended solids provide food for opportunistic benthic animals in the receiving water area, food and shelter for microorganisms (Sec. 3.2.1), sites for adsorption and transport of heavy metals and toxic hydrocarbons (Sec. 2.2.5), measures of the areal extent of the biological effects of discharges (Section 10.5 and 10.6), and provisional bases for regulatory limits to sedimentation rates or accumulations per unit area. Results are presented as contour plots for accumulation in terms in terms of particles/m^2/year which, again, could be pro-rated to g/m^2/year according to the suspended solids mass emission rate. The model is conservative in that resuspension was not included.

11.5.6 Construction

Before construction a thorough geophysical/geotechnical study of the outfall alignment was conducted. Consisting of subbottom profiling and many bottom corings, this study resulted in a detailed mapping of not only the seabed but of the bedrock surface below the unconsolidated sediments.

The tunnel was bored from a single vertical shaft drilled and blasted to a depth of 142 m below sea level. After the first few tens of metres of the tunnel had been excavated by drill-and-blast methods, the tunnel boring machine (TBM) was installed in the tunnel. The TBM then started the 14 km bore, with a very slight adverse (uphill) gradient; enough so that seepage would not collect at the cutting face, but naturally drain back to the original and only access shaft; yet not so steep that during operation salt water cannot be easily purged from the tunnel. The whole tunnel is sufficiently deep that at no point along the alignment is there less than three tunnel diameters of competent rock between the top of the tunnel and the unconsolidated sediments on the sea floor. The TBM discharged the tunnel muck to a continuous conveyor reaching back to the tunnel adit. The TBM included machinery to install precast concrete liner sections just a few metres behind the cutting head. The TBM used a precise laser-beam navigation system.

Meanwhile, each of the 55 risers was drilled from a rig on the sea surface, with the 660-mm diameter liner installed and capped with an 8-port rose diffuser. Finally, each riser was filled with red-dyed water.

At this writing (December 1995) the bored tunnel is about to be brought alongside the 2000-m long row of 55 risers. A pilot bore from the tunnel will be aimed to reach each of the risers. Success will be indicated by a copious leakage of the red-dyed water from the pilot bore. Drilling and blasting will then be used to enlarge the pilot hole to walk-in size so that 660-mm pipe can be placed to connect the riser with the main tunnel, tangent to the tunnel invert; after this, the excess space around the connector pipe will be filled with grout.

When its work is complete, the TBM will be driven forward beyond the diffuser a sufficient distance that it can be abandoned in place, and the tunnel finished behind it. The diffuser tunnel diameter, 8 m like the rest of the tunnel, will be reduced in stages by concrete benching as one proceeds offshore along the diffuser.

Seawater will be allowed to fill the tunnel--slowly, so that any trapped air pockets will not adiabatically heat to excessive temperatures. After the tunnel is filled with seawater, all caps will be removed from the seabed rose diffusers at the top of the risers. Then effluent will be led into the outfall. A transition period of several days will be necessary to ensure that the seawater is expelled from all parts of the outfall tunnel and from all diffuser risers.

11.5.7   Public Awareness, Community Participation

When the Facilities Planning phase began in the mid-1980s, "polluted Boston Harbor" was a notorious issue, with widespread media coverage, and was even a potent, although spurious, derogation. in the 1988 US Presidential elections

The Facilities Planning Process was undertaken with due inclusion of the local community. A Citizens' Advisory Council (CAC) was formed, consisting of 15 to 20 members chosen from the affected geographic communities (such as the shore towns of Quincy, Winthrop, and Nahant), economic communities (e.g. fishermen, condominium realtors, and other commercial interests), recreational interests, and ratepayers from inland communities. In regular and frequaent meetings, the CAC were briefed on the technical progress of the Facilities Plan, and were permitted to advise and question the technical staff.

Because the Greater Boston area has so many institutions of higher learning with oceanic research programs, a special Technical Advisory Group (TAG) was also formed, composed of interested and concerned scientists and engineers in oceanography, water resources management, and environmental chemistry from

Harvard, MIT, Woods Hole Oceanographic Institution, U.S. Geological Survey, the University of Massachusetts, the Audubon Society, and so on, who could at once be briefed and asked to contribute their insights and opinions to the process.

With such public and expert participation and endorsement, the Facilities Plan preparation went forward, relatively smoothly and within the prescribed time, to finalize the Plan and estimated costs that would be supported by all concerned. The Plan for a treatment plant and outfall went forward to detailed design and construction without significant public incident, with two notable exceptions:

(1) Well after plans had been made, debated, and endorsed, and construction well begun on the 14-km outfall, some residents on Cape Cod suddenly raised objections to the project. These people, 70 or more km southeast of Boaton and not represented on the CAC, were alarmed that the outfall would release effluent almost 14 km closer to their shores than the present system. They formed pressure groups (e.g. Stop the Outfall Pipe, or "STOP"), hired consultants, and published letters and articles. Construction halted, and further studies were made.

Responsible authorities were not convinced by any studies subsequent to the Facilities Plan that the outfall site selection was fatally flawed Indeed, important computations showed that the long outfall would impact Cape Cod less than the present outfalls as shown on Figure 11.21 by removing effluent from the alongshore current system. With a promise to halt operations should adverse impacts be evident, construction resumed.

This awkward and costly episode could have been minimized, and useful input from Cape Codders acquired, if Cape Cod had been represented in the CAC. The assumption that the Cape was too far from Boston to be interested was wrong.

(2) The issue of whether conventional secondary treatment is necessary or appropriate for Boston did not rest with the USEPA denial of the 301(h)waiver applications. Senior experts in the Boston area and elsewhere in the US have long maintained that other technologies, such as chemically enhanced primary treatment, better address coastal effluent treatment needs, at far less cost. The debate resulted in a valuable National Research Council study of appropriate and necessary treatment needs for coastal cities such as Boston, San Diego, and elsewhere. USEPA did not relax its secondary treatment requirement for Boston.

11.5.8 Recommended Reading

- Civil Engineering Practice. 1994. Special issue: The Boston Harbor Project. Volume 9 (1). Boston Society of Civil Engineers.
- Oceanus. 1981. Special issue: The Ocean as Waste Space. Volume 24 (1). Woods Hole Oceanographic Institution, Woods Hole, Massachusetts.
- Oceanus 1990. Special issue: Ocean Disposal Reconsidered. Volume 33 (2) Woods Hole Oceanographic Institution, Woods Hole, Massachusetts.
- Oceanus 1995. Special issue: Coastyal Science & Policy. Volume 36 (1) 91. Woods Hole Oceanographic Institution, Woods Hole, Massachusetts

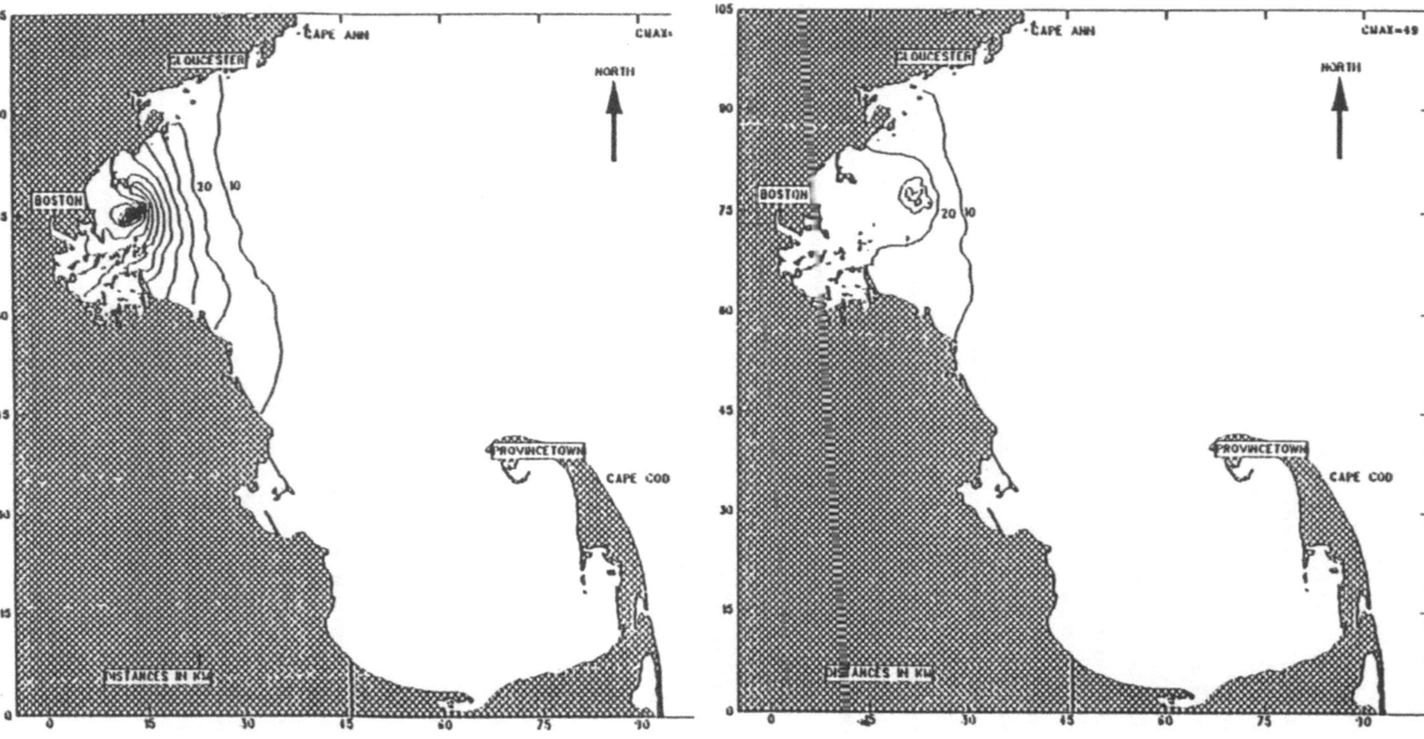

Figure 11.23. Relative concentrations, in percent of that after initial dilution, of constituent in primary effluent discharged from (A) near present ourtall locations near the mouth of Boston Harbor and from (B) the project site 14 km offshoree. The figures are based on pollutants with a 60-day half-life.

11.6 Southern California Bight

The Southern California Bight extends seaward of the coastline from the U.S.-Mexico border to Point Conception west of Santa Barbara (Figure 11.24). Drainage from 22,500 km^2 of mountains and coastal plains enters the Bight.   Some 15 million people live in this sami-arid area where imported water, climate, and the motion picture industry have combined to transform it into "The Cadillac Desert" (11). The climate is Mediterranean with most of the average 370 mm/y rain falling during December through March in Los Angeles. Average monthly temperatures are from 13 to 22° C. Economic development is based on local ground water and imported water from northern and central California areas and the Colorado River. Per capita demand is high and increasing. Occasional grand schemes for importing more water invoke local engineering and financial enthusiasm. These have ranged from floating chunks of Antarctic ice sheets into Santa Monica Bay to importing water from Canadian and Alaskan ruvers and from the North American Great Lakes.

Cornucopian water demand projections are based on historical availability of water that can be imported and so to assign less importance to conservation than to wastewater reclamation that is even more expensive than disposal (see Chapter 12). Except for golf courses, recovery of the marginal costs of treating and distributing reclaimed wastewater over the costs of ocean disposal has received little attention.   Proposals for direct reuse in landscape irrigation or even industrial use with dual supply systems are deterred by the certainty of regulatory vigor and by preemptory litigation by "green" activists. Interestingly, cemeteries are reported as offering premium prices for reclaimed water in order to be assured of a constant supply during drought periods when rationing is applied to conventional municipal and industrial supplies. Indirect reuse through spreading or injection into ground water is promoted by labeling sewage treatment plants as wastewater relamation plants, but most wastewater goes to the sea. As befits a wealthy urban area, total demand is rapidly increasing, particularly in the commercial and residential sectors while supplies are being limited by litigation from other riparian users. The dilemma is underscored by the numbers in Table 11.10

11.6.1 Ocean Disposal of Southern California Wastewaters

Some historical factors. The interactions of neighborhood ("not in my back yard"), municipal (health and sanitation costs and benefits), State (regulatory), and more recent Federal (macroeconomic) priorities in environmental management have been played out in Santa Monica Bay. In 1894, the City of Los Angeles constructed its first ocean outfall to discharge sewage previously used on a farm at the southwestern edge (now in the south-central part) of the city. The outfall was placed near a shoreline streetcar terminal known as Hyperion During the

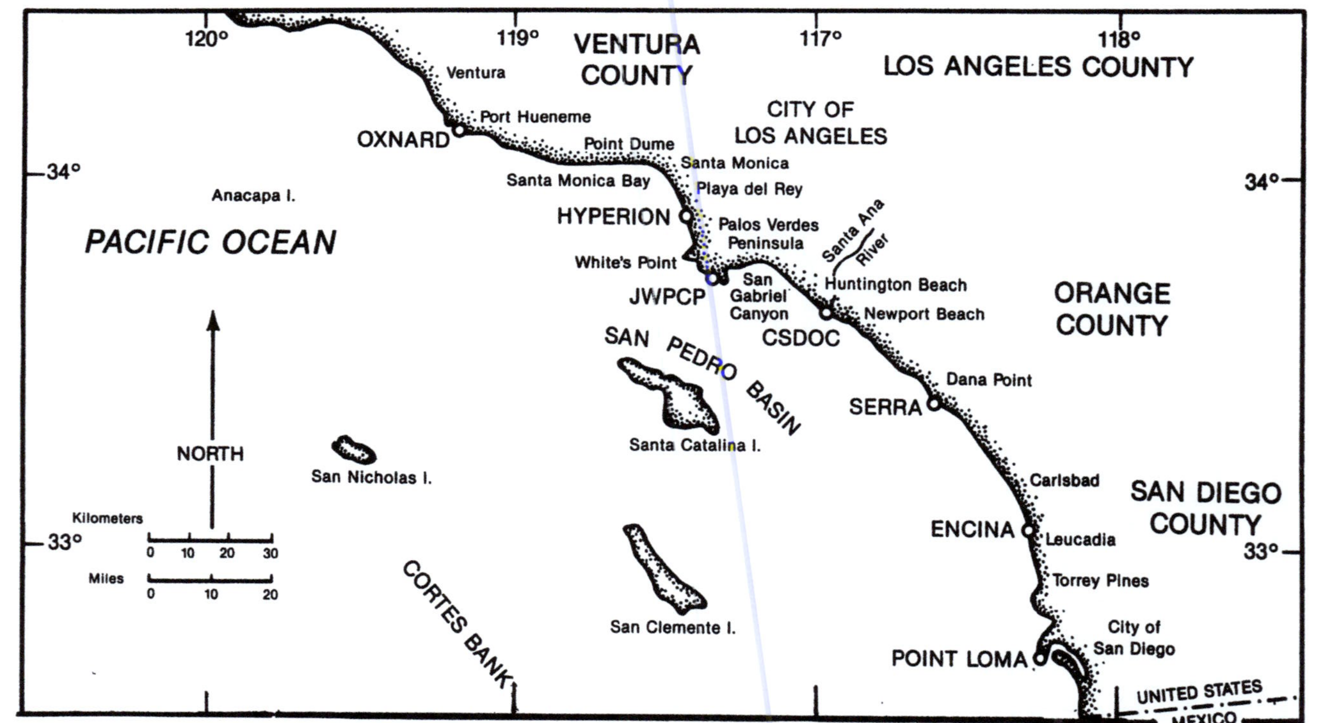

Figure 11.24. The Southern California Bight

Table 11.10. Water supplies and demands in southern California.

Parameter	1990	2010
Estimated population, millions	15.29	17.75
Water supply (mllions of m^3/day)		
Local [1]	7.40	7.40
Reclaimed [2]	0.54	0.58
Los Angeles Aqaueduct [3]	1.42	1.42
Metropolitan Water District [4]	2.70	1.59
State Water Project [5]	4.35	4.90
Total supply	16.39	15.88
Water demand (millions of m^3/day)		
Residential	6.08	7.91
Commercial	1.79	2.43
Industrial	1.25	1.55
Public [6]	1.56	1.76
Agricultural	2.98	2.20
Total demand [7]	13.68	15.85
Supply minus demand	2.71	-0.16

Notes: (1) includes infiltration from Los Angeles Aqueduct spreading operations in San Fernando Valley. (2) includes both direct and indirect utilization. (3) Owens Valley and Mono Basin water. (4) Colorado River reductions to comply with adjudication of riparian rights. (5) transfers across Sacramento-San Joaquin Delta of flows from State Feather River project. (6) includes unaccounted for water. (7) aoorixunately half for population growth and half for increased per capita demand to 8.5 m^3/cap/y. Sources: California Department of Water Resources 1987 Annual Report; Los Angeles Department of Water and Power, 1985-86 Annual Report; Sate Water Contractors, Bay-Delta Hearings, Exhibit Numbers 3,6,13,17,76; June 1977.

following 50 years, the city grew more rapidly than its willingness to invest in sanitation except to remove problems from their sources and transport them downstream to Santa Monica Bay where increasing fecal nuisances and local opposition to the pollution both fluorished (4).

After World War II, attention reverted to Hyperion and Santa Monica Bay, first with engineering and then with baseline oceanographic studies. Hyperion has become a continuing case study in the design, expansion, and monitoring of ocean outfall systems. Early-on decisions by the consultants, Kennedy Engineers, San Francisco, and Pomeroy and Montgomery, Pasadena, were based on observations of the maximum measurable extent of coliform bacteria along the shore line. Coliform pollution from the "1-mile" (1.6 km) outfall with leaky near-shore joints ended 5 miles (8 km) northerly to Santa Monica. Flotables were beached by afternoon sea breezes. If flotables were removed, an outfall discharging five-mile (8 km) offshore from Hyperion would eliminate beach pollution. The empirical solution was obvious.. The "5-mile" (8 km) outfall was carefully aligned to stay just north of the extended boundary between the Cities of Los Angeles and El Segundo, and , for engineering symmetry, a Y-diffuser was added.

Having arrived at a solution, it was necessary to scientifically validate it. The City bought a second-hand 65-foot northern trawler, the PROWLER, and added oceanographic and navigation equipment. The University of Southern California Allan Hancock Foundation and its VELERO IV were commissioned for the more exacting observations and interpretations. Throughout 1954 -1958, currents, water column characteristics, baseline benthos and plankton characteristics, and coliform survival studies were conducted in and around Santa Monica Bay. A WWII British navy frog man who held the world record for aqualung diving and a casual regard for schedules sampled the sea floor. To see if bottom pressures would collapse cell structures in sludge, he released a bag of prime quality grapefruit at 100m depth. They floated. It would have taken another 50 to 100 meters for them to collapse. Aerial observations of fluorescein dye diffusion were made of intermittent and continuous releases from skiffs anchored in the 5-mile discharge area just under the flight path from Los Angeles International Airport. This caused great alarm to former military pilots who recognized the WWII distress marker and whose signals brought City and County lifeguards, the County aero-squad and at one time, the U.S. Coast Guard to the rescue.

In situ comparisons of persistence of coliform bacteria for effluents from Los Angeles City, Los Angeles County, and Orange County outfalls rdemonstrated that effluent characteristics and local dilution factors determined the rates of disappearance . For protection of the surf zone bathers, primary effluent with its more rapid coagulation and sedimentation of effluent solids and attached coliform bacteria was found to be superior to secondary (see Chapter 3). We were attributed with setting sanitary engineering back fifty years.

Meanwhile on shore, a 245 mgd (10.7 m³/s) "high-rate" activated sludge plant was being constructed. Heat-dried sludge was to be sold as a fertilizer at about $20/ton, a price that was being received by San Diego. The idea was not new. After all, Milwaukee had for years been solving its sludge disposal problem by packaging and shipping it all over the country and marketing it for horticulture. The sudden impact of an additional 100 tons per day caused the price to fall to about $4/ton. This was far from the $22 to 24/ton needed to pay fot state-of-the-art, operationally unstable high-temperature flash drying that produced a highly flammable flour-sized product that then had to be pelletized   San Diego's price dropped to $4/ton.

Having adopted the 5-mile offshore effluent disposal, it was logical to consider adding the sludge to the effluent. However, the small-diameter "7-mile" (11 km) sludge outfall could be installed in much less time (actual construction time was 7 days). This was important because local residents, the courts, and State Health authorities were becoming impatient with delays and continuing beach quarantines, and were threatening litigation. Although the sludge could have been discharged along with the effluent through the new effluent outfall scheduled for completion in 1960, earlier discharge into 100m of water would be proof that the City was acting in good faith and with dispatch. The sludge outfall was placed in service in 1957, three years ahead of the effluent outfall, and continued until 1987. With the construction of the onshore and offshore facilities, the basic design goal of meeting coliform standards for ocean bathing waters (see Chapter 3).was achieved

Since then, the combination of Federally funded employment opportunities for local design and construction of state-of-the-art or otherwise innovative sewage treatment technologies, increases in private funding for non-governmental environmental activism, a corresponding although grudging municipal acceptance of Federally mandated operation and maintenance costs, and marketable optimism for attaining zero-discharge systems (see Chapter 10) have continued. By definition, innovation carries the risk of failure. This is particularly true for high temperature (and pressure) processes for sewage sludge processing.   For example, the theoretically attractive gennerally pilot--scale successes of the Porteous process in Chicago in the 1940s and its daughter Zimmerman process (8, 10) for wet oxidation of sludge can be made to work, although not easily. Their O&M costs are high. Neither has been a conventional option at a municipal scale

The most recent manifestation of a high-temperature technology is the Carver-Greenfield multiple-effect evaporation process with pyrolysis and incineration options (8). To approach their theoretical potential, pyrolysis and similar systems require predictable flows and physical and chemical properties in their feed stocks This is possible for, say, brewery wastes but not for municipal--scale sewage or solid wastes systems; too many individuals and enterprises are constantly changing quantities and characteristics of their wastes. Examples of chemical and mechanical failures include the coating of reactor and valve surfaces with an insulating varnish, and inefficient recovery of the solvent used to transport the off-gases intended as an energy source.

Along with other high-temperature schemes for municipal wastes, the Carver-Greenfield process is certainly innovative, but it has been dependent on secondary benefits (research, training, employment, financial disbursement schedules, etc.) to justify its costs. Meanwhile, municipal-scale operations on sewage sludge have led to its regretfully being declared by both Los Angeles County and City authorities as a failed technology.

Some recent operations. Annual discharges to the Southern California Bight are reported by the Southern California Coastal Water Research Project, SCCWRP,   from four major outfalls, San Diego, Los Angeles, Los Angeles County, and Orange County (12)   Total flows during 1971 through 1993 increased monotonically from 1284 to 1656 $x10^6$ m^3/year in 1989, decreasing to 1485 $x10^6$ m^3/year in 1993 principally as a result of increasing block tariffs imposed in response to 1986-92 drought. Another 186 $x10^6$ m^3/year were discharged in 1993 by 15 smaller plants.

Total wastewater discharges into the Bight increased during 1971-93 by about 15 percent. In 1972, the original Clean Water Act began the introduction of much Federal money into pollution control that had previously gone into construction of interstate transportation systems. There were corresponding increases in Federal and State environmental oversight, public and private litigation, and local design and construction employment. Upgraded municipal treatment works reduced mass loadings to the sea. Suspended solids and BOD decreased by about 15 and 10 percent, respectively. In November 1987, the discharge from Hyperion of about 100,000 tons/year (dry weight) was approximately halved when the 11 km disgested sludge outfall discharge of some 50,000 tons/year (dry weight) was changed by court order to land disposal. New facilities with an estimated replacement cost of $1.2 billion had been constructed and another $2 billion has been projected for the Los Angeles metropolitan area (5).

Meanwhile, improved source controls have reduced the mass discharges of DDT by 99 percent and PCB's by 80 percent. Arsenic, cadmium, chromium, copper, mercury, nickel, and lead are down by 50 to 70 percent. The effects of waste discharges into the Bight have been characterized by SCCWRP since 1971 and reported on in a series of annual reports edited by Bascom and his successors (12).

The infaunal trophic index (ITI) defined in Chapter 10 is considered sufficiently diagnostic to be used for predicting changes resulting from increasing or decreasing annual loadings of wastes. The index was developed from 1987 and 1988 field surveys when about 82,000 t/y of suspended solids were being discharged into Santa Monica Bay and another 103,000 t/y were discharged from the Palos Verdes Peninsula wherecurrents are generally up-coast towards Santa Monica Bay. Values for the benthic biomass and its corresponding ITI are shown on Figures 11.25 and 11.26 (1,14, 15). The dependence of impacted areas upon the solids loading is shown on Figure 10.1 along with a discussion of the ITI. In addition, high biomass colonies of *Listriolobus* have been found in shallow water near Oxnard and in deep water in Santa Monica Bay that have no measureable geogrAphic or oceanographic relation to wastewater discharges.

11.6.2   Recovery of Damaged Ecosystems

Throughout the Bight, there have been ecosystem improvements since about 1960. Bathing beaches are no longer quarantined for obvious aesthetic reasons or for precautionary microbiolobical standards. These quarantines have been temporarily reimposed when physical or biological processes have failed at treatment plants, or when there are large amounts of urban runoff Fish stocks in the Southern California Bight have remained within historic ranges. Kelp beds in outfall areas are more affected by marine climate changes than by wastes (9).

One of the problems in assigning health benefits to improvements in southern California coastal cities wastewater management practices is the lack of credible epidemiological background and monitoring data. Even Gillespies' (4) pioneering study "...made no particular effort to discover cases of disease which could have reasonably been reasonably contacted from this beach." Investigation of his single sample case (for which findings would have had a standard deviation of infinity) of lifeguard paratyphoid was terminated when the patient joined the U.S. Navy (4). Negative proofs of zero incidence are in any event rejected by zero-discharge advocates who contend that people are sick without knowing it. A more important issue is a growing number of people who ascribe transcendental properties to all natural waters (6,13). Here we argue that valid long-term pollution abatement planning requires well-designed and completely transparent prospective and retrospective epidemiological studies in which both heaalth and aesthetic issues are considered. These studies should include Santa Monica Bay, San Pedro Channel, and the San Diego-Tiajuana area. These investigations would ensure rational setting of priorities, and woujld require much less time and less expense than the

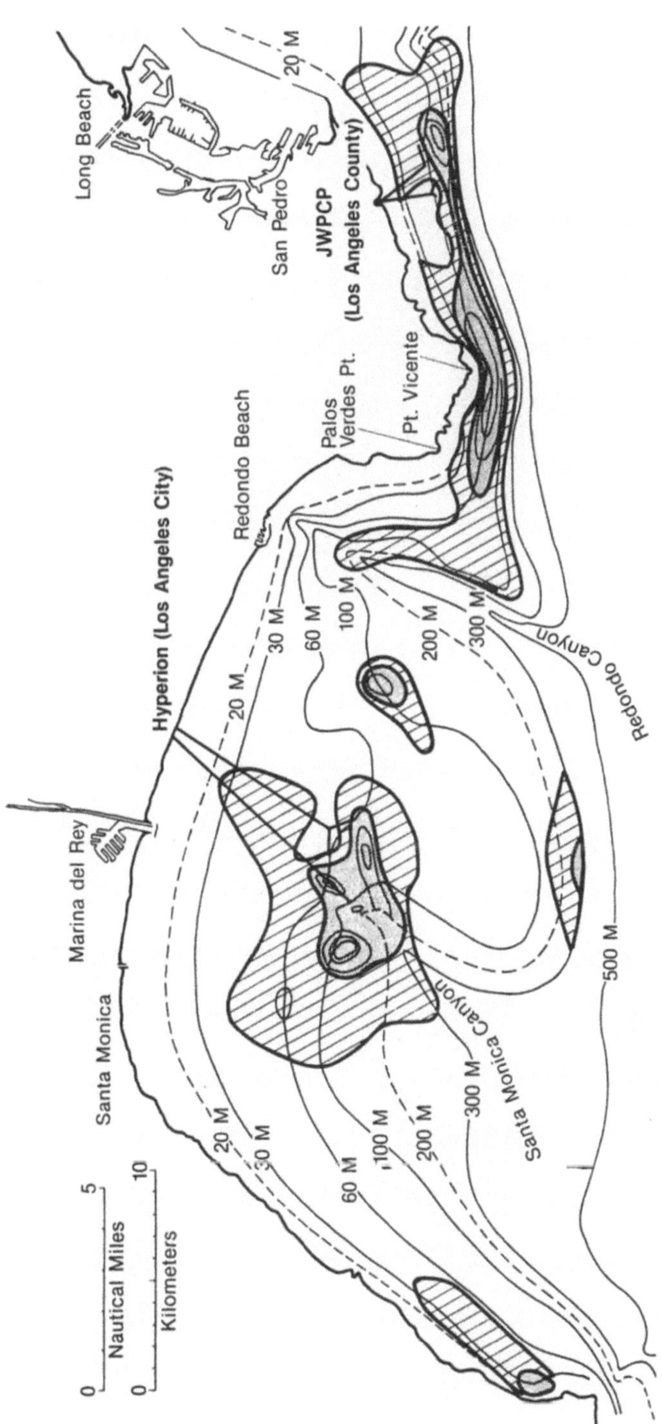

Figure 11.25. Benthic biomass in Santa Monica Bay and San Pedro Channel in 1967-68. Shaded areas indicate areas with >200 g/m^2, and 100-200 g/m^2. Background was $\approx$ 5 -40 at < 20m depth and 20-100 at >20 m depth except near JWPCP outfalls. Source: Bascom (1).

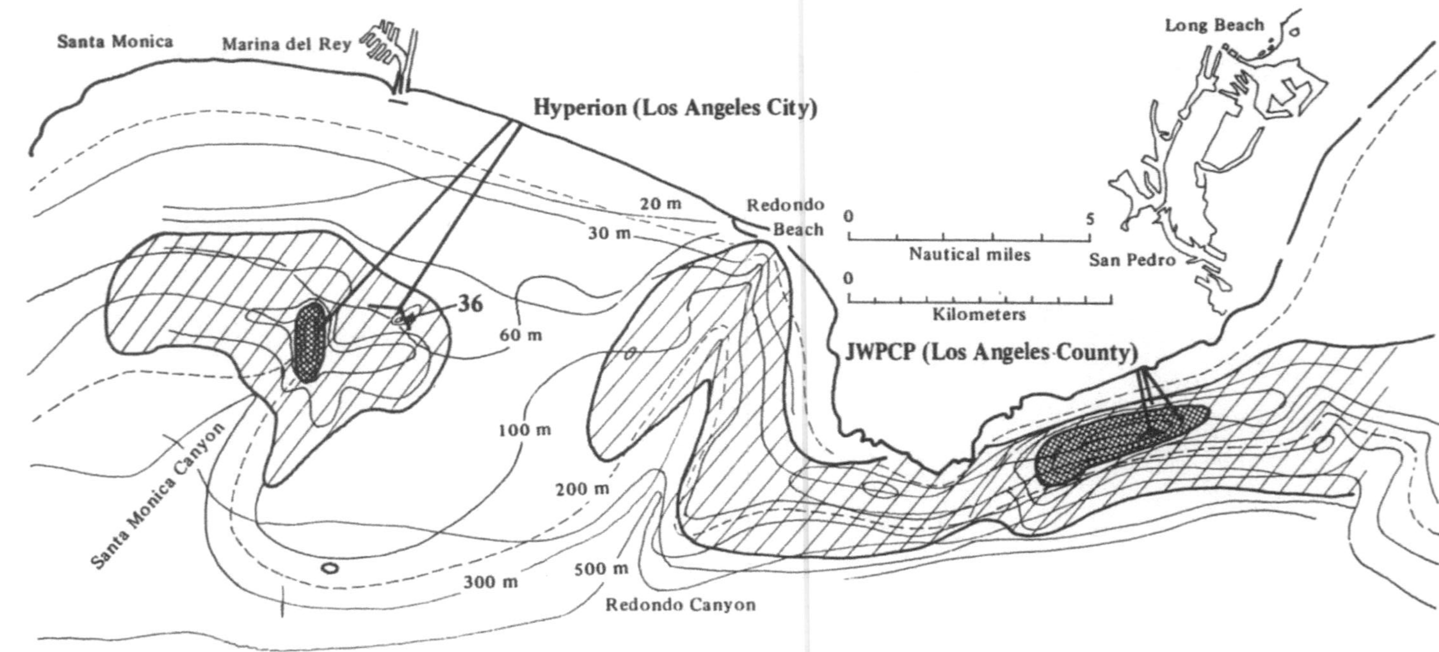

Note: Degraded area (I.T.I., 0 to 30); damaged area (I.T.I., 30 to 60). Depths are in meters.

Figure 11.26. Infaunal Trophic Index (ITI)) in Santa Monica Bay and San Pedro Channel, 1967-68. Shading indicates degraded area with ITI = 0 to 30, damaged areas ti ITI = 30 to 60. Sources: Bascom (1), Word (15), Word and Mearns (16).

feasibility studies and physical works promoted in the name of health benefits.

Long-term monitoring in the Southern California Bight supports the conclusion that pollution effects of treated domestic sewage effluents are not permanent. Adverse changes have been have been followed by recoveries once the source was identified and control measures were put in place. Still, there are residuals in the sediments of DDT, PCBs, and other chemicals whose stability led to their sale in the first place and whose persistance in the environment arguably should have come as no surprise to their manufacturers and distributers.

Recovery of the Santa Monica Canyon Area. In November, 1987, the annual discharge of some 50,000 tons (dry weight) of digested slude at a depth of 100 m at the head of Santa Monica Canyon stopped. Since then, the City of Los Angeles, SCCWRP, USEPA, and local univeersity scientists have been defining and monitoring rates of the area's recovery. (2,12,14). The area is also affected, although to a much lesser extent, by the continuing discharge of a similar amount of suspended solids through the nearby effluent outfall. As noted in Chapter 10, shallower areas with higher energy due to wave and sediment movement recover and reach steady-state reach more quickly than deeper, lowerr energy areas. By 1992, the impaccted area was were reduced to about a third of its 1989 value as measured by ordination analyses of infaunal assemblages. This was accompanied by the return of benthic species distributions toward those of control areas. These findings led Thompson (14) to estimate that full recovery to normal deep (>100 m) benthic conditions would not be reached until about 2002. Boundaries for measuring gross biological impacts using either ITI or ordination indices are based on subjective assignment of limiting values (2,15,16). Here, we argue that both are more suitable than species composition becaus of their dependence on waste discharges that dominate oceanographic variables and successions of species.

Note Added in Proof. Termination of ocean discharge of 50,000 tons per year (dry weight) of digested sludge at 100 m depth in Santa Monica Bay was terminated in December 1987 far overshadowed by ending ocean dumping of 8,000,000 tons wet weight of raw sludge at about five percent solids (400,000 T/y dry weight) from barges (nominally) 19 km offshore in the New York Bight into about 15 m depth near the head of the Hudson Canyon. Much of the sludge accumulated in the adjacent low-energy Christiansen Basin at depths between 15 and 3o meters. In 1972, 380 km^2, later increased to about 800 km^2, had been closed to shellfishing because of bacterial contamination. Also, 8he sludgc discharge had contributed to occasional anoxia and fish kills in the bottom waters.

A 1989 to 1991 study by the National Marine Fisheries Service reported in 1995 on residual effects in sediments and fish. Pathologies were reduced affter dumping was stopped. The survey was too short to evaluate recovery. In contrast, the higher energy site at about the same depth and distance from Delaware Bay into which Philadelphia had dumped up to 700,000 wet tons of per year had recovered within about four years. New York's shelfish quarantine remains in effect. Stocks are too low for commercial interest following years of heavy expoitation. However, there is hope that the area will serve as a nursey for future stocks (17).

11.6.3 References

1. Bascom, W. .1978. Life in the bottom. Annual Report, Southern California Coastal Water Research Project, 1978. Fountain Valley, CA. 57-80.
2. Dorsey, J.M., Phillips, C.A., Dalkey, A., Roney, J.D., and Deets, G.G. 1995. Changers in assemblages of infaunal organisms around wastewater outfalls in Santa Monica Bay, Ca. Bull. So. Calif. Acad. Sci. 94, 1, 46-74.
3. Freysinius, W., Schneider, W., Böhnke, and Poppinghaus, K..1989. Waste Water Technology. Springer Verlag, Berlin. 563.
4. Gillespie, C.G. 1943. Report on a Pollution Survey of Santa Monica Bay Beaches. California State Department of Public Health. Berkeley.
5. Garber, W.F. 1995. Personal communicatrion.
6. Garber, W.F., and Garrison, W.E. 1989. Marine pollution assessment and abatement in California.in Gunnerson, C.G., editor. Post-audits of Environmental Programs and Projects. Amer. Soc. Civil Engrs. New York.73-104.
7. Mearns, A.J., and O'Connor, T.P. 1984. Biological effects versus pollutant inputs: the scale of things. In White, H.H., editor. Concepts in Marine
8. Metcalf and Eddy. 1991. 3rd edition, G. Tchobanoglous, editor. Wastewater Engineering: Treatrment, Disposal, Reuse. McGraw Hill, New York. 879.
9. National Research Council. 1990. Monitoring Southern California's Cioastal Waters. National Academy Press, Washington.
10. Ramalho, 1977. Introduction to Wastewater Treatment Processes. Academic Press, New York. 336-339.
11. Reisner, M. 1986. The Cadillac Desert. Viking Press, New York..
12. SCCWRP. 1974 to 1994. Annual Reports. Southern California Coastal Water Research Program. Westmiinster, CA.
13. (Fourth) Stockholm Water Symposium 1994. Education Workshop. Stockhom Water Company, Vaten.. Stockholm.
14 Thompson, B. 1991. Recovery of Santa Monica Bay from Sludge Discharge. Technical Report C-349. Southern California Coastal Water Research Project, Fountain Valley, CA.
15. Word, J.C. 1978. The Infaunal Trophic Index. Annual Report. Southern California Coastal Water Research Project. Fountain Valley, CA. 19-40.
16. Word, J.C., and Mearns, A.J. 1979. The 60-meter Control Survey of Southern California. Southern California Coastal Water Research Project, 1978. Fountain Valley, CA.
17. Studholm, A.I., O"Reilly, J.E., and Ingham, J.C., Editors, October, 1995. Effects of the Cessation of Sewage Sludge Dumping at the 12-Mile Sitge. NOAA Technical Report, NMFS 124U.S. Department of Commerce, Seattle.

12   Cost and Sustainability Factors

Earlier chapters focused mostly on the scientific and engineering aspects of ocean disposal of coastal city wastewaters. This chapter looks at costs and other non-structural engineering determinants in the urban water and sanitation sector. The final selection of technologies and instruments for a coastal city depends upon four interrelated sets of planning and implementation factors. These include: (1) what costs? who benefits? who pays? (counterpart and public participation, cost recovery, relative costs of water and sanitation, outfall costs, and shared benefits), (2) system options (scales in water supply, disposal, reclamation, and conservation, comparative costing), (3) water, sanitation and public health service levels, and (4) cooperation, competition, and water conflict issues from local to international scales (capacity building, information and technology transfer, community participation, the prisoners dilemma). Chapters 3, 10, and 11 include narrative introductions to some of these factors. Selected empirical and/or theoretical models for each of the four categories follow, beginning with an historical development of the unity of water supply and sanitation.   Much of the background for this chapter lies in operational information exchanges between industrial country expatriate engineers and their developing country counterparts (13). Institutional and political policy issues such as potential roles of market forces, "unbundling" and privatization of components that are natural monopolies within transportation, water, and other sectors are documented elsewhere (36, 37).

Background.   Financial and environmental costs and technologies of water supply and wastewater disposal are interdependent at all scales Nearby latrines pollute shallow wells. American midwest and British power plant stack discharges cause acid rain in Canada and Europe.   Historically, costs of water were reduced in ancient times by slave labor. By 2500 BC. in Egypt, Mesopotamia, and the Indus Valley slaves were building wells, qanats, aqueducts and interior plumbing discharging outside the walls for palaces and temples. Irrigation driving the first agricultural revolution and the civilizing influences of a warm bath were discovered
Imperial Rome built monumental aqueducts, house connections, continuously flowing public baths, and latrines that remain to this day. Eventually, the costs of maintaining an expanding perimeter exceeded the benefits from trade and tribute from within it, and Empire collapsed. Meanwhile, they built the 2nd century BC. Cloaca Maxima to drain the Forum that is now a pilgrimage for environmental engineers and a bête noire for some historians. Looking back from flush toilets and chemical fertilizers, Louis Mumford (23) viewed the Imperial City as one of deprivation and dung farming, and the Cloaca as the paradigm of "Roman engineering and decadence". Engineers never know how their works will be judged.

Nothing matched Rome's enthusiasm for sanitation until late 19th century England.. In 1842, Edwin Chadwick in 1842 (3) proposed community water supplies not for drinking but for flushing. This was reinforced by John Snow's 1864 (31) classic work on cholera transmission through wells polluted by nearby cesspools. Fortuitously, revenues from the East India Company reverted to the Crown and government financing of sewers became possible. These trends have continued with projects for sewerage in Europe, North American and, as a symbol of enterprise and modernity, former colonies. Programs mushroomed after about 1950 when Middle Eastern oil came onto the market at 10 cents a barrel at the wellhead, only a small fraction of its real value. Since then, declining revenues due to oil price shocks of the 1970s and early 1980s shifted priorities from public health benefits to secondary participatory and distributive benefits (employment, environmental reverence, etc.) measured at the ballot box. Exogenous financing has almost become the rule with little consideration of opportunity costs.

12.1 Who Benefits? Who Pays?

People move to cities where they can find the goods and services they want at less economic cost than in the villages and farms from which they came. Their priorities, demonstrated by their loan repayment records, are jobs for themselves and education for their children (11). They need and will pay whatever it takes for water to live. At all levels, people want sanitation and will pay for privacy, then modernity, convenience, comfort, ambiance, and finally health. At some level of population density, the city or other central government assumes responsibility.

12.1.1   Limits to Scale in Water Supply and Sanitation

Relationships of water costs to sanitation costs were studied during 1976-1978 World Bank field studies of alternative urban water supply and sanitation systems throughout the world (19, 35). It was concluded that in developing country capitol cities with moderate water service levels, it cost five or six times as much to get rid of water as it did to supply it in the first place. This work was extended in Jakarta (11) and to industrial countries (12) Figure 12.1 shows economies of scale on the supply (diverging) side with complementary diseconomies of scale on the disposal (converging) side. The same is found in freeway systems. One flat tire causes minor problems during evening rush hours when diverging traffic takes advantage of alternative routes.. Diseconomies occur during the morning rush hour when everything is converging and there is no redundancy. Here, a minor accident or even one flat tire can cause a thousand people to be ten minutes late for work. The aggregated cost data are from urban, urban fringe, and rural sources.

The cost and service level data on Figure 12.1 show the need to use the saame demand projections and coordinated construction schedules for both water and sanitation without which sewers have been built without water to flush them in cities as far apart as Accra and Jakarta (38). For Kyoto and developing countries, they provide full marginal costing including that for flushing water (19). A constraint to the model was the use of published revenues, average and marginal

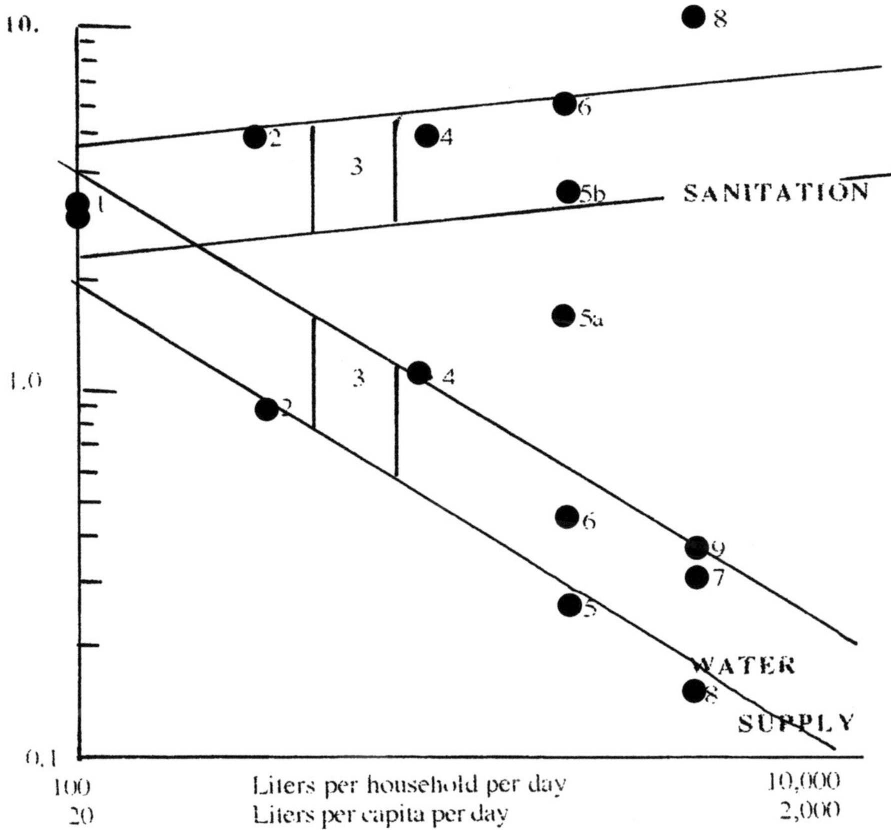

Figure 12.1. Economies and diseconomies of scale in water and sanitation systems Source: Gunnerson (12).

Note: Costs are normalized to 1988 U.S. dollars per cubic meter. Sources: 1- World Bank Research Project 671-46, 1976-78 for village scale hand pump and household pit latrines. 2- Jakarta 1988. 3- Wolrd Bank Basic Needs 1980. 4- Malacca, Malaysia, 1978, 5- Kyoto, Japan, 1978: 5a - household vault and vacuum truck to trunk sewer, 5b- conventional sewerage, both with activated sludge treatment and sludge incineration (Kalbermatten, et al., 1982), 6- Washington, DC, 1982, World Bank Data. 7: Boulder, Colorado, 1989. 8: Chicago, 1986. 9: Los Angeles, 1988.

prices, partial sewer service charges, and other surrogates for water costs. Sanitation costs are conventionally undercounted by neglecting grants, subventions, burden and overhead. Supply costs are for transmission, treatment of known quality sources, and distribution through small, shallow pipelines under pressure. Disposal costs are for deep, gravity flow large-diameter pipes and for treatment of variable water quality. Land costs are included for on-site systems only. Costs of household plumbing are excluded for all systems.

Efficiencies for sewage treatment are usually stated in terms of removing solids from the water, where Figure 12.1 clearly indicates the role of dilution that complements higher service levels. It is argued that a better estimate of efficiency would come from measuring the water removed from the solids as practiced by process chemists and metallurgical engineers but overlooked by sanitary engineers.

We conclude that both water and sanitation costs are understated, so that cost ratios are more reliable than the dollar costs shown. Working ratios of disposal to supply costs increase monotonically from about 1.3:1 for 20-40 ℓcd and on-site disposal to 15:1 at some 600-700 ℓcd and large central systems disposal (12).

12.1.2   Allocating Costs of Water and Sanitation Benefits

Criteria for allocating costs commensurate with benefits of developing country water and sanitation systems are evolved by the community, municipal, and central government entities responsible for cost recovery.. One such model was prepared according to the following criteria for sequential World Bank urban environmental development projects in Jakarta (11).   Table 12.1 is based on 1988 discussions among expatriate colleagues, counterpart professionals, municipal officials, academics, and ministers in support of recommendations for formalizing information and technology transfer (Section 12.5.3) These exchanges have been followed by on-going documentary research and theoretical analyses. The table is a framework that can be adapted by counterparts and expatriates to other cross-cultural realities.

1. As service levels increase, so do direct and indirect costs to the household.
2. As service levels increase, costs to the municipality decrease so long as environmental impacts are effectively restricted to communities of origin.
3. As service levels increase, proportionate amounts to be covered by World Bank loans decrease. There is abundant evidence that poor communities will pay for higher service levels when they participate in the decisions (24,37,38)
4. As environmental impacts of higher service levels extend beyond source community boundaries, provincial and central government costs increase. For example, increasing amounts of wastewater require increasing levels of treatment by the municipality and increasing oversight by provincial and central governments.
5. As experience, skills, and knowledge derive from specific projects and programs that can be transferred to other entities through the state, the central government assumes an increasing allocation of the costs.
6. Beneficiary community contributions to costs can be monetary or in-kind services (labor, supervision, etc.).

Table 12.1. Worked example of allocating costs according to benefits in water and sanitation, from a Jakarta case study.

(Column)	Construction costs (percent)					Operation and maintenance		
	Government to Municipality	World Bank Loan to Government	Rev. Fund	Community Currency	Labor	Munici-pality	Community Currency	Labor
	(1)	(2)	(3)	(4)	(5)	(6)	(7)	(8)
1. Water supply (asterisk indicates included in water service charge)								
1.1 Rehabilitate existing system	20	40	20	10	10	---	---	---
1.2 MCK with pump and inside tap	20	40	20	10	10	10	90	---
1.3 MCK with outside standporft	20	30	30	20	—	10	90*	---
1.4 Neighborhood standpost	10	20	40	30	—	10	90*	---
1.5 Household handpump	—	10	70	10	10	---	100	---
1.6 Household yard tap	—	---	70	30	—	---	100*	---
1.7 Household full plumbing	---	---	30	70	—	---	100*	---
2. Sanitation or sewerage								
2.1 Rehabilitate existing system	20	40	20	10	10	---	—	---
2.2 MCK (mandi-cuci-kakus)	20	40	20	10	10	10	90	---
2.3 Shared leaching pit	10	30	40	10	10	---	100	---
2.4 Household leaching pit	10	30	40	10	10	---	100	---
2.5 Household pour-flush toilet	—	---	—	100	—	---	—	---
2.5.1 Household septic tank	—	10	30	30	30	---	100	---
2.5.2 Houseconnection to sewer	—	10	30	30	30	---	100	---
2.5.3 Microsewer	50	25	25	---	—	50	50	---
2.5.4 Microtreatment (septic tank)	50	25	25	---	—	50	50	---
2.5.5 Macrosewer	50	25	25	---	—	100	—	---
2.5.6 Macrotreatment (imhoff tank0	50	25	25	---	—	100	---	---
2.5.7 Colluminity sludge collection	30	60	—	10	—	100	—	---
2.5.8 Household sludge collection	30	50	—	20	—	100	---	---
2.6.9 Sludge treatment and disposal	50	50	—	---	—	100	—	---

Table 12.1, contd.

(Column)	(1)	(2)	(3)	(4)	(5)	(6)	(7)	(8)
2.7 Conventional megasewer	30	70	—	---	—	100	—	---
2.8 Conventional megatreatment	30	70	—	---	—	100	—	---
2.9 Ocean disposal	30	70	—	---	—	100	—	---
2.10 Wastewater reclamation	50	50	—	---	—	50	50	---
3. Drainage								
3.1 Build/rehabilitate microdrains	---	30	30	---	40	50	—	50
3.2 Clean microdrains	—	30	30	---	40	50	—	50
4. Solid Waste Management								
4.1 Community collection	—	---	99	1	—	---	100	---
4.2 Community storage	—	---	100	---	—	---	—	---
4.3 Municipal collection	50	50	—	---	—	---	—	
4.4 Regional transfer	50	50	—	---	—	100	—	---
4.5 Transport	50	50	—	---	—	100	—	---
4.6 Disposal	50	50	—	---	—	100	—	---
4.7 Resource recovery	25	25	25	---	25	50	50	--
5. Advisory and supervisory services								
5.1 Water supply and sanitation	40	60	—	---	—	---	—	---
5.2 Drainage	30	70	—	---	---	---	---	---
5.3 Solid waste management	20	80	—	---	---	---	---	---
5.4 Research and development	30	70	—	---	---	---	---	---
5.5 Strategic planning	50	50	—	---	—	---	—	---
6. Training and technology transfer								
6.1 Water supply and sanitation	15	85	—	---	—	---	----	---
6.2 Drainage	15	85	—	---	—	---	—	---
5.3 Solid waste management								
5.4 R&D and demonstration								
5.5 Strategic planning								

* - asterisk indicates included in water service charge

Notes to Table 12.1. The Indonesian MCK (mandi, cuci, kakus) indicates community facility for laundry, bath, and latrine, respectively. A leaching pit serves from 1 to 4 households and is usually about 1 m diameter, 2 m deep, and lined with open brickwork. The term septic tank is applied to a variety of structures, from leaching pits to 2-compartment septic tanks with 1 to 3 days detention discharging to reticulated leaching lines on the property. Common practice is for blackwater from lavatories to discharge to leaching pits or septic tanks. Greywater from the kitchen, bath, or laundry is discharged to small, lined surface drains (microdrains). Microsewers are small-diameter (small-bore) sewers 50 to 100 mm in diameter. Microtreatment means on-site leaching pits or septic tanks with a minimum of 1 day detention to remove settleable solids. Macrosewers are generally about 150 mm diameter. Macrotreatment refers to community anaerobic ponds, Imhoff tanks, or septic tank treatment systems that discharge to surface waters for aerobic stabilization or aquaculture. Separate nightsoil or sludge treatment, reclamation (conservancy), and disposal is required for all systems. Megasewers and megatreatment refer to conventional North American and northwest European practices for water service levels of 300 to 1,000 liters per capita per day, adjusted for infiltration of groundwater and in some cases for much greater storm water into the gravity portions of the systems.

7. Complete recovery of beneficiary shares of costs is based on revolving loan funds supplied by central government contributions or World Bank loans.
8. Unless otherwise defined, annual O&M costs are assumed to be 15 percent of initial capital costs. This holds for pond systems and most primary and secondary systems. Numerical values for sharing investment and O&M costs were assigned as listed in Table 12.1.

12.2 Costs of Ocean Outfalls

The initial costs of submarine pipeline systems include those for (1) engineering, land, rents, licenses, and extraordinary legal costs and institutional, financing, and environmental services; (2) mobilization and demobilization; and (3) materials, equipment, and construction. Reported costs for municipal wastewater outfalls and cooling water discharge lines are listed on Table 12.2. Since the mid-1980s, and increasing number have been constructed in tunnels with diffuser risers for the end sections. Unit construction costs per meter have been normalized to an Engineering News Record (ENR) construction index of 4300. Most costs were reported in U.S. dollars. Those that were originally reported in other currencies were converted according to IMF exchange rates.

Unit cost data listed in Table 12.2 include diffuser sections account that usually amount to from about 15 to 50 percent of the total. Outfall lengths have not been found to be a consistent factor for a given diameter (4,9,16,34). Mobilization costs amount to from about 10 to 60 percent of the total. Figure 12.2 reveals two populations of costs, one for diameters of up to about two meters diameter where costs quadruple for each doubling of the diameter. For larger diameters, the costs

Table 12.2. Reported construction costs for selected submarine outfalls normalized to ENR Index 5500

No.	Location	Year	ENR index	Diameter meters	Length meters[1]	ENR Factor	Original millions	1995 millions	1995 per meter	Remarks
		(1)	(2)	(3)	(4)	(5)	(6)	(7)	(8)	
1	Kolta, DK	1995	- - - -	0.4	150	- - - -	0.45	0.45	3000	HDPE, 75m depth
2	Dubrovnik, Yugoslavia	1974	2020	0.45	2 @ 1500	2.72	5.5	15.0	5000	HDPE, 80m drop
3	Vallo, DK	1985	4220	0.5	800	1.30	0.3	0.39	488	HDPE
4	Coos Bay, OR	1972	1753	0.52	1440	3.14	1.8	5.6	3890	
5	Toledo, OR	1965	1000	0.53	1130	5.66	0.96	5.4	4780	
6	Los Angeles, CA	1957	724	0.56	11260	7.6	2.6	19.7	1750	Santa Monica Bay[2]
7	Passaic, NJ	1920	357	0.60	460	21.9	0.1	2.2	4780	Raritan Bay
8	Carmel, CA	1971	1581	0.61	272	3.48	0.41	1.4	5150	
9	Ketaminde, DK	1989	4800	0.71	650	1.15	0.50	0.58	885	HDPE (Fig. 12.)
10	Guldborg, DK	1976	2401	0.8	700	2.29	0.2	0.46	643	HDPE
11	San Elijo, CA	1965	1000	0.76	820	5.76	0.96	5.5	6710	cooling water
12	Seattle, WA	1962	872	0.76	640	6.31	0.27	1.7	2660	Puget Sound
13	Seattle, WA	1962	872	0.84	490	6.31	0.15	0.95	1940	Puget Sound
14	San Mateo, CA	1962	872	0.84	490	6.31	0.84	5.3	10820	San Francisco Bay
15	Oceanside, CA	1972	1753	0.91	2500	3.14	1.9	6.0	2400	extend outfall
16	Watsonville, CA	1959	797	1.0	1170	6.91	0.47	3.2	2740	
17	Encina, CA	1964	926	1.2	1370	5.88	0.35	2.1	1530	cooling water
18	San Francisco, CA	1974	2020	1.2	180	2.73	0.57	1.6	8890	
19	Istanbul, Üsküdar	1994	5504	1.2	2 @ 270	1.01	5.1	5.2	9630	Bosporus[3]
20	Mokapu, Hawaii	1971	2577	1.2	1547	2.14	6.2	3.3	2130	Mamala Bay
21	San Francisco, CA	1966	1019	1.4	250	5.90	0.46	2.7	10800	San Francisco Bay
22	Bellingham, WA	1973	1895	1.5	850	2.90	9.44	1.3	1530	Puget Sound
23	Istanbul, Ahirkapı	1989	4800	1.6	2 @ 1100	1.15	13.7	15.8	7180	Sea of Marmara[4]
24	Hampton, VA	1981	3600	1.7	2930	1.53	12.3	18.8	6420	Chesapeake Bay
25	Istanbul, Baltalimanı	1994	5504	1.7	2 @ 270	1.01	9.5	9.6	16000	Bosporus[5]
26	Contra Costa County, CA	1959	797	1.8	520	6.91	0.17	1.2	2210	San Francisco Bay
27	Ponce, Puerto Rico	1972	1753	1.8	1550	3.14	3.3	10.4	6710	
28	Encina, CA	1973	1895	1.8	700	2.90	1.05	3.0	4290	cooling water
29	Suffiolk County, NY	1981	3600	1.8	5777	1.53	28.8	44.1	7630	New York Bight
30	Manila, Philippines	1985	4220	1.8	3600	1.28	13.0	16.6	4610	Manila Bsy
31	Sand Island, Hawaii	2401	2401	2.1	3816	2.29	13.6	31.1	8150	Marmala Bay

Table 12.2 Continued	(1)	(2)	(3)	(4)	(5)	(6)	(7)	(8)	
32 Los Angeles County, CA	1954	628	2.3	3170	8.77	2.2	19.3	6090	Catalina Channel[6]
33 Seattle, WA	1964	986	2.4	1110	5.87	1.2	7.0	6300	Puget Sound
34 Rio de Janiero, Brazil	1975	2212	2.4	4325	2.48	22.0	54.6	12620	Ipenema Beach
35 Bombay, India	1985	4220	2.4	6000	1.28	65.0	83.3	13880	first estimate
36 Aberdeen, Scotland	1978	2600	2.5	2700	2.11	17.6	37.1	13740	
37 San Diego, CA	1962	872	2.7	4340	6.31	10.5	66.2	15250	Point Loma
38 San Diego, CA	1992	5200	2.7	5330	1.06	90	95	15670	Pt Loma estension[7]
39 San Francisco, CA	1970	1385	2.7	90	3.97	0.40	1.6	17770	San Francisco Bay
40 Los Angeles County, CA	1964	986	3.0	3620	5.57	4.5	25.1	6930	Catalina Channel[8]
41 Orange County, CA	1969	1269	3.0	8350	4.22	9.0	38.0	4550	Catalina Channel
42 San Diego, CA	1995	5500	3.1	1430	1.00	36.0	36.0	25180	South Bay[9]
43 Redondo Beach, CA	1947	413	3.2	2650	13.3	2.2	29.3	11050	cooling water
44 El Segundo, CA	1954	629	3.3	2620	8.77	2.3	22.8	7630	cooling water
45 San Diego	1995	5500	3.4	5490	1.00	90.0	90.0	16405	South Bay tunnel[10]
46 Los Angeles, CA	1950	570	3.6	1610	10.8	2.0	21.6	13230	Santa Monica Bay[11]
47 Huntington Beach, CA	1957	724	3.6	820	7.60	1.5	11.4	13900	cooling water
48 Los Angeles, CA	1957	724	3.6	8046	7.60	20.2	153.	19080	Santa Monica Bay[12]
49 San Onofre, CA	1965	971	3.6	1740	5.67	3.3	17.0	10750	cooling water
50 Sydney, Austr.- Bondi St	1986	4300	2.5	2300	1.28	55.	70.	30610	gravity g;petunnel
51 Sydney - North Head	1986	4300	3.5	3400	1.28	68.	87	25580	pumped flow tunnel
52 Sydney -- Malabar	1986	4300	3.3	3500	1.28	55.	70.	20000	gravity flow tunnel
53 Bermuda	1992	5000	0.49	180	1.10	0.50	0.55	3055	
54 Vahcouver, BrC, Canada	1993	5300	1.17	250	1.04	1.26	1.31	5240	Puget Sound
55 EBMUD, Oakland, CA	1995	5500	1.37	146	1.00	0.55	0.55	3770	San Francisco Bay
56 Los Angeles Harbor	1995	5500	1.8	1585	1.00	20.0	20.0	12620	
57 Barbers Point, Hawaii	1976	2401	1.98	3200	2.28	12.0	27.3	8530	Mamala Bay
58 San Diego, CA	1995	5500	3.05	2652	1.00	50.0	50.0	18850	outfall and diffuser
59 Hong Kong	1995	5500	3.25	1200	1.00	12.5	112.5	13330	diffuser section
60 Seattle, WA	1995	5500	3.66	168	1.00	3.2	3.2	19050	tunnel
61 Shanghai	1984	- - - -	4.2	2 @1460	- - -	19.6	34.5	11830	Huang Pu River[13]
62 Shanghai	1995	- - -	4.2	1600	-- - -	23	23	14375	Yangtze tunnel [14]
63 Boston, MA	1995	5500	4.3	7650	1.00	151.1	151.1	19751	tunnel [15]
64 Hong Kong	1995	5500	5.0	1800	1.00	45.0	45.0	25000	tunnel section
65 Boston, MA	1995	5500	8.1	15200	1.00	334.2	334.2	17150	tunnel [16]

Footnotes to Table 12.2.

1 Includes diffuser sections
2 Hyperion 7-mile (11 km) sludge outfall
3 Into lower northerly current layer
4 Into lower layer at southerly entrance to Bosporus
5 Into lower layer of Bosporus
6 Costs of force account work using agency staff
7 5330m extension of 4340m outfall increased diffuser depth from about 65m near edge of continental shelf to about 300m further down the continental slope
8 Force account
9 Proposed South Bay outfall
10 Proposed South Bay tunnel including drop structure
11 Hyperion 1-mile (1.6 km) outfall
12 Hyperion 5-mile (8 km) outfall
13 Tunnel to Yangtze River estuary. Costs normalized to reflect 6.5% local inflation rate and 3.5¥ / US$
14 Tender for first of two parallel tunnels to Chiangjiang (Yangtze) estuary, not including head tank. Estimated unit costs increase linearly with pipe diameter. Construction by municipal corporation force account.
15 Inter-island tunnel costs include prorated shares of facilities, spoil disposal, and service costs
16 Effluent tunnel and diffuser costs include prorated shares of marine and other facilities, spoil disposal, and other service costs.

Sources: Clancy and Carroll (4), , Grace (9), Hennessy (16), Sayınlı and Yiğit (29), Wallis (33), , CDM International, M. Laginestyra, Mott-McDonald Engineers, and COWI Engineers (personal communications (1995),

increase by a factor of about 6 as the diameter is doubled. The same discontinuity using a smaller data set was reported in 1968 edition of this book

The higher costs for larger diameter outfalls presumably reflect costs of materials, constraints on construction methods (see Chapters 5 through 9), greater sophistication in diffuser systems, increased demand for scarce professional and technical skills, and higher mobilization and infrastructure costs.

Figure 12.3 shows the costs of each cubic meter per second capacity for 100, 500, and 1,000-meter long outfalls with an initial (elevation) head of 10 meters, based on the Manning formula. The Hazen-Williams formula is generally used for 100 to 200 mm diameter smooth pipes. The Manning formula is more appropriate for pipes with diameters over 200 mm and velocities of about 1 m/s and is used here:

$$Q = \frac{d^{8/3}s^{1/2}}{n} \tag{12.1}$$

where Q = flow in m^3/s, d= pipe diameter in meters, s = hydraulic slope, and n = roughness coefficient. Incremental capacities in m^3/sec as functions of pipe diameter (6, 11). Unlined steel pipe has an "n" value of about 0.012. are shown for "n" values of 0.011 (plastic or smooth lined steel pipe) and 0.015 (concrete pipe). Figure 12.3 shows the higher marginal costs of additional outfall capacities that are associated with smaller diameters, greater lengths, and higher "n" values. The discontinuity in unit costs is found in the 1.2 to 3.2 diameter range. Within this range, additional capacities may be obtained at the same cost with either smaller or larger pipes. The significance of this choice is that staged construction of smaller pipes is more adapted to closer matching of capacity to demand and complementary costs savings rather than a single construction period.

12.3 Estimating Marginal Costs and Benefits

Although there is a large body of information on costs of wastewater treatment, local accounting differences makes comparisons difficult. Standard engineering works such as Degrémont (6), Freysinius, et al (7), and Metcalf and Eddy (22) concentrate on how to remove pollutants rather than how to pay for their removal although the latter makes brief reference to "engineering economics."

This is changing. World Bank and other multilateral funding agencies have always dealt with economic and environmental development needs that are many times greater than resources to satisfy them, and have developed analytical tools to appraise and compare cross-sectoral costs and benefits in economic development (2,24,32). Similar tools are used in bilateral development assistance that respond to national political and financial interests where general cost-benefit comparisons are not applicable. U.S. interests in marginal costs of removals are beginning to respond to the combination of the economic downturn beginning in the late 1980s, the phasing out of the USEPA (sewerage) Construction Grants Program, the birth of a uniquely defined Superfund Program of remediation projects for land and groundwater pollution, and of supporting creative ventures for full-scale innovative wastewater treatment projects. These evolving interests in costs are demonstrated in the pioneering study by Murcott and Harleman (25) summarized in a National Research Council report on managing wastewaters in coastal urban areas (26).

12.3.1   Simulating Sewage Treatment Costs and Benefits

Here we focus on unit costs of removal in dollars per kilogram. Murcott and Harleman (25) used data from two surveys of over 100 treatment plants with widely varying processes and practices. They identified ten alternative combinations of treatment processes, four of which are listed below. Costs are undercounted by excluding the opportunity costs of land in accordance with conventional U.S environmental engineering practice, so that valid comparisons of, say, mechanical and pond systems cannot be made. Costs were synthesized from field data and the literature such as USEPA design tables and cost curves, and

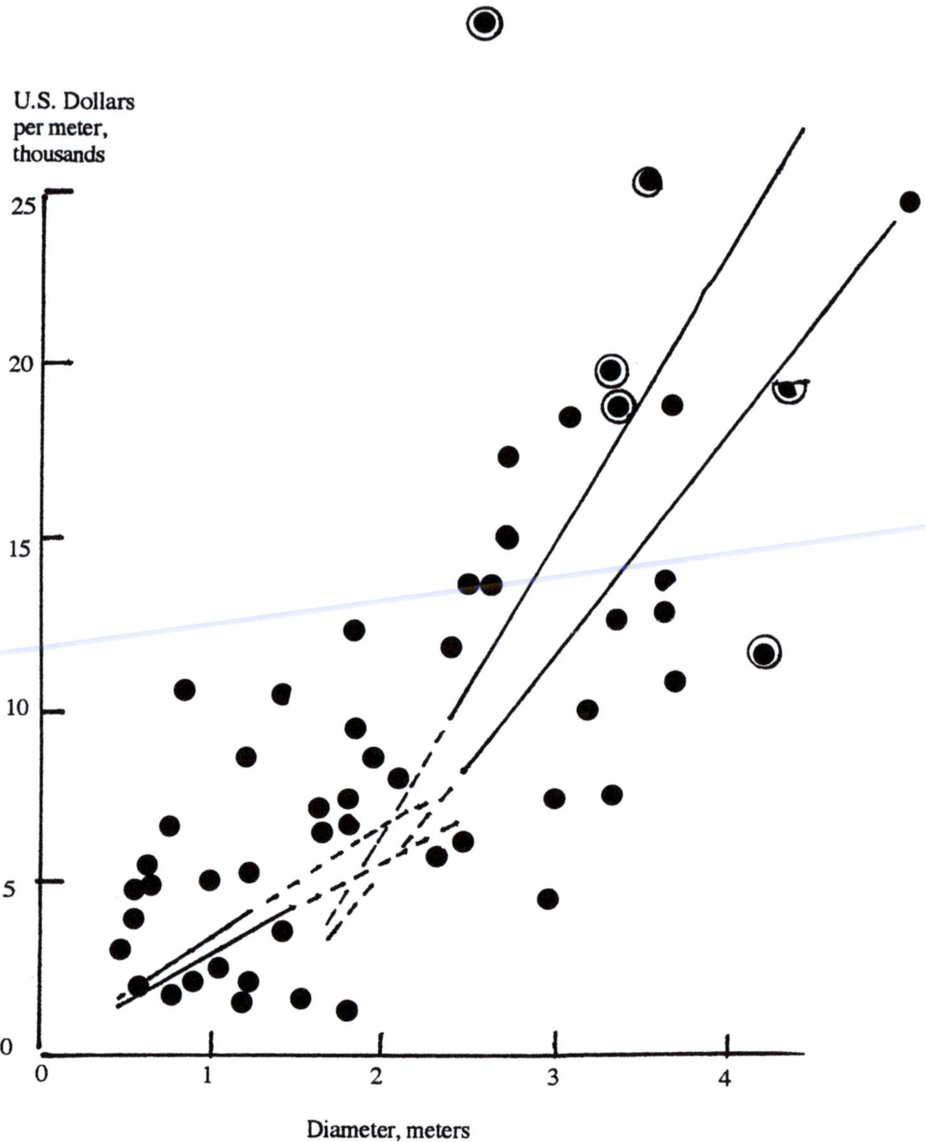

Figure 12.2. Unit costs of ocean outfalls in 1995 U.S. dollars per meter (Table 12.2). Boston Harbor not included. Medians shown for 1995 (upper, ENR 5500) and 1988 (lower, ENR 4300 used in Figure 12.3). Tunnels indicated by open circles.

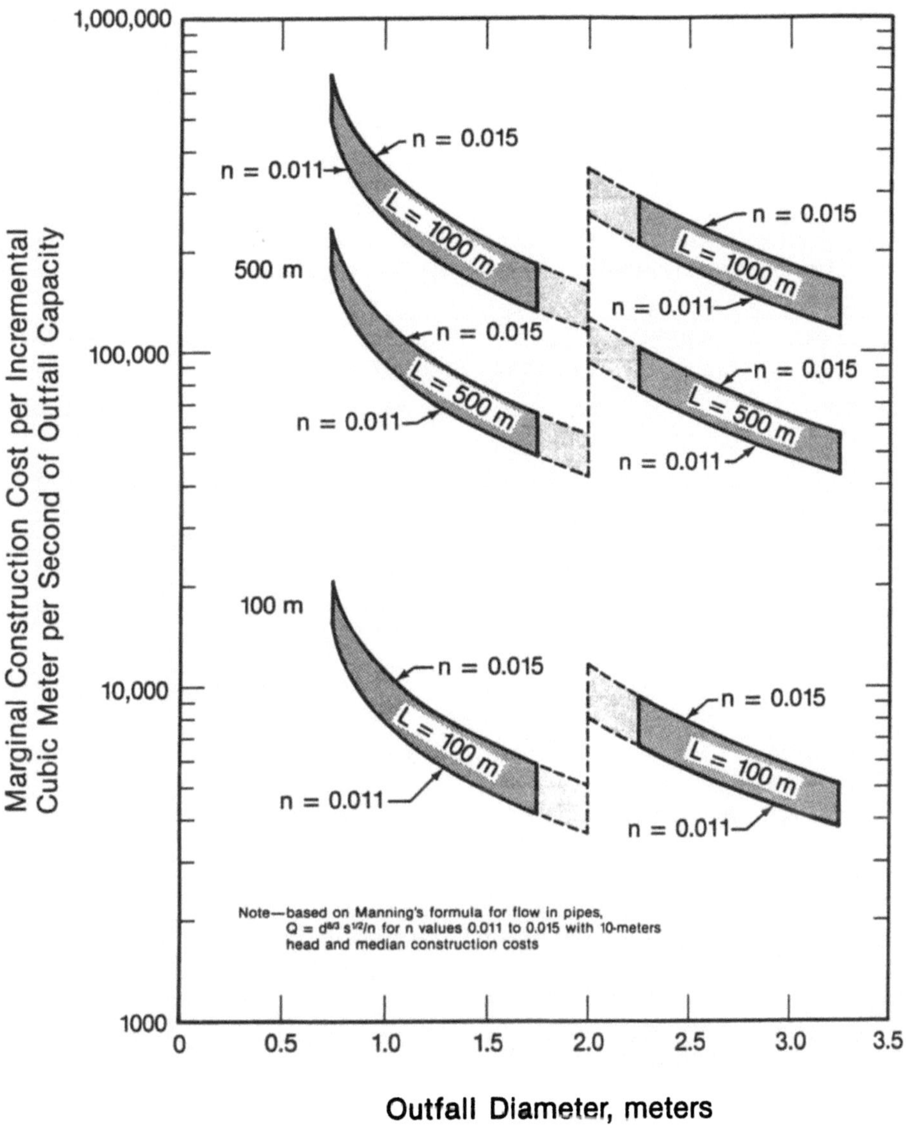

Figure 12.3. Marginal costs for ENR index 4300 of ocean outfall capacities achieved by increasing diameters of 100, 500, and 1000 meter outfalls based on Manning's formula where $Q=d^{8/3}\,s^{1/2}/n$ for n values 0.011 to 0.015.

then normalized to 20 mgd (880 ℓ/s) by a professional engineering firm. Removals were aggregated for suspended solids, BOD_5, total P, total N, NH_4, and oil and grease. Their findings, in terms of single-purpose costs and suspended solids removals, are here further normalized to influent concentrations of 200 mg/ℓ and summarized in Table 12.3.

Table 12.3. Simulated marginal costs of primary, chemically enhanced, and secondary treatment for influent suspended solids concentrations of 200 mg/ℓ.

Wastewater treatment system	Effluent mg/ℓ	Marginal benefit as reduction in mg/ℓ	Ranges of costs in $/million gallons			Median costs per million gallons	Costs per kilogram suspended solids removed
			Capital	O&M	Total		
Primary	87	87	245 to 310	205 to 400	450 to 550	500	$1,046
Primary + low-dose chemical	57	57	320 to 400	230 to 280	550 to 680	615	$2,617
Primary + high dose chemical	15	104	400	200 to 350	650 to 750	700	$1.778
Secondary (biological)	12	75	610 to 720	320 to 410	930 to 1130	1030	$3,640
Low-dose primary + secondary	10	47	750 to 870	350 t0 450	1050 to 1150	1100	$6,179

Source: National Research Council (26). Low-dose is 5 - 100mg/ℓ, high-dose is 100 to ~250 mg/ℓ.

Operation, maintenance, and although to a lesser degree, capital costs incline to be undercounted (Sec. 12.1.1 and 12.10). Most of this undercounting is due to excluding some or all of full burden and overhead, and of opportunity costs of land. Normalizing and averaging unit costs loses information on their ranges. Table 12.3 confirms expected trends. Costs increased monotonically from about $1,000 per kilogram suspended solids removed by primary treatment to $6,000/kg for chemical enhanced primary plus biological treatment. Annual O&M costs of 5 to 8 percent of original capital costs include direct costs of labor and materials.. The study confirmed that chemically enhanced primary is as effective at much less cost than USEPA-mandated secondary treatment for ocean discharges.

12.3.2 Measuring Sewage Treatment Costs and Benefits

Results of a definitive study of incremental costs for nine wastewater reclamation plants of the Los Angeles County Sanitation Districts (LACSD) where the data are both more exacting and homogeneous (21) are shown on Figure 12.4. The

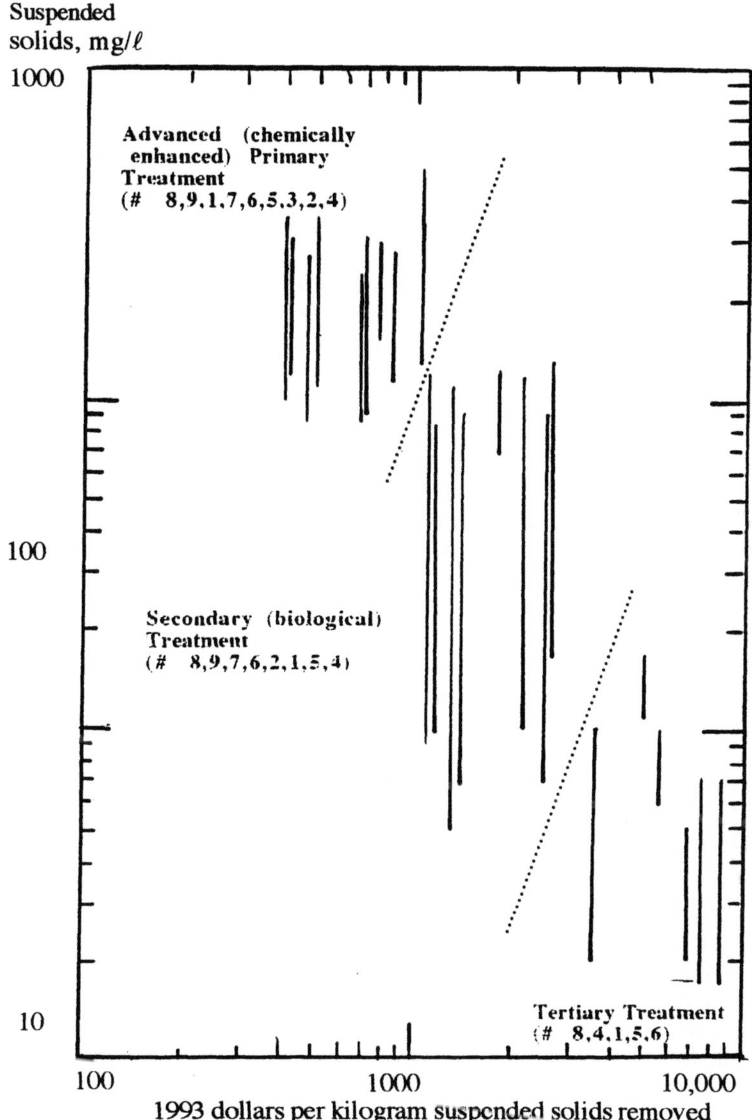

Figure 12.4. Measured marginal costs and benefits of primary, secondary, and tertiary treatment processes at Los Angeles County wastewater reclamation plants. Vertical lines indicate influent and effluent concentrations for each stage at each plant. Water reclamation plant average flows in million gallons/day (liters/sec): #1- Saugus , 5.8 (250); #2 - Palmdale; 7.1 (310); #3 - Lancaster, 8.5 (370); #4 - Valencia, 8.9 (370); #5 - Whittier Narrows, 12.1 (530); #6- Pomona, 12.6 (550); #7 - Long Beach, 17.0 (750); #8 - Los Coyotes, 31.3 (1380); #9 San Jose Creek, 67.1 (2950). Plants #2 and #3 use ponds systems. Source: adapted from Chiu, et al (21)

capital recovery factor of 10% per year was based on the same 8% interest over 20 years used by Murcott and Harleman.   The greater credibility of LACSD data is that O&M costs explicitly include all chargeable supervision, burden, and overhead items, ranging from general management to plant security.   Land costs are excluded because real estate taxes and public records in California have had little relation to subsequent land values since a dubious 1968 "tax revolt" when citizens voted for a constitutional base property tax limit of 1.25% of the last sale price.

Suspended solids were selected for purposes of this comparison because of the robustness of the data and because concentration changes provide a credible measure of the removals of the pathogenic, toxic, and nutrient constituents of urban wastewaters.   Figure 12.4 quantifies on logarithmic scales the costs and benefits (measured as suspended solids removals) at each level of treatment as indicated by the vertical lines.. Costs include sludge treatment and disposal.

Advanced (chemically enhanced) primary treatment removes suspended solids at costs of about $500 to $1,000 per kilogram removed. Related costs are $500 to $3000 and $3000 to almost $9000 for secondary and tertiary treatment, respectively. Annual O&M costs are 10 to 30 percent of original capital costs. These results extend the exploratory conclusions drawn from the Murcott and Harleman data listed in Table 12.3.   Although economies of scale are evident, their aggregate is much too small to offset the hydraulic water-wastewater system diseconomies of scale shown on Figure 12.1.

Reclamation operations.   The greater public acceptance of the term "wastewater reclamation plant" over "sewage treatment plant" is matched by political, financial, and technical recognition that reclamation will become increasingly important as raw water sources become more distant and more expensive. Average daily flows listed on Figure 12.4 are much higher than the amounts actually reclaimed and sold. The total capacity of the nine plants listed is 223 mgd (9,800 ℓ/s).   Treatment costs ranged from $542 to $2,954 per million gallons. 55.4 mgd or 25% of the total were reclaimed while the balance was returned to the interceptor and trunk systems for another treatment and ultimate discharge to the ocean. Prices for reclaimed water excluding wastewater collection and product water distribution costs are less than their costs except for an occasional golf course (21). Utilization of reclaimed wastewater is generally assumed to increase in the future, particularly where the price of water reflects its quality where the first user pays the highest price and when full marginal costing, environmental attributes, opportunity costs, and community values for conservation are included.   Decisions on reclamation depend on salt balances and the capacity of the soil to accept and transmit water. Earlier World Bank research on appropriate technology found that loamy soils can accept about $10m^3$/ha/d (19,20). This figure needs to be confirmed with empirical data on local septic tank or small irrigation systems. Standard short-term infiltration tests are insufficient.

12.4   Principles of Comparative Costing

Economic costing of water supply and wastewater management requires that all casts of all feasible alternatives be compared across different systems (19). In addition to conventional listing of capital and O&M costs to the utility providing the service, costs include those to central governments for grants, concessionary loans, and subventions; costs to householders for plumbing fixtures and flushing water; and costs to the municipality for transportation, security, administration and personnel , and other services. Comparisons are more than those of with or without the project. Conventional engineering estimates of costs based upon full utilization of the capacity at the end of the design life. Full costing includes that for the unused capacity during the early years of the project. The per capita average incremental cost over the life of the project is calculated by dividing the sum of the present value of capital (C) and operation and maintenance costs (O) by the sum of the present value of the incremental persons served the quantities of wastewater disposed of (N). Note that the rationale for discounting the wastewater service level in the same manner as that for capitol and operating costs is that it is the monetary value rather than the physical quantity that is being considered (24). The formal statement for the average incremental cost is:

$$AIC = \frac{\sum_{t=1}^{t=T} (C_t + O_y) / (1 + r)^{t-1}}{\sum_{t=1}^{t=T} N_t (1 + r)^{t-1}} \tag{12.2}$$

where C = present value of construction costs

   O = incremental value of operation and maintenance costs

   N = present value of benefit, either number of persons served or volume of service, e.g., m3. of water supplied or disposed of, or for an irrigation project, quantities of crops produced.

   T = life of facility

   r = opportunity cost of capital

Because they are intended to show costs to the economy, utility costs are shadow-priced to exclude taxes, duties, rents and licenses, differences between actual and official minimum wage standards, block water tariffs imposed to ensure minimum (lifeline) service, and for international projects, the real and official currency exchange rates. The black market price of currency is a shadow price.

Other exclusions are those of estimated inflation during the service life, of sunk costs and their benefits, and of politically important secondary benefits arising from employment (this would be an example of double-counting).

Average incremental costing provides for comparing costs of projects which have different system capacities, service lives, construction schedules, disbursement targets, and/or different imputed economies of scale. Thus, the AIC

(1) quantifies the costs of unused capacity during the early years of a project.

(2) demonstrates that the most economical system, program or project is the one that provides the greatest efficiency. This requires that incremental increases in capacity most closely match the incremental increases in demand (e.g., well fields and conjunctive use of ground water inherently match demand better than high dams and large reservoirs, on-site wastewater disposal systems inherently match demand more efficiently than do large sewage collection, treatment, reclamation, and disposal systems) and

328

inherently match demand more efficiently than do large sewage collection, treatment, reclamation, and disposal systems) and

(3) removes the conventional bias toward large, centralized systems with large, lumpy investments. This bias favors capital-intensive, high technology systems with minimum redundancy (stability) whose costs and benefits are optimized for the end of the design period.

12.4.1   Worked Example of Average Incremental Costing

Compare the average incremental costs of two-stage vs. one-stage construction of a sewerage system and compare the costs and benefits of more closely matching capacity to demand during the first 20 years of a continuing sanitation improvement program following Kalbermatten, et al, (20).

Assume 6.7 percent annual population increase beginning with 88,340 in year 1, reaching 125,000 in year 6 and 250,000 in year 16. Corresponding wastewater flows at 200 liters per capita per day are 6450, 9125, and 18250 thousands of cubic meters per year, respectively. The 6.7 percent value is higher than most, but it is not unheard of. Herewe use a doubling interval based on 72 years.

The project period is taken at 20 years to demonstrate a reasonable matching of capacity to demand. Staged construction of both the collection (sewer) system and treatment works, either primary sedimentation or 20-day pond systems, are scheduled as shown. Costs of household plumbing, reclamation options and final disposal costs are excluded from the analysis.

Construction costs of additional capacity to avoid overloading would be required after year 16; these are excluded for the purpose of this comparison. Construction periods are 5 years for single-phase construction, 2 1/2 years for the first of the staged construction and 2 years for the second. No economies of scale for mobilization costs are assumed; they are a fixed portion of the construction costs. The opportunity cost of capital is 12 percent in the sample calculation.

Annual operation and maintenance costs vary according to the population served. Construction costs are 100 times those for the 18,250 m3/day capacity system used in the worked example. The higher figures more closely approximate recent costs in US dollars for similar systems. O&M costs are assumed to be directly proportional to flows, based on an estimated 5 percent of construction costs for the collection component and 15% for treatment when operating at capacity. The cost stream calculated as follows is listed on Tables 12.4 and 12.5.

(1) Determine the present values for recurring costs and wastewater volumes for the first 20 years of the program at interest rates of 8 and 16 percent Convert these costs to their present values using a financial calculator, a set of financial tables, or the following equation where C_t is the annual cost for collection and treatment: $PV = \quad C_t / (1 + r)^{t-1}$

(2) Compare the Average Incremental Costs for the first 20 years under the two construction schedules. Costs are in millions of units of national currency (UNCs). Flows are in thousands of cubic meters. In this example, the total economic cost for a single construction period is 4% greater than that for the two-phase construction.

Table 12.4 Costs in millions and flows in thousands for sewerage scheme

Year	Single Construction Period				Staged Construction				Flows m^3/y
	Collection		Treatment		Collection		Treatment		
	Capital	O&M	Capital	O&M	Capital	O&M	Capital	O&M	
1	90	0	2	0	40	0	18.5	0	0
2	90	0	0	0	95	0	37	0	0
3	80	4.67	32.5	0	95	4.67	37	1.92	* 3704
4	80	10.00	30	4.11	0	10.00	0	4.11	7940
5	120	10.72	30	4.41	0	10.72	0	4.41	8511
6	0	11.49	0	4.72	0	11.49	0	4.72	9122
7	0	12.32	0	5.06	0	12.32	0	5.06	9777
8	0	13.20	0	5.43	0	13.20	0	5.43	10478
9	0	14.15	0	5.82	0	14.15	0	5.82	11231
10	0	15.17	0	6.23	0	15.17	0	6.23	12036
11	0	16.26	0	6.68	115	16.26	46.25	6.68	12902
12	0	17.42	0	7.16	115	17.42	46.25	7.67	14822
13	0	18.68	0	7.67	0	18.68	0	7.48	14822
14	0	20.02	0	8.23	0	20.02	0	8.23	15886
15	0	21.44	0	8.84	0	21.44	0	8.83	17027
16	0	23.00	0	9.45	0	23.00	0	9.45	18250
17	0	24.53	0	10.08	0	24.53	0	10.08	19473
18	0	26.18	0	10.76	0	26.18	0	10.76	20777
19	0	27.93	0	11.48	0	27.93	0	11.48	22170
20	0	29.81	0	12.25	0	29.81	0	12.25	23655

Table 12.5. Present values at 12% in millions and flows in thousands.

Year	Single Construction Period				Staged Construction				Flows
	Collection		Treatment		Collection		Treatment		
	Capital	O&M	Capital	O&M	Capital	O&M	Capital	O&M	m^3/y
1	90.0	0	2	0	40.0	0	18.5	0	0
2	80.4	0	0	0	84.8	0	33.0	0	0
3	63.7	3.72	25.9	1.92	75.7	3.72	29.5	1.92	2953
4	56.9	7.12	21.3	2.92	0	7.12	0	2.92	5651
5	76.3	6.81	19.1	2.80	0	6.81	0	2.80	5408
6	0	6.52	0	2.68	0	6.52	0	2.68	5176
7	0	6.24	0	2.56	0	6.24	0	2.56	4953
8	0	6.02	0	2.45	0	6.02	0	2.45	4739
9	0	5.71	0	2.35	0	5.71	0	2.35	4536
10	0	5.47	0	2.24	0	5.47	0	2.24	4340
11	0	5.23	0	2.15	37.0	5 23	14.6	2.15	4154
12	0	5.01	0	2.05	33.1	5.01	13.3	2.05	3975
13	0	4.79	0	1.97	0	4.79	0	1.97	3804
14	0	4.58	0	1.89	0	4.58	0	1.89	3641
15	0	4.39	0	1.81	0	4.39	0	1.81	3484
16	0	4.20	0	1.73	0	4.20	0	1.73	3334
17	0	4.00	0	1.64	0	4.00	0	1.64	3176
18	0	3.81	0	1.57	0	3.81	0	1.57	3026
19	0	3.63	0	1.49	0	3.63	0	1.49	2883
20	0	3.46	0	1.42	0	3.46	0	1.42	2746
Column	367.3	90.71	68.3	37.64	270.6	90.71	108.9	37.64	71,979

AIC for single stage construction = 563.95/71.979 = 7.83 UNCs per cubic meter

AIC for two stage construction = 507.85/71.979 = 7.05 UNCs per cubic meter

12.5 Sustainable Water and Wastewater Management

There is a large and authoritative literature on how ancient models of survival and growth were revived under the name of sustainability during the second half of the 20th century. Concepts of urban and rural interactions, limits to scale in water and other natural resource exploitation, and comparative costing have been presented in previous chapters and paragraphs. All are determinants of defining and developing strategies in wastewater management for coastal cities. Following are three unifying models that support the thesis that cooperation is more efficient than competition in the water sector. These include water conflict identification, the prisoner's dilemma as a paradigm for cooperation, and information and technology transfer as means for capacity building and system sustainability.

12.5.1   Water Conflict Identification

Water is a tenacious source of competition and conflict. Water conflicts arrive first over quantity or efficiency in allocation. Seldom is there just the right amount for everybody in the places where they want it. In 1957, Karl Wittvogel (34) applied Marco Polo's term, oriental despotism, to rulers or peoples whose schemes for water development at least temporarily interfered with others' territorial or riparian rights. Water conflicts have ancient roots in cultural, historical, and economic disparities, and new sources in needs for institutional legitimacy or technological modernity that generate realpolitik and funding agency disbursement targets. They would be vastly exacerbated by sea level increases due to global warming. (Sec. 2.1.3) and further declines in marine fisheries (Sec. 2.2.7). Effective coastal city water conservation, reclamation, and wastewater management is embewdded in the interdependence of upstream and local water economies.

Identifyingspecific points of potential conflicts over water quantity and water quality is a first step towards cooperation and sustainability, especially in allocatring the costs and benefits of multipurpose river basin development projects Conflicts have arisen in areas of both water scarcity and abundance in cll climates, usually   by overestimating supplies and the abilities of technologies for meeting and using them, and waste assimilative capacities.

Figures 12.5 and 12.6 show intersections of demand sectors and supply technologies with respect to water quantity and water quality, respectively. The models identify 53 flash points in water quantity, and 44 in water quality whose sustainable resolution requires cooperation, transparency, and trust (see Section 12.6.2). Expatriate scientific and engineering contributions to this end include cross-cultural information and technology transfer, creation of agricultural policies that lead to water compacts to support them, and the identification of inter communal single and multipurpose water projects.

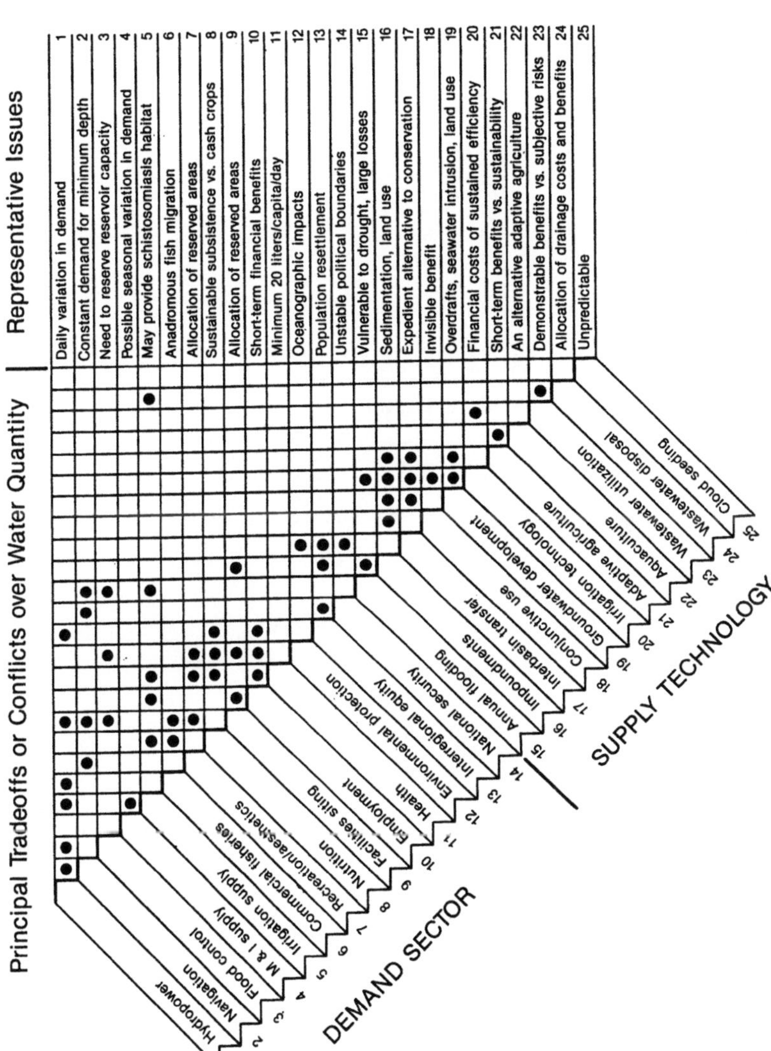

Figure 12.5. Principal tradeoffs or conflicts over water quantity in river basin development from tributary areas to coastal cities and coastal zones. Source: Gunnerson (14)

332

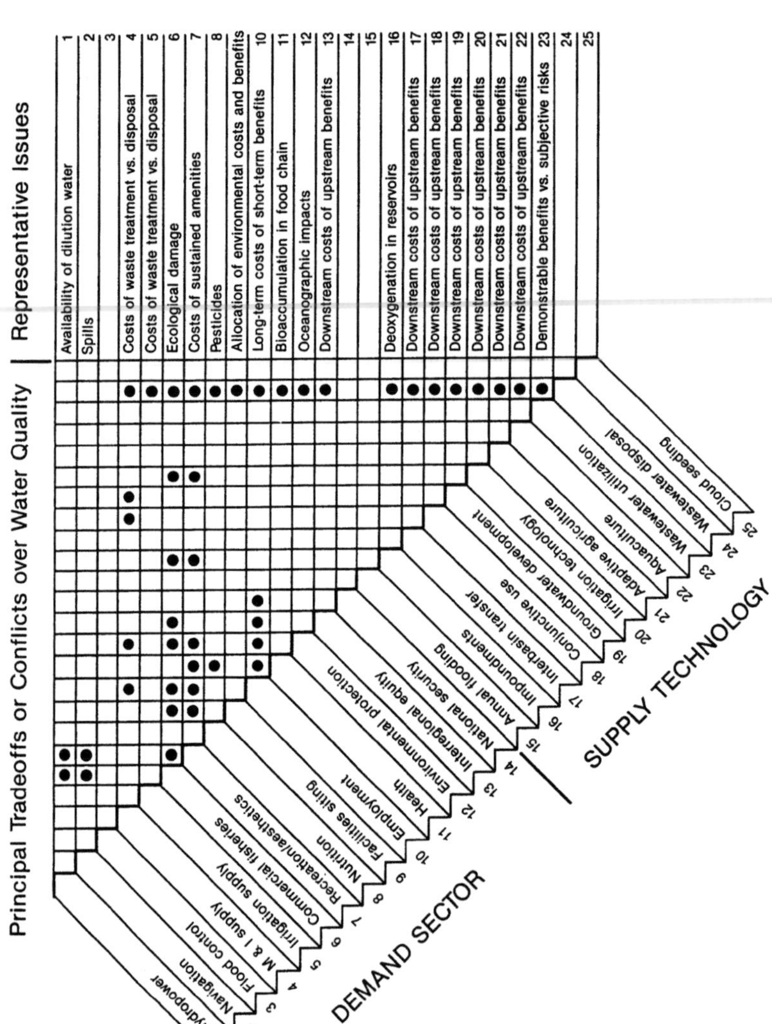

Figure 12.6. Principal tradeoffs or conflicts over water quality in river basin development from tributary areas to coastal cities and coastal zones. Source: Gunnerson (14)

12.5.2   Water Conflict Resolution

Supply-side engineers and economists have long worked on development of water and other resources and in the negotiations that are part of sustainable project identification, preparation, and implementation. Resource wars continue to be one of the options, but their successes tend to be temporary (10,34). Although no solutions are likely to be permanent, the design lives of water and wastewater projects are commensurate with the argument that cooperation is more efficient than competition and conflict.   In this and the following sections includes theoretical bases for cooperation and a practically approach for supporting it.

The prisoner's dilemma.   This model provides a rigorous test of the thesis that cooperation is more efficient than competition. The story begins (28) with two men caught with stolen goods whose possession means six months in prison. They are suspected of burglary where conviction results in a minimum two year sentences but the evidence is weak. The prisoners are in separate cells and cannot communicate. The prosecutor wants convictions. So he offers each of the prisoners a deal. If one confesses to burglary and the other does not, the first one goes free and the second one gets the maximum of six years. If both confess, each gets two years.   If neither confesses, each gets six months, so the total cooperative prison time would be one year which would be the optimum solution. Neoclassical economic theory holds that it is to the advantage of each prisoner to confess, for if he does not and the other does, he will receive six years. Without a credible previous commitment or on-going communication, neither can trust or cooperate with the other, so they compete. Both confess and between them serve four years in prison. This inefficient, suboptimal solution is a special case of Garret Hardin's tragedy of the commons (15). It is expressed in the 2x2 matrix:

Table 12. 6.   Prisoner's dilemma. How many years in prison.?

	Prisoner #2	
	Cooperate (deny)	Compete (confess)
Prisoner #1		
Cooperate (deny)	0.5, 0.5	0.0, 6.0
Compete (confess)	6.0, 0.0	2.0, 2.0

Decision theory research confirms that suboptimal choices abound in single-play binary games. The choices become more efficient during iterative decisions that characterize extended negotiations and compromises between individuals and nations. Classical and neoclassical economics hold that individual and collective "rational man" who always strives for personal gain and (in a perfect market) magically benefits all others in the community.   Empirical and theoretical research both demonstrate that in a two-player iterative contest, the best strategy is for the first player to begin by cooperating, and from then on to follow a tit-for-tat strategy. Barrett (1) has recently adapted the model as an adjunct to negotiation in a variety of international cooperation and conflict issues in the water sector, an example of which defined is in Table 12.7.

Table 12.7. Benefits from unilateral national water projects

	Country B	
	Build	Don't build
Country A		
Build	3, -5	1, 9
Don't build	2, 0	-1, 1

This is a single binary decision in an interdependent market like the prisoner's dilemma. Should a country build or not build a project within its boundaries by claiming unlimited territorial sovereignty. On a river, the downstream country would receive less water (due to consumptive use) and more pollution. Both would impact on its unlimited territorial (riparian) integrity. In an estuary or coastal embayment, both countries would be faced with polluted nearshore waters. With the payoffs indicated, The worst would be for Both A and B to build. The best solution is for A to build and B to not build.. Here B would improve its payoff by 14 and A would loose by two. The optimum could be reached only by negotiation over how to share the net gain of 12. Transparency, a cooperative first step, a tit-for-tat approach, and trust are required for the most efficient projects and programs and the sustainability that results. Means to this end in international environmental development projects include information and technology transfer from expatriate consultants and suppliers to their host country counterparts for which the prisoner's dilemma is a model. Clearly, decisions that take advantage of situations where both cooperative and individual decisions are possible are more efficient than decisions based on primitive self-interest under free competition (28).

12.5.3 Information and Technology Transfer

Technological, environmental, and cultural intervention and innovation are among the principal instruments of economic and environmental development. Post-audits measure results of new, often intrusive, technologies and the technology transfers to make them work.   At the operational level, most technology transfer occurs between external consultants and resident practitioners. Meanwhile, as global populations increase, there are signals that regional and local limits to scale are being reached in developing country resource exploitation, wastes disposal, conventionally defined economic development, and in technical and social ingenuity at all levels. (17)

An operational definition of technology transfer is the two-way exchange across cultural thresholds of information, skills, and practice between foreign consultants and their resident counterparts.

Here we argue that the most successful international consultants will be those who best apply the essential resources to meet both explicit and implicit complementary contractual obligations for training services that accompany those for advisory and supervisory services. The premise that cooperation is more efficient than competition is extended to include *de facto* competition between consultant and client rooted in the idea of a continuing monopoly ("competitive edge") and the hope of future gain from the same client. The cooperative alternative is to transfer information and training so that clients will prepare the terms of reference for similar projects while the consultants are taking advantage of their lead in technical and organizational ingenuity by developing new and innovative project technologies (see Sec. 12.5.3).

Technology Transfer in Context. Technology transfer includes information, skills, and practice. These take place in historical, cultural, economic, environmental, and technological contexts. Practice is easy. Counterparts participate in field surveys, computer-assisted design programs, or laboratory analyses. The minimum skills needed to sustain that practice are also easily transferred. Seldom is there time, funding, or expertise under conventional engineering services contracts for adequate cross-cultural explanations to the counterparts why things are done this way in the first place.

This means that client enterprises and governments are not getting their money's worth from expensive foreign loans, advisors and consultants. National, institutional, and personal patron-client bonds are strengthened, dual economies rooted in colonialism are embedded, and full economic and technological partnerships remain foreclosed. There is a unique utility for technology transfer among developing countries through UN agencies as the UNDP Special Unit for Technical Cooperation among Developing Countries (TCDC).

Cross-cultural thresholds abound in international work. In the first place, engineering is not a universal language because language is culture-bound. Financial and monetary thresholds are easily identified. Others are subtle, such as conceptions of roles and status, and of problem solving. Comparisons are confounded by the mystique of money from development banks, agencies, and industries. The point is that other cultures have their norms, and none are congruent with the American one. Sustainable development projects depend on accommodating these differences.

Table 12.8 is an introductory model for comparing two cultures, and for working within them. The table reflects experiences of resident expatriate and counterpart engineers (13,30). It could be expanded to include such things as the affordability of cultural mistrust and the prisoner's dilemma, particularly in business. Cultural norms are categorized to reveal the internal consistency of specific attributes, but norms are only that. Certainly, there are Americans who ascribe evil to its disclosure and Javanese who place self-interest above mutual self-help (*gotong royong*). A table prepared in another language would differ in detail because language is culture-bound.. In any event, European sources of and differences from American norms are recognizable, especially to Europeans. Table 12.8 is designed to be adaptable for operational rather than academic comparisons. For example, the Chinese "yes" is like the Javanese, and "no" is the same as in English while the Japanese way of saying "no" is more like the Javanese.

Table 12.8. An approach to comparing cultural norms.

	Javanese Views	American Views
Premises		
Emphasis	Syncretic, Islamic, Javanese	Calvinist, messianic [1]
Environment	Man with nature	Man over nature
Development	Colonialism, invasive trade	Raw materials, markets, manifest destiny
Orientation	Past to present to future	Present to near future Indifference to long-term
Means	Intuitive, deductive	Analytical, inductive, deductive
Roles and Status		
Emphasis	Being, hierarchical	Doing, arguably egalitarian
Means	Ancestry, cultural heritage	Position, possessions, fluidity
Ideal Behavior		
Constitutional emphasis	*Pancasila* (religion, nationalism, guided democracy, humanitarianism just and prosperous society	Individual freedom, human rights Separation, balance, and competition between executive, legislative, and judicial powers
Unifying principle	Command and control	External threat, (Godzilla, and the Andromeda Strain)
Means	Cooperative, *gotong royong* (community self-help)	Individual, competitive, winner takes all
Governance		
Emphasis	Protect the community at expense of the individual	Protect the individual at the expense of the community[3]
Authority	Hierarchical, parental	Functional, adversarial, high-minded
Evil	Shame (tact of exposure)	Guilt (the deed itself)
Foundation	Law and government are good, good, a governor may be bad	Law is good, government is not
Means	Consensus	Majority rule, litigation, gridlock
Knowledge		
Emphasis	Basic morality in science, humanities, etc.	Neutral, amoral, applied, convenient, marketable
Education	Islamic, social, rote Form, theory, formula	Civic, critical examination Content, results, marketability
Means	Mechanical, mystical,	Analytical, practical
Communication		
Emphasis	Social harmony	Exchange information
Accept invitation	Honor to host	Honor to guest
Context	"Yes" means yes, I hear you, or yes, I understand you, or yes, I agree with you "Not yet" (*belum*) or "none" (*tidak*) can convey "no".	Yes means I agree with you, (at least for now) No means not ever (for now)
Means	Indirect, temporizing, harmonious	Direct, word of honor
Problem Solving		
Emphasis	Consensus, trust is assumed	Trust is a competitive advantage
Resolution	Mediation	Litigation, conflict management
Negotiating	Principles first, then details	Get the details out of the way first
Means	Harmonious, moral aspects	Confrontational, neutral, amoral
Work Ethic		
Emphasis	Social	Personal
Means	Cooperative, stratified	Competitive, hands-on

Notes to Table 12.8.

(1) The epithet, "Calvinist," is from Max Weber's (1904) argument in "The Protestant Ethic and the Spirit of Capitalism." Calvinist predestination by God (election) meant that no one who really believed in it could ever be sure of personal salvation. The psychological need for some certainty in life was met by worldly hard work with resulting abundance and status . Devolution of central authority from Rome to individual opportunity was enhanced by the Industrial Revolution. Christianity entered the market, and the (predestined) individual was more important than the community. This belief was enshrined in the U.S. constitution, made affordable by conquest of the American West, and given expression in comparisons of native patriotism and foreign nationalism. Its destiny is that, "All those people out there are just like us, or want to be like us, or *should* be like us" or overhead at a 1995 international conference in Washington "How come it's taking you (Europeans) so long to be like us?" Their most accurate response is that they're different. Reciprocity in the public trust, except for campaign contributions, is seen as bribery that leads to a greater egalitarianism than can be tolerated under Calvinist doctrine and its secular expression in American culture. The community guarantees to all equal rights under the law, although, as elsewhere, some people are more equal than others.

(2) Godzilla and the Andromeda Strain are from 1960s cinema in which an identifiable malevolent, budget-busting monster (an evil empire, oil price shock, etc..) or an unseen, diffuse pathogen (HIV) that is ignored as long as possible. These alien threats temporarily suspend competition and gridlock.

(3) The community guarantees to all equal rights under the law although, as elsewhere, some people are more equal than others.

(4) The dichotomy between law and government reaches new heights during political national elections every four years when both incumbents (who have been there for years) and challengers promise to "clean up the mess in Washington." This is difficult to explain to people who accept the proposition that government is good, even while a governor may be bad.

The significance of all this in international work is that, sooner or later, cultural diversity and relativism are driven home to expatriate engineers and others, some of <u>whom will want explanations and understanding. A sense of humor helps.</u>

Sources: Boeke, J..H. 1953. Economics and Economic Policy of Dual Societies. Institute of Pacific Relations, New York; Geertz, C. 1962. Peddlers and Princes: Social Development and Economic Change in Two Indonesian Towns. Chicago University Press, Chicago; Gould, D.J., and Amaro-Reyes, J.A. 1983. The Effects of Corruption on Administrative Performance: Illustrations from Developing Countries. World Bank Staff Working Paper No. 580. Washington. Gunnerson, C.G., et al, 1991 and 1994. International information and technology transfer. J. Professional Issues in Engineering Education and Practice. 117,4,336-350, and 120,1,99-118. Am. Soc. Civil Engrs, New York; Scott-Stevens, S. 1987. Foreign Consultants and Counterparts: Problems in Technology Transfer, Westview Press, Boulder; Sievers, A. M. 1974. The Mystical World of Indonesia: Culture and Economic Development in Conflict. Johns Hopkins, Baltimore.

12.5.4 Terms of Reference for Information and Technology Transfer

It is concluded that the most successful engineering feasibility studies are those that are updated by trained counterparts, and that expatriate firms who provide this training will be favored over those who don't. Needs for on-site training throughout project preparation for international engineering services are readily acknowledged and sometimes detailed during negotiations with ministers and senior staffs of public works, manpower, and environment. Doubts arise when owners and consultants see information and technology transfers as threatening to established patron-client relationships and benefits, and to future consulting services. Development and training needs apply equally to expatriate and counterpart staffs; only the details are different. Here, we conclude that successful transfers in economic and environmental development projects will, instead of people working themselves out of their jobs, result in more demands for their services than ever. To this end, we present the following requirements for terms of reference:

- insistence by the host government that it take place,
- commitment to a goal of full economic and technological partnerships between and industrial and developing countries, and among developing countries,
- commitment to the objective that the counterpart staff will prepare the terms of reference for the next project
- secondment of the most competent and motivated professionals as counterparts, recognizing that they will be sorely missed by their agencies.
- an additional 15 to 20 percent in time and money for training during project identification and preparation; this should include time during working hours for language instruction, for both expatriates and counterparts, designed to promote transparency and recognition of the two cultures,
- Assignments for individual expatriate and counterpart managers should be for the life of the project preparation; with a corresponding minimum for senior counterpart and expatriate personnel, and a minimum of one year for short-term consultants and advisors, and,
- recognition by individuals and firms that international engineering has many non-structural attributes, and that host-country staff training is almost certainly in their best strategic interest.

12.6   References

1. Barrett, S. 1994. Conflict and Cooperation in Managing International Water Resources. Policy Research Working Paper 1303. World Bank, Washington.
2. Baum, W.C., and Tolbert, S.M. 1985, Investing in Development: Lessons of World Bank Experience. Oxford, New York
3. Chadwick, E.   1841. Report on the Sanitary Condition of the Labouring Population in Great Britain. Poor Land Commission. Reprinted 1985, Edinburgh University Press, Scotland.
4. Clancy, K. G., and Carroll, D. J. 1986. Key Issues in Planning Submarine Outfalls for Sydney, Australia. In Proceedings, Marine Disposal Seminar, Rio de Janeiro, August 1986. International Association for Water Pollution Research and Control, London, pp. 159-170.
5. California State Water Resources Control Board. 1990. California Municipal Wastewater Reclamation in 1987. Sacramento.
6. Degrémont. 1991. Water Treatment Handbook, 6th Edition, 2 volumes. Lavoisier Publishing, Paris.
7. Fresinius, W., Schneider, W., Böhnke, B., and Pöppinghaus, K.   1989. Waste Water Technology. Springer-Verlag. Berlin, New York.
8. Gould, D. J., and Amaro-Reyes. 1983. The Effects of Corruption on Administrative Performance: Illustrations from Developing Countries. Staff Working Paper No. 580, World Bank, Washington, D.C.
9. Grace, R.A. 1978. Marine Outfall Systems. Prentice-Hall, Inc. Englewood Cliffs, NJ.
10. Gruen, G. 1991. The Next Middle East Conflict: The Water Crisis. A Simon Wiesenthal Center Report. Simon Wiesenthal Center, Los Angeles.
11. Gunnerson, C.G. 1988. Report to Government of Indonesia on water supply and sanitation for Jakarta. Motor-Columbus Engineers, Baden/Jakarta.
12. Gunnerson, C.G. 1991. Costs of water supply and wastewater disposal: forging the missing link.   In Rosen, H., and Durkin-Keating, A., editors. Water and the City, American Public Works Association, 185-202.
!3. Gunnerson, C.G. et al. 1991, 1994. International information and technology transfer. J., Professional Issues in Engineering Education and Practice, 117, 4, 336-350. and discussions 120, 1, 99-118. Am.Soc. Civ. Engrs. New York.
14. Gunnerson, C.G. 1992. Water management as an instrument for cooperation and reconciliation. In Karamouz, M., editor. Water Resources Planning and Management. American Society of Civil Engineers, New York. 731-737.
15 Hardin, G. 1268. The tragedy of the commons. Science, 162, 13 December 1968, 1243-1248.
16. Hennessy, Paul V. 1986. Personal communication. J. M. Montgomery Co., Inc. Pasadena, California.
17. Homer-Dixon, T.F., 1991. Environmental changes as causes of acute conflict. International Security, 16 (2) 76-117, and (1994) Environmental scarcities and violent conflict, International Security, 19 (1) 3-40.
18. Homer-Dixon, T.F. 1995. The ingenuity gap: can poor countries adapt to resource scarcity? Population and Development Review. 21, 5, 587-612.

19. Kalbermatten, J. M., Julius, D. S., and Gunnerson, C. G. 1982. Appropriate Sanitation Alternatives: a Technological and Economic Appraisal. World Bank Studies in Water Supply and Sanitation 1. Johns Hopkins, Baltimore..

20. Kalbermatten, J. M., Julius, D. S., and Gunnerson, C. G. and Mara, DD.. 1982. Appropriate Sanitation Alternatives: a Planning and Design Manual World Bank Studies in Water Supply and Sanitation 2. Johns Hopkins, Baltimore.

21. Los Angeles County Sanitation Districts data courtesy of C.C.Carey and F. Gaarrett, summarized in Chiu, L., Doughman, P., and Lyon, S. 1994. Marginal costs and benefits of treatment processes for wastewater reclamation. Project report for E278, Wastewater Management for Coastal Cities, Spring 1994, University of California at Irvine.

22. Metcalf and Eddy. 1991. Wastewater Engineering, 3rd edition. G. Tchobanoglous, editor.. McGraw-Hill, Inc. New York.

23. Mumford, L. 1961. The City in History. Harcourt, Brace/ New York.

24. Munasinghe, M. 1992. Water Supply and Environmental Management: Developing World Applications. Westview Press, Boulder.

25. Murcott, S., and Harleman, D. 1992. Performance and innovation in wastewater Treatment, Technical Note #36, Parsons Laboratory, Massachusetts Institute of Technology, Cambridge.(summarized in Ref. No. 26).

26. National Research Council. 1993. Managing Wastewater in Coastal Urban Areas. National Academy Press, Washington.

27. Operations Evaluation Department. 1992. Water Supply and Sanitation Projects: The Bank's Experience -- 1967-1989. World Bank, Washington.

28. Rapoport, A. 1987. Prisoner's dilemma. In Eatwell, J., Milgate, M., and Newman, P., editors. Palgrave Dictionary of Economics. Stockton Press, New York. Vol. 3, 973-976.

29. Sayınlı, T., and Yiğit, S. 1994. Üskudar and Baltalimanı Sea Outfalls of the İstanbul sewerage project, construction case history, Proc. International Specialized Conference on Marine Disposal Systems, İstanbul International Association on Water Quality, London. 237-244.

30. Scott-Stevens, S. 1987. Foreign Consultants and Counterparts: Problems in Technology Transfer. Westview Press, Boulder.

31. Snow, J. 1855. On the Role of Well Water in the Communication of Cholera. London. Reprinted Oxford University Press, 1936.

32. Squire, L., and van der Tak, H.G. 1975. Economic Analysis of Projects International Bank for Reconstruction and Development, Washington.

33 Wallis, I. G. 1979. "Ocean Outfall Construction Costs. Journal, Water Pollution Control Federation , 31,5,951-957.

34. Wittvogel, K. 1957. Oriental Despotism. Yale, New Haven

35. World Bank. 1980. Water Supply and Waste Disposal. Poverty and Basic Needs Series. Washington.

36. World Bank. 1992. World Development Report: Development and the Environment. Oxford, New York.

37. World Bank. 1994. World Development Report: Infrastructure for Development. Oxford, New York.

38. World Bank data. 1982, 1992.

13   Index

I

J, K, L

M

N

O

Springer-Verlag and the Environment

We at Springer-Verlag firmly believe that an international science publisher has a special obligation to the environment, and our corporate policies consistently reflect this conviction.

We also expect our business partners – paper mills, printers, packaging manufacturers, etc. – to commit themselves to using environmentally friendly materials and production processes.

The paper in this book is made from low- or no-chlorine pulp and is acid free, in conformance with international standards for paper permanency.